| IB | IIB | IIIA | IVA | VA | VIA | VIIA | O |
|---|---|---|---|---|---|---|---|
| | | | | | | | 2 · 2 He 4.0026 |
| | | 2 3 · 5 B 10.811 | 2 4 · 6 C a 12.01115 | 2 5 · 7 N 14.0067 | 2 6 · 8 O 15.9994 | 2 7 · 9 F 18.9984 | 2 8 · 10 Ne 20.183 |
| | | 2 8 3 · 13 Al 26.9815 | 2 8 4 · 14 Si 28.086 | 2 8 5 · 15 P 30.9738 | 2 8 6 · 16 S 32.064 | 2 8 7 · 17 Cl 35.453 | 2 8 8 · 18 Ar 39.948 |
| 2 8 18 1 · 29 Cu 63.546 | 2 8 18 2 · 30 Zn 65.37 | 2 8 18 3 · 31 Ga 69.72 | 2 8 18 4 · 32 Ge 72.59 | 2 8 18 5 · 33 As 74.9216 | 2 8 18 6 · 34 Se 78.96 | 2 8 18 7 · 35 Br 79.904 | 2 8 18 8 · 36 Kr 83.80 |
| 2 8 18 18 1 · 47 Ag 107.868 | 2 8 18 18 2 · 48 Cd 112.40 | 2 8 18 18 3 · 49 In 114.82 | 2 8 18 18 4 · 50 Sn 118.69 | 2 8 18 18 5 · 51 Sb 121.75 | 2 8 18 18 6 · 52 Te 127.60 | 2 8 18 18 7 · 53 I 126.9044 | 2 8 18 18 8 · 54 Xe 131.30 |
| 2 8 18 32 18 1 · 79 Au 196.967 | 2 8 18 32 18 2 · 80 Hg 200.59 | 2 8 18 32 18 3 · 81 Tl 204.37 | 2 8 18 32 18 4 · 82 Pb 207.19 | 2 8 18 32 18 5 · 83 Bi 208.980 | 2 8 18 32 18 6 · 84 Po (210) | 2 8 18 32 18 7 · 85 At (210) | 2 8 18 32 18 8 · 86 Rn (222) |

(Left partial entries: 28 Ni ...71 · 46 Pd ...6.4 · 78 Pt ...5.09)

| | | | | | | | |
|---|---|---|---|---|---|---|---|
| 64 Gd 7.25 (2 8 18 26 9 2) | 2 8 18 28 8 2 · 65 Tb 158.924 | 2 8 18 28 8 2 · 66 Dy 162.50 | 2 8 18 29 8 2 · 67 Ho 164.930 | 2 8 18 30 8 2 · 68 Er 167.26 | 2 8 18 31 8 2 · 69 Tm 168.934 | 2 8 18 32 8 2 · 70 Yb 173.04 | 2 8 18 32 9 2 · 71 Lu 174.97 |
| 96 ...m (47) | 2 8 18 32 26 9 2 · 97 Bk (247) | 2 8 18 32 27 9 2 · 98 Cf (249) | 2 8 18 32 28 9 2 · 99 Es (254) | 2 8 18 32 29 9 2 · 100 Fm (253) | 2 8 18 32 30 9 2 · 101 Md (256) | 102 No (254) | 103 Lr (257) |

# Charles W. Spangler

Department of Chemistry, Northern Illinois University

# Organic Chemistry

## A Brief Contemporary Perspective

Prentice-Hall, Inc., Englewood Cliffs, New Jersey 07632

*Library of Congress Cataloging in Publication Data*

SPANGLER, CHARLES W
  Organic chemistry.

  Includes index.
  1. Chemistry, Organic.  I. Title.
QD251.2.S68      547      79-10494
ISBN 0-13-640318-2

Acquisition Editor: Fred Henry
Editorial/production supervision: Eleanor Henshaw Hiatt
Interior design and cover: Virginia M. Soulé
Manufacturing buyer: Edmund Leone

The cover painting by Olympia Shahbaz is of the paper models
of the asymmetric carbon atom, made by J. H. van't Hoff in 1875.
The models are in the National Museum for the History of Science,
Leyden, The Netherlands.

To my mother for her early encouragement,
To my wife, Brenda, for her love, patience, and understanding,
To my sons, David and Daniel, my hope for the future,
And to Gandalf, for just being a dog.

Printed in the United States of America

10  9  8  7  6  5  4  3  2  1

PRENTICE-HALL INTERNATIONAL, INC., *London*
PRENTICE-HALL OF AUSTRALIA PTY. LIMITED, *Sydney*
PRENTICE-HALL OF CANADA, LTD., *Toronto*
PRENTICE-HALL OF INDIA PRIVATE LIMITED, *New Delhi*
PRENTICE-HALL OF JAPAN, INC., *Tokyo*
PRENTICE-HALL OF SOUTHEAST ASIA PTE. LTD., *Singapore*
WHITEHALL BOOKS LIMITED, *Wellington, New Zealand*

# Contents

## PART I

# 3

## ALKANES: The "Saturated" Hydrocarbons  39

# 4

## STEREOCHEMISTRY  57

# 5

## ALKENES, ALKYNES, AND DIENES: The "Unsaturated" Hydrocarbons  75

# 6

## BENZENE AND THE CONCEPT OF AROMATICITY  101

# 11

## ACYL COMPOUNDS: Carboxylic Acids and Their Derivatives    206

# 12

## CARBOHYDRATES    234

# 13

## AMINES AND AMINE DERIVATIVES    259

# 14

## AMINO ACIDS AND PROTEINS    282

# PART II
# Core Enrichment

# 5

## AROMATIC SUBSTITUTION   369

# 6

## FATS, OILS, WAXES, DETERGENTS, AND OTHER LIPIDS   383

# 7

## POLYMERS: Backbone of Present-Day Life   397

# 8

## DRUGS: Boon and Bane   410

## INDEX   425

# Preface

Look around you. The life you lead is very different from that lived by your parents. Before World War II, clothing was made only of such natural fibers as wool, cotton, and silk. Synthetic drugs and antibiotics were virtually unknown, and many diseases we now consider minor inconveniences were deadly killers, especially of the very young. Furniture and household goods were made from wood or metal, and the most widely used polymer was natural rubber. Oddly enough, some items manufactured from natural materials today are quite expensive or unobtainable, but their synthetic substitutes are relatively cheap—a trend that will undoubtedly continue.

Technology produces change, and the incredible pace and growth of our technological society since World War II has changed our lives forever. One can argue interminably whether such change is good or bad, but it is irreversible, particularly in a world of declining natural resources and expanding population. Organic chemistry has provided much of the raw material for our changing life-style and will probably continue to do so. The purpose of this text, therefore, is twofold. First and foremost, the "guts" of organic chemistry must be mastered in order for many people to function in today's complex world and in their own chosen professions. By "guts" we mean the fundamentals of structure, reactions, synthesis, and language of organic chemistry. You will not be a practicing organic chemist when you finish this course of study—you won't even be close, but you will have an appreciation of where organic chemistry has been, where it is going, and, to borrow from the language of today, where it's at.

Why study organic chemistry at all? Why is it important enough to your chosen field of study to be a requirement for graduation? These are favorite questions of nonchemistry majors, which brings us to the second reason for writing this book. Very few texts attempt to relate organic chemistry to life as it is lived or, in other words, to place organic chemistry in perspective. It is my hope that when you finish your study, you will have a better feeling of why people study organic chemistry and why our present society could no longer exist or progress without it.

I would like to thank the many people who participated in the genesis of this book—particularly the reviewers for their many useful comments during the early stages of writing and my colleagues at Northern Illinois University. It has been a distinct pleasure to be associated with the staff of Prentice-Hall, and I would like to extend my sincere appreciation to David Hildebrand, who twisted my arm in the first place; to Fred Henry, who never lost his sense of humor; and especially to Eleanor Henshaw Hiatt, who was unflappable even when it seemed I was lost forever in the Selway-Bitterroot Wilderness.

C.W.S.

# To the Student
# and the Instructor

This book is an attempt to present organic chemistry in an organized, rational manner to nonchemistry majors. This is a difficult task at best, and at its worst, can be a very frustrating experience for the student. Most nonchemistry majors, with the exception of those in the premedical track, take a series of one-semester courses, usually the familiar freshman and organic chemistry sequence followed occasionally by biochemistry. In many universities, laboratories associated with these courses are not required, and it is these nonlaboratory courses many students elect. In general, this is a grave error because chemistry is first and foremost an experimental science, and the instructor is left with the unpleasant task of trying to teach an experimental science without the experimentation. Thus, after taking several such courses, many students are left with the impression that chemistry is a very cold, cut-and-dried field far removed from trial and error and human emotion. Facts and experimental results give the appearance of flowing from abstract theory, and some instructors fall into the fatal (to the student!) habit of teaching theory alone, because it is so much more intellectually satisfying (to the instructor!). This text tries to avoid such pitfalls.

The organization of this book is unusual for a brief organic course. The material is divided into a "core" section and a "core enrichment" section. The rationale behind such a division is the heterogeneity of the student interests and needs in various colleges and universities. The makeup of each class can vary widely from school to school, and each instructor may have different ideas as to how the course should be taught and what his or her particular class really needs. I have attempted to identify the essential core material that really *must* be taught in any modern course in organic chemistry and to present it in a logical fashion. The optional enrichment topics, for the most part, are self-contained and relatively independent from each other. Some instructors may argue about some inclusions in an enrichment designation, but such pedagogic arguments should be of little concern to the student. In essence, each instructor can make up his or her own course. The optional topics can be introduced at any point in the core material and almost in any order. They are *not* special topics, as they are in some organic texts, that can be totally ignored. Rather, they are an attempt to given organic chemistry instructors at junior colleges, colleges, and universities some degree of control over the type of course they wish to present to their students. Although this idea is not new in education, it is new for a subject as complex as organic chemistry.

A solutions manual for this text is available and recommended to allow the student to gain as much as possible from the course and the text. The manual contains answers, with full explanations as necessary, for each exercise and problem.

Structure, stereochemistry, and the third dimension are inseparable in organic chemistry. I cannot recommend the use of organic models too strong—without them, students tend to think in two dimensions, a mistake for many topics. Buy the models and use them. I especially recommend instructors to allow their use in exams. There are several types of models on the market, all with inherent advantages and disadvantages, which are discussed on page xiii. If you find the models too expensive for your budget, reasonable models can be made with toothpicks and either gum drops or jelly beans. They have the advantage of both ready availability and edibility.

# The Use of Molecular Models

In order to understand the structure of organic molecules an appreciation of all three dimensions will be required. No amount of paper-and-pencil chemistry will ever replace the actual building and study of molecular models whenever stereochemistry is essential to the solution of a problem. There are many different kinds of models available in kit form today, and as in structure writing, they all have inherent advantages and disadvantages. Some of the more common types and their sources are described below.

1. Framework Molecular Models (Prentice-Hall, Inc.)—a low-cost kit composed of metal valence "clusters" and plastic tubing. The metal units represent tetrahedral, octahedral, and trigonal bipyramidal arrangements. The plastic tubing is color-coded to represent the various elements. The tubing can be cut in lengths to represent the covalent atom radius, resulting in models that have relatively accurate internal dimensions.

2. HGS Molecular Models—Organic Chemistry Kit (Benjamin–Maruzen)—another low-cost kit, which comes in both freshman (general) and organic chemistry versions. The organic chemistry versions has a preponderance of black plastic tetrahedral carbon atoms and small hydrogen atoms with a few oxygen, nitrogen, and halogen atoms (red, blue, and green). Bonds are plastic connectors in two different straight lengths (C—C and C—H). There are curved pieces for multiple bonds.

3. Orbit Molecular Building System (Science Related Materials, Inc.)—a more flexible kit modeled on the Prentice-Hall system, with plastic valence "centers", green and white plastic straws and small plastic "pegs" to represent atoms or atomic centers. As in the Prentice-Hall kit, the plastic straws may be cut to length in order to give relatively accurate internal scale models. Some space-filling plastic units that depict true atomic volumes are available as a separate kit.

4. Ealing Scale Atom Molecular Models (Ealing Corporation)—a space-filling molecular model kit composed of hollow spheres, which snap together to give models that duplicate the actual size and shape of various molecules. These molecules are particularly good for predicting steric interference of accessibility of various functional groups to reagents.

Methyl chloride                                    2-Butanol

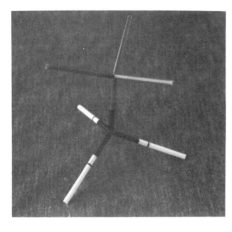                               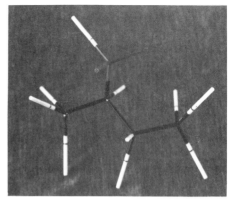

                        Prentice-Hall, Inc.

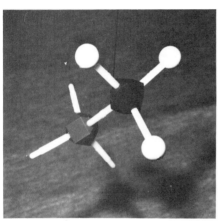                               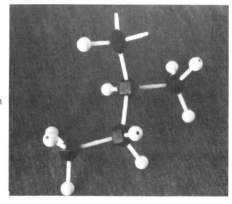

                        Benjamin-Maruzen

                            Ealing

These four kits and systems are by no means the only available materials. Others exist and cover a wide range of application and price, but the preceding descriptions cover the basic types commercially available. The point of these comparisons is to show you that you do have a choice as to the kinds of models you use. Buy a set and *use them, use them, use them*!

A comparison of three of the kits is shown in the accompanying illustration. Models of methyl chloride ($CH_3Cl$) and 2-butanol ($CH_3CH(OH)CH_2CH_3$) are shown as examples of each type of kit.

PART I

Core

# Introduction

## THE NATURE OF ORGANIC CHEMISTRY

Organic chemistry has always been associated with life. In the earliest days of chemistry, even the terms *organic* and *inorganic* emphasized this distinction. Organic compounds were thought to require a "vital force" or life principle to be actively engaged in their preparation or synthesis. An "organic compound" was thus loosely defined as a compound isolated from a natural source (plant or animal) and made *by* the natural source. Inorganic compounds, on the other hand, came from nonliving sources such as rocks and mineral deposits. In 1828 Friedrich Wöhler shattered this concept when he announced that he had been able to prepare urea, a compound normally excreted in urine, from ammonium cyanate, an inorganic salt, without requiring an animal kidney or liver. Further experimentation by other German chemists produced other synthetic examples, and the "vital force" theory died a quiet death.

$$NH_4^{\oplus}, \overset{\ominus}{O}CN \xrightarrow{\Delta} H_2N-\overset{\overset{\displaystyle O}{\|}}{C}-NH_2$$

Ammonium          Urea
isocyanate

We still obtain large quantities of organic chemicals from living sources. Fossil fuels such as petroleum and coal provide not only energy but also a storehouse of petrochemicals, primarily hydrocarbons. Many drugs, both licit and illicit, are still derived from plants, such as morphine from the opium poppy. Antibiotics are obtained from the controlled growth of fungi by the pharmaceutical industry, and enzymes are normally isolated from plant, animal, and bacterial sources. The simple fact is that nature, and natural biosynthetic processes, can produce many chemicals important to our modern way of life much more cheaply than the organic chemist can synthesize them in the laboratory. However, these same chemists have produced hundreds of thousands of new compounds never thought of by Mother Nature, and every bit as useful as hers. In every study of organic chemistry, then, we must learn to appreciate both the beauty and diversity of natural biosynthesis and the ingenuity of humans in producing new and better materials through laboratory synthesis.

## THE NATURE OF ORGANIC COMPOUNDS

Organic chemistry may be loosely defined as the study of the compounds of carbon. In your study of organic chemistry you will soon realize that only a few elements in the periodic table are actually responsible for the tremendous variety of structure and functionality found in organic compounds. Carbon's role in organic structure is to provide the *backbone*, or *framework*, of the molecule. This framework determines both the size and, to a large extent, the shape of the molecule in three-dimensional space. Hydrogen is present in almost every organic compound, and the covalency of the C—H and C—C bonds is responsible for many of the unique physical properties of organic molecules. But even though the C—H framework of an organic molecule determines its gross structure, it actually has little to do with the chemical reactions of the molecule. This is reserved for the *functional group.*

## THE CONCEPT OF THE FUNCTIONAL GROUP

A small number of elements attached to the carbon framework in an equally small number of ways are responsible for the chemical reactions of organic molecules. The more important of these are the first-row elements oxygen and nitrogen. A few second-row elements, such as P and S, are important biochemically. The halogens (F, Cl, Br, and I) are useful reactive sites in organic molecules used as synthetic intermediates, but they are also potentially lethal to living systems. The hydrocarbon portion of the molecule in an organic reaction essentially "goes along for the ride." In fact, it is common practice when writing reaction equations to designate this portion by the letter R, and indicate what occurs only to the reactive part of the molecule. We refer to these reactive parts as *functional groups*. Consider, for example, the —OH group. Simple organic molecules containing this grouping are referred to as *alcohols.* The chemistry of this class of compounds is determined by the reactions of the —OH functional group. Thus we can systematize organic reactions by recognizing and learning the reactions of the individual functional groups. Although this does not make organic chemistry easy, it does greatly simplify our approach to its study.

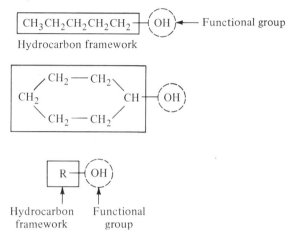

Table I-1 indicates the functional groups covered in the core material in this text.

Table I-1.   Common Functional Groups in Organic Chemistry

| Compound | Example | Functional group | Class of compound |
|---|---|---|---|
| RH | $CH_3CH_2CH_2CH_3$ | None | Hydrocarbon–alkane |
| $RCH=CH_2$ | $CH_3CH_2CH=CH_2$ | $C=C$ | Hydrocarbon–alkene |
| $RC\equiv CH$ | $CH_3CH_2C\equiv CH$ | $C\equiv C$ | Hydrocarbon–alkyne |
| R—X (X = F, Cl, Br, I) | $CH_3CH_2CH_2CH_2Cl$ | C—Cl | Alkyl halide |
| R—OH | $CH_3CH_2CH_2CH_2OH$ | OH | Alcohol |
| R—O—R | $CH_3CH_2OCH_2CH_3$ | —O— | Ether |
| $R-\overset{\overset{O}{\|\|}}{C}-H$ | $CH_3CH_2CH_2\overset{\overset{O}{\|\|}}{C}H$ | $-\overset{\overset{O}{\|\|}}{C}-H$ | Carbonyl–aldehyde |
| $R-\overset{\overset{O}{\|\|}}{C}-R$ | $CH_3CH_2\overset{\overset{O}{\|\|}}{C}CH_3$ | $-\overset{\overset{O}{\|\|}}{C}-$ | Carbonyl–ketone |
| $R-\overset{\overset{O}{\|\|}}{C}-OH$ | $CH_3CH_2\overset{\overset{O}{\|\|}}{C}-OH$ | $-\overset{\overset{O}{\|\|}}{C}-OH$ | Carboxylic acid |
| $R-\overset{\overset{O}{\|\|}}{C}-OR$ | $CH_3CH_2\overset{\overset{O}{\|\|}}{C}-OCH_3$ | $-\overset{\overset{O}{\|\|}}{C}-OR$ | Ester |
| $R-\overset{\overset{O}{\|\|}}{C}-NH_2$ | $CH_3CH_2\overset{\overset{O}{\|\|}}{C}-NH_2$ | $-\overset{\overset{O}{\|\|}}{C}-NH_2$ | Amide |
| $R-NH_2$ | $CH_3CH_2CH_2NH_2$ | $-NH_2$ | Amine |

It is interesting to note that some of the more complicated functional groups are composed of two simpler groups. Thus the carboxylic acid group is composed of both a carbonyl ($-\overset{\overset{\textstyle O}{\|}}{C}-$) and an alcohol ($-OH$) grouping. It should not be surprising, then, to learn that the carboxyl group has some of the characteristics of both an alcohol and a carbonyl. The chemistry of single functional groups is relatively uncomplicated. However, as the number of functional groups in a molecule increases, reaction complexity increases markedly, and it is this complexity and multiplicity of reaction that makes the study and synthesis of biochemically active molecules so difficult.

## THE STUDY OF ORGANIC CHEMISTRY

In order to make some sense out of the diversity of organic reactions, three topics must be understood:

*Structure.* In essence, this topic is the fundamental building block for any further study of organic chemistry. We will learn how carbon is able to form such a wide variety of different compounds and the different ways carbon can bind, both to itself and to other elements such as H, O, and N. We will see how bond polarity helps us predict compound reactivity and how the nature of the intermediates formed by carbon can influence product formation. An intimate understanding of structure will enable you to make many predictions about the potential behavior of organic molecules in many different reaction situations.

*Stereochemistry.* Organic chemistry is a three-dimensional science. We will learn that this aspect leads to many new, and at first bewildering, types of molecular isomerism. The arrangement of the atoms of a molecule in space affects both how it reacts and also, to a large extent, what products can be formed in any particular transformation. You will learn to think in three dimensions, to apply this knowledge to the understanding of biochemical transformations as well as simple organic reactions, and finally, to appreciate that stereochemistry governs virtually all important life functions.

*Synthesis.* The conversion of one compound into another, whether in a test tube or in your own body, is *the* fundamental problem in organic chemistry. The importance of all other topics can eventually be judged by how well they extend our knowledge and understanding of how a conversion can best be performed. We will first learn how one functional group can be transformed into another, and when a sufficient background has been developed, that a rational synthesis of a compound can be planned from readily available starting materials. From this study of laboratory synthesis, we can progress to elementary biosynthesis and examine how a living system solves the same problems from *its* store of starting materials, usually quite different from what is found on laboratory shelves. Finally we will discuss how molecules can be "tailor-made" to solve specific wants and needs.

These three topics follow one another naturally. One cannot study stereo-chemistry without first understanding structure, and a rational synthesis cannot be planned without an appreciation of both structure and stereochemistry. And certainly an appreciation and understanding of biochemistry is impossible without a firm grounding in all three.

## PROBLEMS: INTRODUCTION

**1.** In each of the following compounds, identify the functional groups present:

(a)  $CH_3CHCH_2CH_3$
       |
       OH

(b)  $CH_2{=}CHCH_2CH_2Cl$

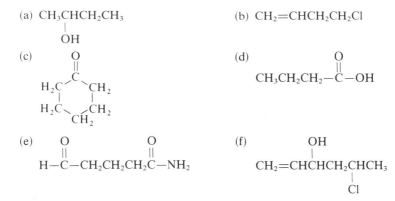

(c)

(d)      O
         ||
    $CH_3CH_2CH_2{-}C{-}OH$

(e)     O            O
        ||           ||
    $H{-}C{-}CH_2CH_2CH_2C{-}NH_2$

(f)           OH
              |
    $CH_2{=}CHCHCH_2CHCH_3$
                      |
                      Cl

**2.** How many functional groups are there in each of the following naturally occurring compounds?

(a)  COOH
     |
     CHOH
     |
     CHOH
     |
     COOH
     Tartaric acid

(b)      CHO
         |
     H{-}C{-}OH
         |
     HO{-}C{-}H
         |
     H{-}C{-}OH
         |
     H{-}C{-}OH
         |
       CH_2OH
       Glucose

(c)           OH
              |
    $HOOCCH_2CCH_2COOH$
              |
            COOH
         Citric acid

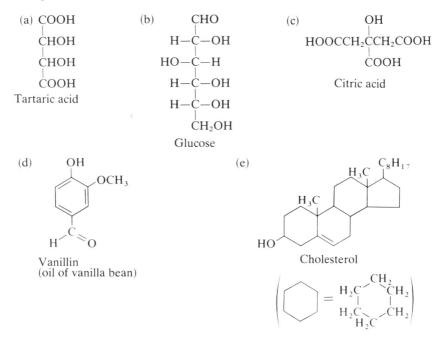

Vanillin
(oil of vanilla bean)

Cholesterol

# 1

# Structure and Physical Properties of Organic Molecules

Organic chemistry is first and foremost a study of covalently bonded molecules. For many students familiar with the "water system" and ionic inorganic reactions, the concept of covalency introduces a number of conceptual problems. One of these is having to consider a molecule's structure as an intact entity rather than simply a crystal lattice which magically falls apart in solution yielding the more readily understood ions. In this chapter we will investigate the nature of *covalent bonding* in organic molecules and how the degree of *polarity* of these bonds can affect both chemical and physical properties. We will learn that organic bonding cannot adequately be explained by simple electron sharing resulting from overlapping atomic orbitals, and that the utilization of *hybrid orbitals* for carbon and certain other first-row elements gives a more accurate explanation of the structure of carbon compounds. Similarly, we will see that the carbon backbone of most organic molecules may be assembled in many different ways, leading to the formation of *structural isomers*. Finally, the language of organic chemistry (nomenclature) will be introduced and the need for a systematic method of describing organic structure emphasized.

## COVALENT BONDING IN SIMPLE MOLECULES

Chemical bonds are formed by electron transfer between atoms, or by electron sharing. Electron transfer results in the formation of *ionic bonds*, while the sharing of electron pairs yields *covalent bonds*. Metallic elements such as sodium, calcium, copper, and iron form positive ions by the loss of one or more electrons from an outer valence shell. Similarly, nonmetallic elements such as oxygen and the halogens tend to gain electrons, yielding negative ions. Ion formation in each case results in a

$$Li\ (1s^22s^1)\ \xrightarrow{-e}\ Li^+\ (1s^22s^0) + e$$

$$Cl\ (1s^22s^22p^63s^23p^5)\ \xrightarrow{+e}\ Cl^-\ (1s^22s^22p^63s^23p^6)$$

$$O\ (1s^22s^22p^4)\ \xrightarrow{+2\ e}\ O^{-2}\ (1s^22s^22p^6)$$

*closed-shell* or "rare-gas" configuration which is more stable than the original elemental electron configuration. First-row elements with 3–5 electrons in their valence shells, such as boron, carbon, and nitrogen, form ions only with difficulty. Carbon, for example, must gain or lose four electrons to achieve a rare-gas configuration. It is easy to imagine carbon gaining or losing one electron, but further gains or losses could only be achieved by overcoming large electrostatic forces.

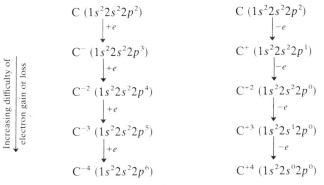

Increasing difficulty of electron gain or loss

| C $(1s^22s^22p^2)$ | C $(1s^22s^22p^2)$ |
|---|---|
| ↓ +e | ↓ −e |
| C$^-$ $(1s^22s^22p^3)$ | C$^+$ $(1s^22s^22p^1)$ |
| ↓ +e | ↓ −e |
| C$^{-2}$ $(1s^22s^22p^4)$ | C$^{+2}$ $(1s^22s^22p^0)$ |
| ↓ +e | ↓ −e |
| C$^{-3}$ $(1s^22s^22p^5)$ | C$^{+3}$ $(1s^22s^12p^0)$ |
| ↓ +e | ↓ −e |
| C$^{-4}$ $(1s^22s^22p^6)$ | C$^{+4}$ $(1s^22s^02p^0)$ |
| Rare gas configuration (Ne) | Rare gas configuration (He) |

Consider the last step in forming either $C^{+4}$ or $C^{-4}$. In the former, one would have to remove an electron (negatively charged) from an entity with a +3 charge, while in the latter the fourth electron must be forced upon an ion already carrying three negative charges. It is not surprising, then, that carbon does not form either ion in normal chemical processes.

Covalent bonds are formed when two elements share one or more electron pairs in order to achieve the stable closed-shell configuration. An *equal sharing*, however,

$$:\ddot{B}r\cdot\ +\ \cdot\ddot{B}r: \longrightarrow\ :\ddot{B}r\boxed{:}\ddot{B}r:$$

$$\uparrow$$
Shared electron pair
$$\downarrow$$

$$H\cdot\ +\ \cdot H \longrightarrow\ H\boxed{:}H$$

is only possible in "pure" covalent bonds between identical atoms. When covalent bonds are formed between atoms of differing electronegativities (a measure of an element's ability to attract electrons toward itself in a molecule), a *polar* bond is formed in which the electron density is skewed toward the more electronegative atom.

H     H            Br     Br

Symmetrical electron clouds — equal sharing

H     Br            H     O

Unsymmetrical electron clouds — unequal sharing

Polar covalent bonds may be indicated in written formulas by approximating the polarized electron cloud as shown above for HBr, or by indicating the polarity by drawing an arrow from the positive end of the dipole toward the negative end or by writing *partial* ($\delta$) charges on the respective atoms.

$$\underset{\leftrightarrow}{\text{H—Br}} \qquad \underset{\delta^+ \quad \delta^-}{\text{H—Br}}$$

Bond polarities may be predicted by considering the electronegativity series for the elements commonly found in organic molecules:

$$F > O > N, Cl > Br > C, H$$

Note that there is actually very little difference in the electronegativities of carbon and hydrogen, which predicts a high degree of covalency for the C—H bond.

    A third type of covalent bond results when both electrons are donated by one atom. Such bonds are commonly formed by elements with nonbonding electron pairs such as nitrogen and oxygen, and are referred to as *coordinate covalent bonds*.

Hydronium ion
(a type of oxonium ion)

Ammonium ion

Oniom-type salts, or ions, produced in these reactions are common intermediates in organic chemistry. Quite often, coordinate covalent bond formation is accompanied by the development of *formal charges* within the molecule. Formal charge on an atom in a covalently bonded molecule may be calculated by comparing the number of electrons surrounding the atom in the free versus the bound state. In the ammonium ion, the nitrogen atom "owns" one-half of each electron pair engaged in bonding with the hydrogen atoms. Since nitrogen has five electrons in its valence shell in the free atom, there has been a *formal* net loss of one electron resulting in a charge of +1 on the nitrogen atom in the ammonium ion.

$$N\ (1s^2 2s^2 2p_x^1 2p_y^1 2p_z^1) \equiv\ :\overset{\cdot}{N}\cdot\ =\ 5\ \text{valence } e\text{'s}$$

$$\begin{array}{c} H \\ | \\ H\!-\!\overset{|}{\underset{|}{N}}\!: \\ | \\ H \end{array} \qquad N \text{ has } \tfrac{1}{2}(3 \ e\text{-pairs}) + 1 \quad \underset{e\text{-pair}}{\text{nonbonded}} = 5\ e\text{'s}$$

$$\left[ \begin{array}{c} H \\ | \\ H\!-\!\overset{|}{\underset{|}{N}}\!:\!H \\ | \\ H \end{array} \right]^{\oplus} \qquad \begin{array}{l} N \text{ has } \tfrac{1}{2}(4\ e\text{-pairs}) = 4\ e\text{'s} \\ 5 - 4 = \text{net formal charge of } +1 \text{ on } N \end{array}$$

In more complicated molecules, such as $HNO_3$, it is necessary to satisfy the octet rule for first-row elements and to establish the most probable electron structure before a calculation of formal charge can be attempted. One should (1) determine the number of electrons available for bonding; (2) join the atoms in the molecule by electron pairs (single covalent bonds); (3) satisfy the octet rule where necessary with multiple bonds (2 or 3 $e$-pairs); and (4) calculate formal charges on the appropriate atoms. The following example will illustrate how this may be accomplished for $HNO_3$:

(1) In $HNO_3$ there are:

$$\begin{array}{rl} 1 \times 1 = & 1\ e \text{ for H} \\ 1 \times 5 = & 5\ e\text{'s for N} \\ 3 \times 6 = & 18\ e\text{'s for O} \\ \hline & 24\ e\text{'s for HNO}_3 \end{array}$$

(2) $H\!:\!O\!:\!\overset{O}{\underset{\ddot{O}}{N}}$    8 $e$'s used for bonding component atoms by single covalent bonds

(3) $H\!:\!\ddot{O}\!:\!\overset{:\ddot{O}:}{\underset{:\ddot{O}:}{N}}$   Octet rule for each atom satisfied for each atom with remaining 16 $e$'s

(4) $H\!:\!\ddot{O}\!:\!\overset{:\ddot{O}: \ -\ 1}{\underset{:\ddot{O}:}{N}} + 1$    Formal charges on N and O are then calculated

Final formula:    $H-\ddot{\text{O}}-\overset{\oplus}{N}\overset{\displaystyle :\ddot{\text{O}}:^{\ominus}}{\underset{:\ddot{\text{O}}:}{\|}}$    or    $H-O-N\overset{\nearrow O}{\underset{\searrow O}{}}$

It is common practice when writing covalent structures to indicate *all* valence electrons, especially unshared pairs. It is a particularly good habit for beginning students in organic chemistry to adhere to this practice until they gain confidence in interpreting covalent structures. Thus chlorine should be written as

$:\ddot{\text{C}}\text{l}:\ddot{\text{C}}\text{l}:$    not    $Cl-Cl$

This type of formula is commonly referred to as an *electron-dot structure.*

$H:\ddot{\text{B}}\text{r}:$                 $:\ddot{\text{O}}::C::\ddot{\text{O}}:$        $:N:::N:$

Hydrogen bromide        Carbon dioxide    Nitrogen

Some typical electron-dot structures

---

**Exercise 1-1:**  Keeping in mind that in writing structural formulas carbon has four electrons in its valence shell, nitrogen five, oxygen six, and halogen seven, write electron-dot formulas for the following compounds:

(a) $CH_4$                     (b) $CCl_4$                     (c) $CHCl_3$
(d) $CH_3OH$                 (e) $CH_3NH_2$              (f) $CH_3OCH_3$
(g) HONO                     (h) $HOClO_3$               (i) $HOSO_2OH$
     (Nitrous acid)               (Perchloric acid)           (Sulfuric acid)

Indicate formal charges where necessary.

---

## POLARITY AND PHYSICAL PROPERTIES

The physical properties of covalently bonded molecules depend, to a large extent, on the molecular weight and the degree of molecular polarity. In simple molecules containing simple functional groups, this polarity may be related to the polar nature of each functionality, and such physical properties as solubility, boiling point, melting point, refractive index, and dielectric constant may all be related to the bond polarities. To illustrate this behavior, let us consider two of these phenomena: solubility and boiling point.

An old rule of thumb in organic chemistry is that "like dissolves like," or in other words, compounds tend to dissolve better in solvents having similar polarity. When a compound dissolves in a solvent, it is normally because of the extra stability derived from one of three different interactions: ion–dipole, dipole–dipole, or induced dipoles, illustrated below:

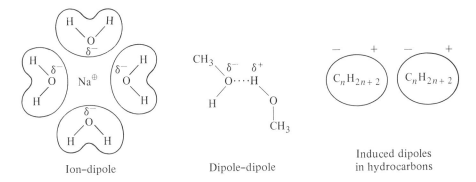

Ion–dipole                    Dipole–dipole                    Induced dipoles in hydrocarbons

In essence, in order for a compound to dissolve in a solvent, the stability derived from solvent–solute interactions must be greater than the intermolecular forces in the solute. In the case of nonpolar molecules such as hydrocarbons ($C_nH_{2n+2}$), the intermolecular attractions are of an induced-dipole nature, the approach of one molecule to another inducing dipoles in both by mutual repulsion of two electron clouds surrounding the molecules. Thus polar molecules tend to be more soluble in

Molecule A          Molecule B          A          B          A          B

polar solvents which have the capability of overcoming the dipole–dipole *intermolecular* attraction between solute molecules by replacing these interactions with new *intermolecular* solute–solvent dipole–dipole interactions.

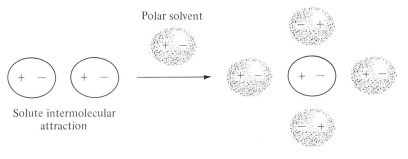

Polar solvent

Solute intermolecular attraction

Solute–solvent intermolecular attraction

In order to convert a liquid into a gas (or to cause individual molecules to escape from a liquid surface) we must overcome the same kind of intermolecular forces discussed above. For ordinary organic molecules, then, we must overcome either dipole–dipole or induced-dipole forces. Since the former are considerably stronger than the latter, it should not be surprising that for compounds of equivalent molecular weight the more-polar compound will have the higher boiling point. For example, water (MW = 18), a highly polar associated liquid, boils at 100°C, while $CH_4$(MW = 16), a nonpolar hydrocarbon held together in the liquid state only by induced-dipole forces, boils at −161°C. Further examples are shown below:

| Compound | MW | B.p.(°C) | Compound | MW | B.p.(°C) |
|---|---|---|---|---|---|
| $CH_3CH_2OCH_2CH_3$ | 74 | 36 | $CH_3(CH_2)_6CH_3$ | 114 | 126 |
| $CH_3CH_2CH_2CH_2OH$ | 74 | 118 | $CH_3(CH_2)_5CH_2OH$ | 116 | 176 |
|  |  |  | $CH_3(CH_2)_4COOH$ | 116 | 205 |
| C—OH more polar than C—O—C |  |  | —COOH more polar than —$CH_2OH$ |  |  |

Other physical properties show similar trends.

**Exercise 1-2:** (a) Arrange the following compounds in order of increasing boiling point. Remember to consider *both* molecular weight and functional group polarity. Look up the actual boiling points in the *Handbook of Chemistry and Physics*. What conclusions can you draw about the relative polarities of the functional groups in these molecules?

pentane ($CH_3CH_2CH_2CH_2CH_3$)
1-pentene ($CH_3CH_2CH_2CH=CH_2$)
*n*-butyl alcohol ($CH_3CH_2CH_2CH_2OH$)
proprionic acid ($CH_3CH_2COOH$)
*n*-butyl chloride ($CH_3CH_2CH_2CH_2Cl$)

(b) Which of the above compounds would have a moderate solubility in water? Explain your choice, then check your answer in the *Handbook*. Were there any surprises?

## COVALENT BONDING IN ORGANIC MOLECULES

If one wanted to predict a structure for the simplest compound between hydrogen and carbon formed by covalent electron sharing, consideration of carbon's outer valence shell would lead one to expect the formation of $CH_2$. One would be able to predict that the HCH bond angle should be approximately 90°. The C—H bonds

$$C \underset{1s}{\overset{\downarrow\uparrow}{\rule{0pt}{0pt}}} \underset{2s}{\overset{\downarrow\uparrow}{\rule{0pt}{0pt}}} \underset{2p_x}{\overset{\downarrow}{\rule{0pt}{0pt}}} \underset{2p_y}{\overset{\downarrow}{\rule{0pt}{0pt}}} \underset{2p_z}{\overline{\phantom{x}}} + 2H \underset{1s}{\overset{\downarrow}{\rule{0pt}{0pt}}}$$

would be formed by the sharing of electrons between a carbon $2p$ orbital and the hydrogen $1s$ orbital ($s-p$ overlap). The $2s$ electron pair would presumably be similar to the unshared pairs in either water or ammonia.

There are several things that are obviously very wrong with the above prediction. $CH_2$, even if formed by this simple mechanism, would still be an *electron deficient* species compared to the closed-shell model. In fact, the simplest hydrocarbon that exists as a stable molecule is methane, $CH_4$. The HCH bond angle in methane is not 90°, but 109°28' and all four C—H bonds are equivalent (bond lengths are equal). How can we explain this discrepancy and our inability to predict a structure for even the simplest of hydrocarbons? In order to unravel this question, it is necessary to consider atomic orbitals and their interaction in more detail before we rationalize the nature of bonding in carbon compounds.

An atomic orbital, we must remember, is really only a conveniently *defined* volume of space in which the probability of finding an electron has been set at some reasonable, but arbitrary, figure (usually 95%). When chemical bonds are formed by the overlap of two atomic orbitals, the bond may also be considered as a region in space where the probability of finding the electron pair is quite high (again, let us say 95%). If one were to take a cross-section of this probability space, one would find it to be circular, thus the bond is cylindrically symmetrical, and we refer to bonds having this symmetry as sigma ($\sigma$) bonds. These three bonds, $s-s$, $s-p$, and $p-p$, although formed from different overlapping atomic orbitals, actually have the same symmetry. Sigma bonds, then, may be formed from a wide variety of orbital combinations.

The formation of a chemical bond releases energy. If we could utilize *all* the available atomic orbitals in the valence shell in bond formation, then we should achieve greater energy release and greater stability. Let us apply these arguments to carbon. Since carbon has four electrons in its valence shell we must ask ourselves how we can effectively utilize them to form the *maximum* number of covalent bonds.

First the electron pair in the 2s carbon orbital must be unpaired in order to free these electrons for bonding. Since the $2p_z$ orbital is empty, one of these unpaired 2s electrons could conceivably be promoted to the 2p level to produce a carbon atom with *four* unpaired electrons. Unfortunately, this does not solve our problem of having to provide four *equivalent* C—H bonds in methane.

$$C \frac{\uparrow\downarrow}{2s} \frac{1}{2p_x} \frac{1}{2p_y} \frac{\phantom{1}}{2p_z}$$

Unpair 2s electrons and "promote" one of them to the vacant $2p_z$ orbital

$$C \frac{1}{2s} \frac{1}{2p_x} \frac{1}{2p_y} \frac{1}{2p_z}$$

4H(1s)

3 C—H bonds (*p–s* overlap) + 1 C—H bond (*s–s* overlap)

This problem is solved on the atomic level by a unique mechanism called *hybridization*. Remember that our atomic orbitals are arbitrarily defined probability spaces. If we rearrange the probability space associated with the carbon 2s and 2p orbitals so that a set of *four equivalent hybrid* orbitals is formed in a three-dimensional array allowing maximum separation of the newly formed hybrids, we obtain four $sp^3$ orbitals directed toward the corners of a tetrahedron. Overlap of the $sp^3$ hybrid carbon orbitals with four hydrogens (1s orbitals) would produce four equivalent C—H bonds whose HCH bond angles are all 109°28'. The C—H bond formed by $sp^3$–s overlap has $\sigma$ symmetry. The tetrahedral nature of the carbon atom is one of the fundamental tenets of structural organic chemistry.

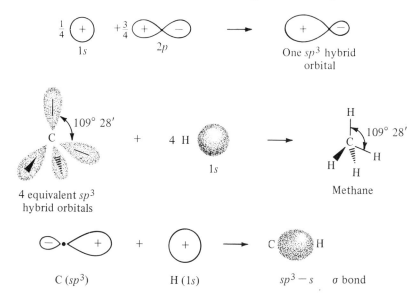

**Exercise 1-3:** The observed H—X—H bond angles in water and ammonia are 105° and 107°, respectively. Do these structures more closely resemble $s$–$p$ or $s$–$sp^3$ X—H bonding? How would you explain the deviations from the theoretical values?

## THE CARBON–CARBON BOND

Organic molecules are composed largely of carbon chains or rings which form the backbone of the structure. Carbon–carbon bonds may be formed from the hybrid orbitals described in the previous section by the overlap of two $sp^3$ orbitals. The resulting $sp^3$–$sp^3$ bond has $\sigma$ symmetry, and the individual carbon atoms maintain their tetrahedral character. The six remaining $sp^3$ orbitals may bond to carbon, hydrogen, or other elements in a similar fashion, forming a backbone of $\sigma$ bonds.

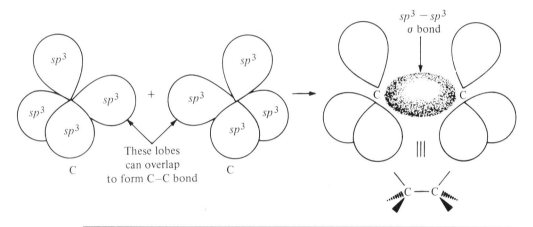

**Exercise 1-4:** The C—H bond results from $sp^3$–$s$ overlap. Predict the makeup of the following bonds:

(a) C—C           (b) C—Cl           (c) C—O
(d) C—N           (e) O—O           (f) N—N

## THE CONCEPT OF STRUCTURAL ISOMERISM

Long carbon chains can be constructed in many different ways; however, once chains contain four or more carbon atoms, branching becomes possible. Such branching produces *structural isomers*, compounds having identical molecular formulas, but whose carbon backbones are arranged differently in three-dimensional space. As the carbon chains increase in length and number of carbon atoms, the number of isomers increases dramatically. For example, there are nine possible isomeric heptanes ($C_7H_{16}$). If another atom (or group) is bonded to the carbon system (say,

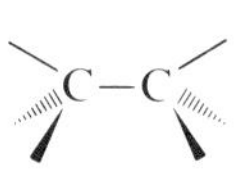

$C_2H_6$

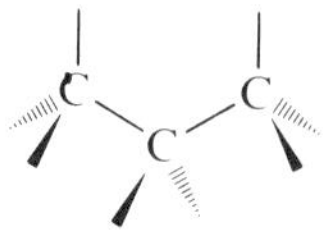

$C_3H_8$

Only one possible carbon backbone

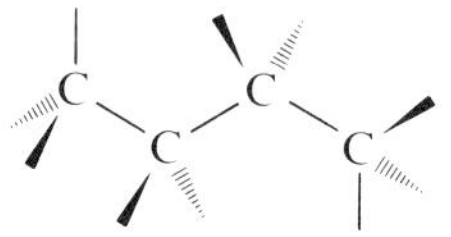

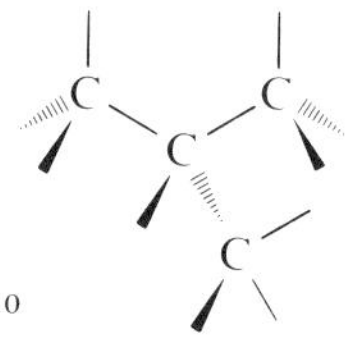

$C_4H_{10}$

Straight chain          Branched chain

Two possible carbon backbones

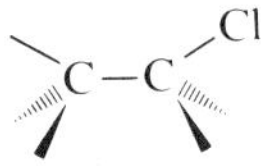

Only one $C_2H_5Cl$ possible

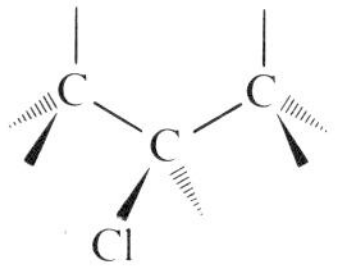

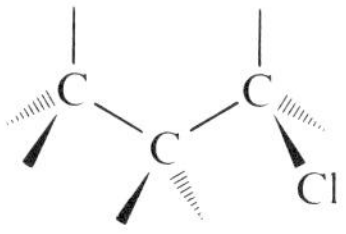

Two $C_3H_7Cl$ molecules possible

chlorine), *positional isomerism* as well as structural isomerism becomes possible. Structural and positional isomers usually differ in physical properties such as boiling point, melting point and solubility. Often they also differ in chemical reactivity.

---

**Exercise 1-5:**   (a) Draw all possible structural isomers for the heptane family ($C_7H_{16}$). (b) Draw all possible isomers corresponding to the formula $C_6H_{13}Br$. Arrange these isomers into groups having identical carbon chains. How many different carbon structural units are there?

---

# THE REPRESENTATION OF ORGANIC
# STRUCTURE WITH PAPER AND PENCIL

There are many different methods of writing structural formulas for organic molecules. Each method has inherent advantages and disadvantages, and there will be times when you will be required to use each and every form. We have already discussed electron-dot formulas and their advantage in showing all valence electrons; however, for large molecules, the display of all valence electrons may prove to be quite cumbersome. In that case, condensed structural formulas give a better "feel" for the total molecular structure and emphasize the presence of functional groups. In the following section, the various common methods of displaying structure are outlined and their advantages and disadvantages briefly enumerated for a typical organic molecule, 2-chloropentane.

(1) **Electron-dot formula—**

$$\begin{array}{c} H \quad H \quad H \quad H \quad H \\ H : C : C : C : C : C : H \\ H \quad H \quad H \quad :\!Cl\!: \quad H \end{array}$$

*Advantages*—Displays all valence electrons, makes calculation of formal charge distribution within functional groups possible.

*Disadvantages*—Makes structure overly complex, difficult to pinpoint functional groups.

(2) **Structural Formulas**—Each bond is represented by a dash; unshared electron pairs may be displayed if desired.

$$\begin{array}{c} H \quad H \quad H \quad H \quad H \\ | \quad | \quad | \quad | \quad | \\ H-C-C-C-C-C-H \\ | \quad | \quad | \quad | \quad | \\ H \quad H \quad H \quad Cl \quad H \end{array} \quad or \quad \begin{array}{c} H \quad H \quad H \quad H \quad H \\ | \quad | \quad | \quad | \quad | \\ H-C-C-C-C-C-H \\ | \quad | \quad | \quad | \quad | \\ H \quad H \quad H \quad :\!Cl\!: \quad H \end{array}$$

*Advantages*—Displays all chemical bonds clearly, appearance less cluttered than dot formulas, functional groups easier to identify.

*Disadvantages*—Requires same amount of space to display as the dot formulas; for large molecules individual structural features are often difficult to spot.

(3) **Condensed Structural Formulas**—Each carbon atom and its attached atoms are condensed to a single line and the total structure read from left to right.

$$\begin{array}{c} H \\ | \\ H-C- \quad \equiv \quad CH_3- \quad or \quad H_3C- \\ | \\ H \end{array}$$

$CH_3CH_2CH_2CHClCH_3$    Functional group included in line

$$CH_3CH_2CH_2CHCH_3$$
$$| \\ Cl$$

Functional group emphasized

If the functional group contains more than one atom, then the group is enclosed in brackets.

$$CH_3 - \underset{\underset{\displaystyle OH}{|}}{\overset{\overset{\displaystyle H}{|}}{C}} - CH_3 \equiv CH_3CH(OH)CH_3$$

*Advantages*—Formula is displayed in the minimum of space, functional groups are easily seen, total molecular structure can be scanned quickly. *Disadvantages*—Most difficult formula for beginning students to read and draw correctly.

Most texts and scientific journals have adopted the condensed structural formula almost exclusively. The beginning student should practice writing all types, but concentrate on mastering the condensed structure. In this section we have not discussed three-dimensional representation. Although this is an extremely important aspect of organic chemistry it is best deferred until Chapter 3, when more familiarity with organic molecules is attained.

---

**Exercise 1-6:**   Draw electron-dot, structural, and condensed structural formulas for each of the following. Include *all* valence electrons and emphasize functional groups in your representation.

(a)  all isomers corresponding to $C_4H_{10}O$
(b)  all isomers corresponding to $C_5H_{10}$
                (*Hint*: Multiple bonds and rings are possible.)

Which class of formula-writing do you find easiest?

---

## NOMENCLATURE—THE LANGUAGE OF ORGANIC CHEMISTRY

It is impossible to discuss organic chemistry without referring to specific organic molecules by name. During the early development of chemistry, the discoverer of a new compound or class of compounds usually had the privilege of naming his discovery. As long as the number and complexity of known organic molecules was relatively small, such haphazard systems of *nomenclature* (literally, "a system of naming") could be tolerated. However, it quickly became apparent to chemists around the world that it was patently ridiculous to have several different names for the same compound, and that a systematic method of naming was required.

In 1892 an international assembly of chemists met in Geneva, Switzerland, to hammer out a rational nomenclature system for organic molecules. The result of their labors is referred to as the *IUPAC* (*I*nternational *U*nion of *P*ure and *A*pplied *C*hemistry) or *Geneva* System, and similar commissions have met periodically since then to modify and extend the system to newly discovered classes of compounds. In essence, the system assigns a *parent* name to a carbon chain according to the number of carbon atoms in a continuous chain which includes all pertinent functional groups. Priorities are also assigned to the various functional groups discussed in the Introduction to determine their order of appearance in the compound name. Groups attached to the chain are named as substituents, and their relative positions indicated by assigning numbers to the carbon atoms in the chain. We will discuss the naming of each class of compounds in the chapter dealing with the relevent functional group chemistry, beginning with Chapter I-3. For alkanes, however, the IUPAC names for the simple straight-chain hydrocarbons are given below. These chains ($C_1$–$C_{10}$) form the basic building units for all future nomenclature and should be committed to memory as soon as possible.

| Formula | No. of carbons in chain | IUPAC name |
|---|---|---|
| $CH_4$ | 1 | Methane |
| $CH_3CH_3$ | 2 | Ethane |
| $CH_3CH_2CH_3$ | 3 | Propane |
| $CH_3(CH_2)_2CH_3$ | 4 | Butane |
| $CH_3(CH_2)_3CH_3$ | 5 | Pentane |
| $CH_3(CH_2)_4CH_3$ | 6 | Hexane |
| $CH_3(CH_2)_5CH_3$ | 7 | Heptane |
| $CH_3(CH_2)_6CH_3$ | 8 | Octane |
| $CH_3(CH_2)_7CH_3$ | 9 | Nonane |
| $CH_3(CH_2)_8CH_3$ | 10 | Decane |

## REVIEW OF NEW TERMS AND CONCEPTS

Define each term and, if possible, give an example of each:

Covalent bond / Closed shell configuration
Polar covalent bond / Coordinate covalent bond
Formal charge / Electron-dot formula
Dipole-dipole interaction / Induced-dipole interaction
Intramolecular interaction / Intermolecular interaction
$\sigma$ bond / Hybrid orbitals
C—C $\sigma$ bond / Structural isomers
Positional isomers / Structural formula
Condensed structural formula / IUPAC nomenclature

1. In each of the following compounds, indicate which bonds would be polar–covalent and which would be nonpolar–covalent.

   (a) $CH_3Br$                           (b) $CH_3CH_2NH_2$

   (c)             $O$
                     $||$
        $CH_3CH_2C{-}OH$             (d) $CH_3CH_2OCH_2CH_3$

   (e) $CH_3CH_2CH_2OH$              (f) $CH_3CH_2CH_2CH_3$

2. In general, as the molecular weight of a series of related compounds increases, the boiling point increases in a more or less regular fashion. Would you expect the melting point behavior to be similar? What forces are overcome when a compound melts? Are these forces similar or dissimilar to those forces holding liquids in the liquid state?

3. Draw electron-dot formulas for the following ions:

   (a) $Cl^-$                   (b) $HO^-$               (c) $AlCl_4^-$

   (d) $NO_3^-$                 (e) $CO_3^{-2}$             (f) $ClO_4^-$

   (g) $NH_2^-$                (h) $Cl_3C^-$             (i) $HCO_3^-$

4. Consider the electron-dot formulas for (d) and (e) in Problem 3 (above). How do you decide which oxygen(s) have negative charge(s)? Draw all possible formulas for the nitrate ion (with the negative charge on different atoms). Is it possible to convert one formula into another by moving electron pairs?

5. Previously, in Exercise 1–3, water and ammonia were said to have H—X—H bond angles of 105° and 107°, respectively. The next members in the Group VA and VIA families, $PH_3$ and $H_2S$, however, have H—X—H bond angles close to 90°. What conclusions can you draw from these data concerning covalent bonding of first-row elements compared to second- (and third-) row family members?

6. Would you expect $Si(CH_3)_4$ and $Ge(CH_3)_4$ to have the same type bonding as $C(CH_3)_4$? The same geometry? Explain your answer in detail in view of the data and questions raised in Problem 5 (above).

7. Build molecular models of the following compounds:

   (a) $CH_3CH_2OH$       (b)        $O$           (c)          $O$
                                    $||$                        $||$
                            $CH_3CH_2C{-}H$           $CH_3CH_2C{-}OH$

   (d)      $CH_2{-}CH_2$     (e) H        H        (f) $H{-}C{\equiv}C{-}CH_3$
            /                    \     /
      $CH_2$          |                $C{=}C$
           \                      /      \
          $CH_2{-}CH_2$         H         $CH_3$

   (g) $CH_3CH_2CH_3$

   Compare the structures of (e), (f), and (g). What conclusions can be drawn concerning the *shapes* of molecules containing C—C, C=C, and C≡C bonds?

**8.** The molecule $CH_2$, which we discussed in the beginning of this chapter, *can* be prepared by thermolysis or photolysis of diazomethane.

$$CH_2N_2 \quad \xrightarrow[\text{or light}]{\Delta} \quad CH_2 + N_2\uparrow$$

Draw possible electron-dot formulas of $CH_2N_2$ and propose structures for $CH_2$. Two different $CH_2$ molecules have been observed experimentally. In what ways do you suppose they differ? Can you propose two different structures for $CH_2$ (*Hint*: You may have to use hybrid orbitals which are *different* than $sp^3$—use your imagination!).

# 2

# The Nature of Organic Reactions

In the previous chapter we learned how covalent molecules in general, and organic molecules in particular, may be assembled from atomic and hybrid orbitals. We also learned that some covalent bonds have intrinsic polarities, while others closely approximate the ideal of equal sharing. In this chapter we shall investigate how these molecules can react with one another and with ionic reagents. One of the more important concepts to be learned in this connection is the nature of the *transition state*, or *intermediate*, in an organic reaction. If one can identify or postulate the most probable intermediaries in a reaction, it is much easier to attain an understanding of product formation. We will also categorize organic transformations rather broadly as *substitutions*, *additions*, *eliminations*, *oxidation–reductions*, *polymerizations*, and *rearrangements*, and show how knowledge of the attacking reagent and the intermediate enables us to discuss the transformation intelligently.

## REACTIONS OF COVALENT MOLECULES— INTERMEDIATES AND TRANSITION STATES

In studying organic reactions, one must realize that the reactions between covalent molecules are quite different from those between ions. When inorganic reagents are dissolved in water, dissociation into ions is followed by reaction, usually precipitation of an insoluble compound or electron transfer between oxidation–reduction couples. In virtually all inorganic reactions there are "spectator" ions which do not participate in the main reaction pathway. For example, if one is studying the reactions of hydroxide ion, it matters little whether KOH, NaOH, or $Ba(OH)_2$ is utilized as the source of hydroxide, as the $K^+$, $Na^+$, and $Ba^{+2}$ ions are not involved in the reaction, but merely "go along for the ride." Similarly, in freshman chemistry one pays little attention to the solvent—water is used predominantly.

Organic reactions are quite different, as very few organic molecules dissociate totally into ions when placed in solution. We have previously indicated that organic chemists rely upon the concept of the functional group to categorize organic reactions, in contrast to the inorganic chemist's reliance upon ions. However, even though the bulk of an organic molecule remains unchanged throughout a functional group transformation, the unchanged part may affect either the course of the reaction, or its rate. There are very few "spectators" in organic chemistry! The choice of solvent also becomes a major factor in covalent chemistry. Water is generally unimportant as a solvent for organic molecules, and the organic chemist must be well versed in choosing the correct solvent (ether, alcohol, hydrocarbon, etc.) for each particular reaction.

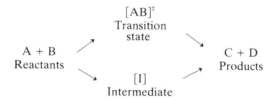

If organic molecules do not dissociate into ions, how do organic reactions occur? Organic chemists have been attempting to answer this question for the past 30 to 40 years with varying degrees of success. It seems fairly clear now that most organic transformations occur through either *transition states* $(AB)^{\ddagger}$, or *intermediates* (I), or both. The concept of an intermediate is readily understood as it is usually a real chemical species with a finite lifetime which results from either a bond-breaking or a bond-making process. Understanding a transition state, however, involves imagining bond-breaking (old bonds) and bond-making (new bonds) taking place simultaneously. The transition state is defined as the point in time when both these ongoing processes attain the maximum potential energy for the system. A more detailed description of the differences between intermediates and transition states may be found in Chapters II-1 and II-2.

## THE NATURE OF ORGANIC INTERMEDIATES

Bond-breaking processes in organic molecules usually involve a C—X bond where X is an atom other than carbon, and they can occur via either heterolytic or homolytic mechanisms:

$$
-\overset{\scriptstyle |}{\underset{\scriptstyle |}{C}}:X
\begin{cases}
\longrightarrow & -\overset{\scriptstyle |}{\underset{\scriptstyle |}{C}}{}^{\oplus} + :X^{\ominus} \quad \text{Heterolytic cleavage} \\[2ex]
\longleftrightarrow & -\overset{\scriptstyle |}{\underset{\scriptstyle |}{C}}\cdot + \cdot X \quad\quad \text{Homolytic cleavage} \\[2ex]
\longrightarrow & -\overset{\scriptstyle |}{\underset{\scriptstyle |}{C}}:^{\ominus} + X^{\oplus} \quad \text{Heterolytic cleavage}
\end{cases}
$$

Bond-breaking processes in organic reactions

In many cases, the type of cleavage is dictated by the inherent bond polarities:

$$-\overset{\scriptstyle |}{\underset{\scriptstyle |}{C}}{}^{\delta^+}-X^{\delta^-} \quad \text{tends to yield} \quad -\overset{\scriptstyle |}{\underset{\scriptstyle |}{C}}{}^{\oplus}, \ :X^{\ominus}$$

$$-\overset{\scriptstyle |}{\underset{\scriptstyle |}{C}}{}^{\delta^-}-X^{\delta^+} \quad \text{tends to yield} \quad -\overset{\scriptstyle |}{\underset{\scriptstyle |}{C}}:^{\ominus}, \ X^{\oplus}$$

In others, the nature of the attacking reagent, say a strong base, predetermines which atom is removed:

$$\text{B:} \quad + -\overset{\scriptstyle |}{\underset{\scriptstyle |}{C}}-H \longrightarrow \text{B:}H^{\oplus} + -\overset{\scriptstyle |}{\underset{\scriptstyle |}{C}}:^{\ominus}$$

Strong base

↑
Most acidic H

The intermediates generated by these processes $C^+$, $C\cdot$, and $C:^-$ are referred to as *carbonium ions (carbocations)*, *free radicals*, and *carbanions*, respectively:

| | | |
|---|---|---|
| $-\overset{\scriptstyle \|}{\underset{\scriptstyle \|}{C}}{}^{\oplus}$ | Carbonium ion ⎱ Carbocation ⎰ | Generated in acidic or neutral media |
| $-\overset{\scriptstyle \|}{\underset{\scriptstyle \|}{C}}\cdot$ | Free radical | Generated in neutral media or gas phase by light, heat, or free-radical generators such as peroxides |
| $-\overset{\scriptstyle \|}{\underset{\scriptstyle \|}{C}}:^{\ominus}$ | Carbanions | Generated in very basic or highly polar neutral media |

If you can develop a feeling for the "how" and "why" of formation of these three classes of intermediates, you will be well on the road to a firm understanding of organic reaction processes.

# THE NATURE OF TRANSITION STATES

The main difference between a transition state and a true intermediate is that the former has no finite existence as a chemical entity. Full separation of charge, as in carbonium ions and carbanions, never occurs. Transition states, in fact, tend to *disperse* charge rather than localize it on a particular atom. Consider, for example, the displacement of the X group by an incoming negatively charged ion such as hydroxide. In the transition state, the hydroxide ion has begun to form a new bond to the central carbon, thereby transferring some of its charge to the X group, whose bond to the central carbon is beginning to break. Ideally, at the point of maximum potential energy (C—OH bond half-formed, C—X bond half-broken) the negative charge is now dispersed almost 50–50 over the incoming and outgoing groups. Thus the transition state in this example is less polar than either the reactants or the products. There are many different types of transition states involving dispersal of the negative and positive charge in organic chemistry, and some where no charge develops whatsoever. What they all have in common, however, is a smooth, concerted redistribution of electron density throughout the course of the reaction.

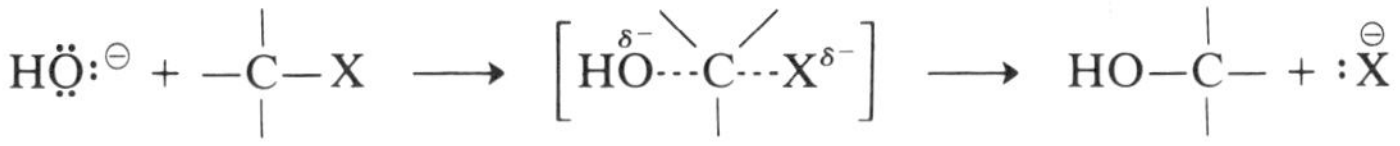

$$HO\!:^{\ominus} + -\overset{|}{\underset{|}{C}}-X \longrightarrow \left[ HO^{\delta-}\!\cdots\overset{|}{C}\cdots X^{\delta-} \right] \longrightarrow HO-\overset{|}{\underset{|}{C}}- + \;:\!\overset{\ominus}{X}$$

Reactants       Transition state       Products

Typical displacement reaction proceeding through a transition state

# ACIDS AND BASES IN ORGANIC CHEMISTRY

In *protic solvents*, such as water, the strongest acid and base that can exist in a given solvent are determined by *autoprotolysis*, or the transfer of a proton from one solvent molecule to another, as shown below for water and liquid ammonia. However, in most organic reactions *aprotic solvents*, such as ethers and hydrocarbons, are utilized and one cannot use the Lowry–Brønsted system of defining an acid as a *proton donor* and a base as a *proton acceptor*.

$$2\,H_2O \;=\; H_3O^{\oplus} \;+\; OH^{\ominus}$$
$$2\,NH_3 \;=\; NH_4^{\oplus} \;+\; NH_2^{\ominus}$$

Solvent      Strongest      Strongest
              acid          base

A more general definition of acids and bases was developed by G. N. Lewis. In the Lewis system, an acid is defined as an *electron-pair acceptor*, a base as an *electron-pair donor*. The advantage of these definitions is that they are neither tied down to a particular type of solvent (protic) nor are they limited to a single species (a proton). In fact, Lewis' approach is as close to a universal system as one could wish. A wide variety of acids and bases that do not fit the Lowry–Brønsted definition may easily be classified by the Lewis system, as shown below:

| Typical Lewis acids | Typical Lewis bases |
|---|---|
| $H^+$, $H_3O^+$ | $H:\ddot{O}:^-$, $R:\ddot{O}:^-$, $H_2\ddot{N}:^-$ |
| $Na^+$, $Ag^+$, $Mg^{+2}$ | $H_2\ddot{O}:$, $R\ddot{O}:$, $H_3N:$ |
| $M^{+n}$ (any metal cation) | $\quad\quad\quad$ H |
| $BF_3$, $AlCl_3$, $FeBr_3$, $SO_3$ | $:\ddot{Cl}:^-$, $H:^-$, $NO_3^-$ |
| (Any electron-deficient species) | (Any negative ion) |
| $NO_2^+$, $Br^+$, $R^+$ | $CH_2{=}CH_2$ |
| $CH_2{=}CH_2$ | |

It is interesting to note that the Lewis system also has its counterpart to the "amphoteric" compounds of the Lowry–Brønsted system, in that $CH_2{=}CH_2$ may function both as an electron pair donor and acceptor.

**Exercise 2-1:**  In the following reactions, identify the Lewis acids and bases in the reactants:

(a)  $Ag^+ + Cl^- = AgCl{\downarrow}$

(b)  $Fe^{+2} + 6\,H_2O = Fe(H_2O)_6^{+2}$

(c)  $H_3N: + BF_3 = H_3\overset{\oplus}{N}:\overset{\ominus}{B}F_3$

(d)  $ROH + H^+ = R\overset{+}{O}H_2$

(e)  $CH_3CH_2^+ + H_2O = CH_3CH_2\overset{+}{O}H_2$

## NUCLEOPHILES AND ELECTROPHILES

Although the concept of Lewis acidity and basicity is fundamental to our understanding of organic reactions, it may be further refined by also considering the *relative abilities* of functional groups to donate or accept electron pairs. Groups which can donate *e*-pairs are referred to as *nucleophiles* (nucleus-loving), while groups which can accept electrons are referred to as *electrophiles* (electron-loving). Although all negative ions are Lewis bases, their nucleophilicity varies over many orders of magnitude, and is not necessarily related to the base strength one is familiar with on the basis of proton chemistry. Thus $RS^-$ is a better nucleophile than $RO^-$, and $I^-$ is better than $Cl^-$. To a large extent, nucleophilicity is related to the *polarizability* of the group. On the other hand, electrophilicity is inversely related to size, the smaller member of a series being the stronger electrophile. The relationship between nucleophilic and electrophilic character and reaction mechanism is discussed in greater detail in Core Enrichment (Chap. II-1), along with an explanation of how an organic chemist utilizes such information in examining a reaction series. For the remainder of this book, we will refer to attacking reagents as nucleophiles or electrophiles, but the student should bear in mind that these terms are also related to the Lewis definitions of acids and bases. All Lewis acids are electrophiles, and all Lewis bases are nucleophiles.

## CLASSIFICATION OF REACTIONS IN ORGANIC CHEMISTRY

Reactions may be divided into categories by considering two facets of the reaction process: (1) what *happens* to the functional groups, and (2) the *nature* of the attacking reagent (nucleophilic or electrophilic). The major reaction types are summarized below:

1. Substitution—the replacement of one group by another:
   (a) Nucleophilic:

$$N\colon + R-X \longrightarrow N\colon R + \colon X$$
$$\colon \ddot{I}\colon^{\ominus} + R-Cl \longrightarrow I-R + \colon \ddot{C}l\colon^{\ominus}$$

   (b) Electrophilic:

2. Addition—$\pi$ bonds add a reagent XY to form two $\sigma$ bonds

Examples of both electrophilic and nucleophilic addition are known, but the former largely predominates.

3. Elimination—$\pi$ bond formed by elimination of XY
   (a) Nucleophilic:

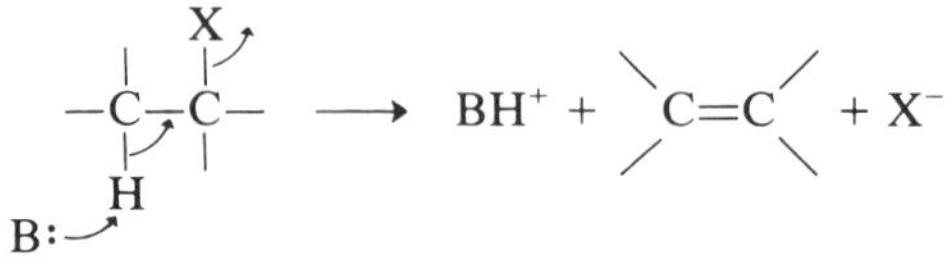

$$-\underset{\underset{\displaystyle B:}{\overset{\displaystyle H}{|}}}{\overset{\overset{\displaystyle X}{|}}{C}}\!-\!\overset{|}{\underset{|}{C}}\!- \longrightarrow BH^+ + \overset{\diagdown}{\diagup}C{=}C\overset{\diagup}{\diagdown} + X^-$$

$$HO^- + CH_3CH_2Br \xrightarrow{\Delta} H_2O + CH_2{=}CH_2 + Br^-$$

   (b) Electrophilic:

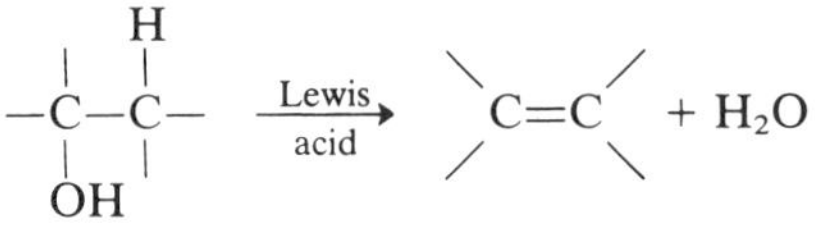

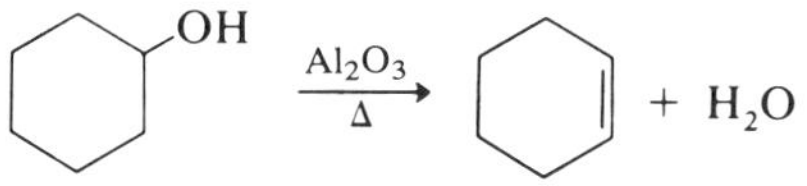

$$-\underset{\underset{\displaystyle OH}{|}}{\overset{\overset{\displaystyle H}{|}}{C}}\!-\!\overset{\overset{\displaystyle H}{|}}{\underset{|}{C}}\!- \xrightarrow[\text{acid}]{\text{Lewis}} \overset{\diagdown}{\diagup}C{=}C\overset{\diagup}{\diagdown} + H_2O$$

4. Oxidation—increase in carbon electrovalence by use of chemical (usually inorganic) reagents

   (a)

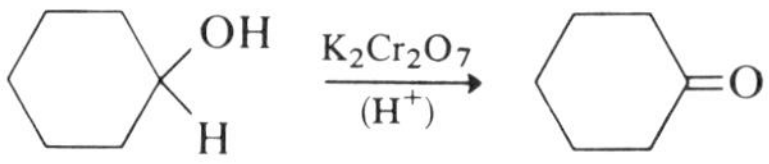

   (b)

$$CH_3CH_2CH_2OH \xrightarrow[\text{(OH}^-)]{KMnO_4} CH_3CH_2\overset{\overset{\displaystyle O}{\|}}{C}{-}OH$$

5. Reduction—decrease in carbon electrovalence by use of either chemical reducing reagents or catalytic hydrogenation

   (a)

$$CH_3CH_2\overset{\overset{\displaystyle O}{\|}}{C}{-}OH \xrightarrow{LiAlH_4} CH_3CH_2CH_2OH$$

   (b) $\quad CH_3CH{=}CHCH_3 \xrightarrow[\text{Pd}]{H_2} CH_3CH_2CH_2CH_3$

   (c)

6. Polymerization—conversion of monomer to polymer by the action of heat, light, free radical, cationic, or anionic pathways

   (a) Monomer

$$n\,A \longrightarrow \text{\textpm}A{-}A{-}A{-}A\text{\textpm}_n \quad \text{Polymer}$$

$$n \ CH_2{=}CHCl \longrightarrow \ {+}CH_2{-}CH{+}_n \quad \text{Polyvinyl chloride (PVC)}$$
$$\text{Vinyl chloride} \qquad\qquad\quad \underset{Cl}{|}$$

(b)  $n \ A + n \ B \longrightarrow \ {+}A{-}B{+}_n$

$$\underset{\text{Monomer}}{Cl{-}\overset{\overset{O}{\|}}{C}{+}CH_2{+}_4\overset{\overset{O}{\|}}{C}{-}Cl} + \underset{\text{Monomer}}{H_2N{+}CH_2{+}_6NH_2} \longrightarrow$$

$$\overset{\overset{O}{\|}}{+}\overset{}{C}{+}CH_2{+}_4\overset{\overset{O}{\|}}{C}{-}NH{+}CH_2{+}_6NH{+}_n$$
$$\text{Copolymer—Nylon 6,6}$$

7. Rearrangement—reorganization of molecular structure by heat, light, or catalysis—no additional chemical reagent is required.

(a)
$$\underset{\underset{CH_3}{|}\ \ \underset{CH_3}{|}}{CH_3{-}\overset{\overset{OH}{|}}{C}{-}\overset{\overset{OH}{|}}{C}{-}CH_3} \ \ \xrightarrow[\text{Catalyst}]{H^+} \ \ \underset{\underset{CH_3}{|}}{CH_3{-}\overset{\overset{CH_3}{|}}{C}{-}\overset{\overset{O}{\|}}{C}{-}CH_3} + H_2O$$

(b)

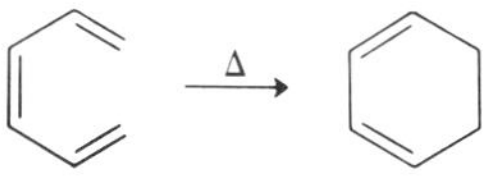

---

**Exercise 2-2:** Classify each of the following reactions as a substitution, elimination, addition, oxidation, or reduction.

(a)  $CH_3CH_2CH_2CH_2I + NaOCH_3 \longrightarrow CH_3CH_2CH_2CH_2OCH_3 + NaI$

(b)  $CH_3CHCH_2CH_2CH_3 + KOH \xrightarrow{\Delta} CH_3CH{=}CHCH_2CH_3 + KBr$
$\quad\ \underset{Br}{|}$

(c)  $CH_3CH_2CH_2CH{=}CH_2 + KMnO_4 \xrightarrow{\Delta} CH_3CH_2CH_2COOH + CO_2 + MnO_2$

(d)  $CH_3CH_2CH_2CH_2NO_2 \xrightarrow[H_2]{Pd} CH_3CH_2CH_2CH_2NH_2$

(e)  $CH_3C{\equiv}CCH_3 + 2 \ Br_2 \longrightarrow \underset{\underset{Br\ \ Br}{|\ \ \ |}}{CH_3\overset{\overset{Br\ \ Br}{|\ \ \ |}}{C}{-}CCH_3}$

(f)  $CH_3CH_2CH_2\overset{\overset{O}{\|}}{C}NH_2 \xrightarrow[\Delta]{P_2O_5} CH_3CH_2CH_2C{\equiv}N$

(g)  $CH_3CH_2CH_2OH \xrightarrow[\Delta]{H^+} CH_3CH{=}CH_2$

# FORMATION OF INTERMEDIATES

## Carbonium Ions

Carbonium ions may be formed from a large number of different organic entities. In general they are formed either by an ionization step or by the addition of an electrophile to a double bond. In the former, the stability of the leaving group

$$(1) \quad RCH=CH_2 \underset{}{\overset{X^{\oplus}}{\rightleftharpoons}} R\overset{\oplus}{C}HCH_2X, \quad X = H^{\oplus}, Br^{\oplus}, Cl^{\oplus}, \text{etc.}$$

$$(2) \quad R-X \rightleftharpoons R^{\oplus} + X^{\ominus}, \quad X = \text{halogen}, -OSO_2-\langle\bigcirc\rangle, \text{etc.}$$

$$(3) \quad R\overset{\oplus}{O}H_2 \longrightarrow R^{\oplus} + H_2O$$

$$(4) \quad R-N\equiv N^{\oplus} \longrightarrow R^{\oplus} + N_2\uparrow$$

Methods of carbonium-ion generation

affects the ease of ionization dramatically. For example, in (2) the position of equilibrium greatly favors RX except in highly polar media, while the reaction in (4) is essentially irreversible. In the addition of an electrophile to an unsymmetrical double bond, two different carbonium ions are possible. As in most equilibria, the addition produces the *more stable* intermediate. One can predict the course of

$$RCH=CH_2 \underset{\overset{X^{\oplus}}{\searrow}}{\overset{X^{\oplus}}{\nearrow}} \begin{array}{l} R\overset{\oplus}{C}HCH_2X \\ \\ R\overset{\oplus}{C}HCH_2 \\ \quad | \\ \quad X \end{array}$$

addition by considering which ion is more highly substituted, as the stability seems to increase as substitution increases. We will discuss this in greater detail in Chapter I-5.

$$\begin{array}{c} R \\ | \\ R-\overset{\oplus}{C} \\ | \\ R \end{array} > \begin{array}{c} H \\ | \\ R-\overset{\oplus}{C} \\ | \\ R \end{array} > \begin{array}{c} H \\ | \\ R-\overset{\oplus}{C} \\ | \\ H \end{array} > CH_3^{\oplus}$$

3° (Tertiary) > 2° (Secondary) > 1° (Primary) > Methyl

Stability of carbonium ions

## Free Radicals

Free radicals may be formed by the action of heat or light, or by interaction with free-radical initiators such as peroxides. When heat or light is utilized to generate radicals, the most labile bond in the molecule tends to dissociate. However, radical generators form organic radicals by removing halogens or the *most reactive hydrogens*. Since radicals are electron deficient, as are carbonium ions, the factors influencing their stability are similar.

(1)   $CH_3CH_3 \xrightarrow[400-500°]{\Delta} 2\ \dot{C}H_3$

(2)   $X\cdot + CH_4 \longrightarrow CH_3\cdot + H{-}X$   X = halogen or other stable radical

(3)   $R{-}N{=}N{-}R \xrightarrow{h\nu} 2\ R\cdot + N_2\uparrow$

$$R_3C\cdot > R_2CH\cdot > RCH_2\cdot > CH_3\cdot$$
$$3°\qquad 2°\qquad 1°$$

Stability of free radicals

$$\underset{\underset{R}{|}}{\overset{\overset{R}{|}}{R{-}C{-}H}} > \underset{\underset{H}{|}}{\overset{\overset{R}{|}}{R{-}C{-}H}} > \underset{\underset{H}{|}}{\overset{\overset{H}{|}}{R{-}C{-}H}} > CH_4$$

Reactivity of hydrogens to free radical formation

Some well-known free-radical generators (initiators) are shown below.

(1)

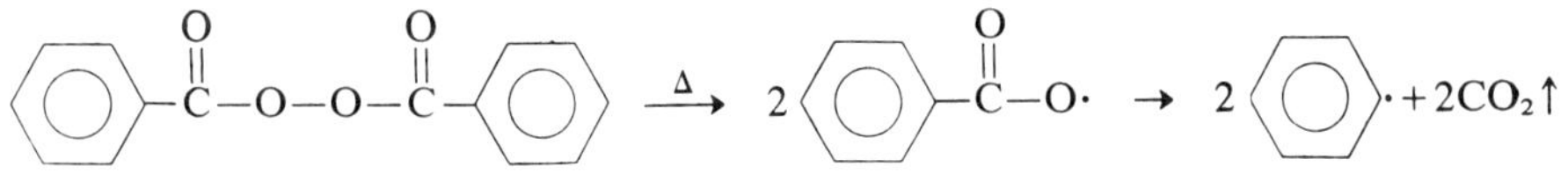

Benzoyl peroxide

(2)   $(CH_3)_3C{-}O{-}O{-}C(CH_3)_3 \xrightarrow{\Delta} 2\ (CH_3)_3CO\cdot$

(3)   $(CH_3)_2\underset{\overset{|}{CN}}{C}{-}N{=}N{-}\underset{\overset{|}{CN}}{C}(CH_3)_2 \xrightarrow[h\nu]{\Delta\ or} 2\ (CH_3)_2\underset{\overset{|}{CN}}{C}\cdot + N_2\uparrow$

AIBN (Azoisobutyronitrile)

## Carbanions

Carbanions are extremely powerful bases and nucleophiles. As such, they are more difficult to form unless other groupings are present in the structure. By far the most common class of carbanionic compounds are organometallic in nature, the alkyllithium reagents being the most stable. These reagents are so reactive, they are normally handled as solutions in inert hydrocarbon reagents such as pentane or

$$R{-}Br + 2\ Li = RLi + LiBr$$

$$CH_3CH_2CH_2CH_2Br + 2\ Li = CH_3CH_2CH_2CH_2Li + LiBr$$

*n*-Butyllithium

hexane. The stability order for carbanions is *opposite* to that for carbonium ions and

$$:CH_3^{\ominus} > \overset{R}{\underset{|}{\overset{}{C}}}H_2^{\ominus} > R_2\overset{\ominus}{C}H > R_3C:^{\ominus}$$

Methide > 1° > 2° > 3°

Stability of carbanions

free radicals. Ease of formation of carbanions increases as electron-withdrawing groups such as halogen are substituted on the carbon.

$$CH_4 \xrightarrow{\text{NaOH}} \text{No reaction}$$

$$Cl_3CH \xrightarrow{\text{NaOH}} [Cl_3C:^{\ominus}] \longrightarrow \text{Further reaction}$$

---

**Exercise 2-3.** In each of the following, arrange the organic intermediates in the order of their stability.

(a)

$$(CH_3)_2CH\overset{\oplus}{C}H_2, \quad CH_3CH_2\overset{\oplus}{C}(CH_3)_2, \quad CH_3\overset{\overset{\textstyle CH_3}{|}}{C}H\overset{\oplus}{C}HCH_3,$$

$$CH_3CH_2CH_2\overset{\oplus}{C}H_2$$

(b)

⬠·· ,  ⬠—CH$_3$, CH$_3$CH$_2$ĊHCH$_3$, CH$_3$CH$_2$CH$_2$ĊH$_2$

(c)

$$(CH_3)_3C:^{\ominus}, \quad CH_3CH_2CH_2CH_2:^{\ominus}, \quad CH_3CH_2\overset{\ominus}{C}HCH_3, \quad CH_3CH_2\overset{\overset{\textstyle Cl}{|}}{\underset{\underset{\textstyle Cl}{|}}{C}}:^{\ominus}$$

---

# THE ROLE OF CATALYSTS IN ORGANIC REACTIONS

Many organic reactions are extremely slow in the absence of a *catalyst*. The mechanism by which some catalysts function is still relatively obscure, but in general they lower the energy barrier to product formation by enabling an intermediate to be formed, rather than a transition state, or by lowering the energy for intermediate formation. By far the most common types of catalyses are those generated by *acids* and *bases*. As you might expect from our previous discussion, acids promote the formation of carbonium ions, and bases increase carbanion concentration, as illustrated below. Catalysts increase the rate of product formation, but they do not alter the *position* of equilibria. Also, they are neither altered nor destroyed in a reaction

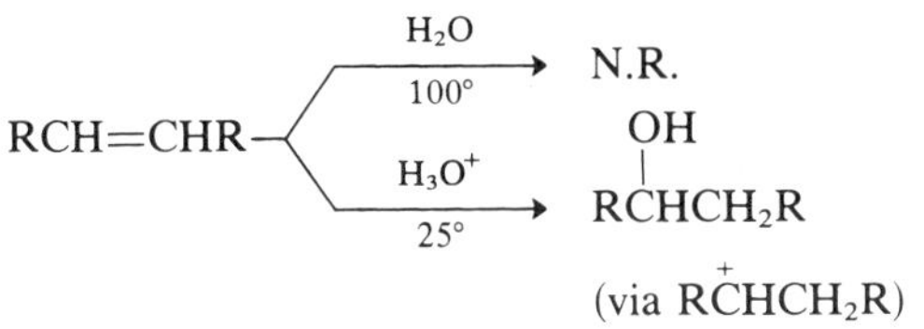

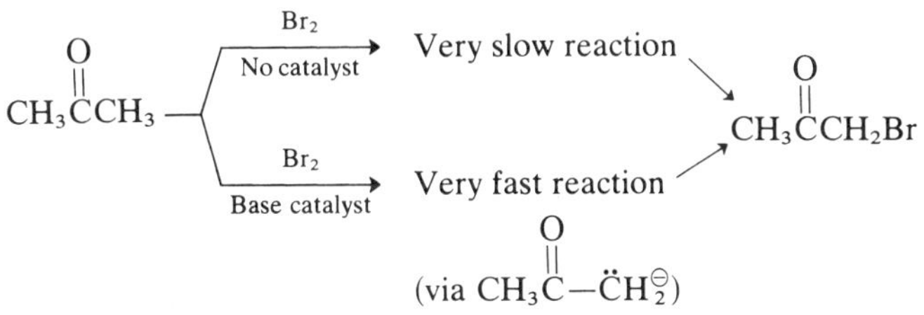

$$RCH{=}CHR \left\langle \begin{array}{l} \xrightarrow[\substack{100°}]{H_2O} \text{N.R.} \\ \\ \xrightarrow[\substack{25°}]{H_3O^+} \underset{\substack{|\\ \text{OH}}}{R\overset{}{CH}CH_2R} \\ \\ \text{(via } R\overset{+}{C}HCH_2R) \end{array} \right.$$

$$\underset{\substack{\|\\O}}{CH_3\overset{}{C}CH_3} \left\langle \begin{array}{l} \xrightarrow[\substack{\text{No catalyst}}]{Br_2} \text{Very slow reaction} \\ \\ \xrightarrow[\substack{\text{Base catalyst}}]{Br_2} \text{Very fast reaction} \end{array} \right. \underset{\substack{\|\\O}}{CH_3\overset{}{C}CH_2Br}$$

$$\text{(via } CH_3\underset{\substack{\|\\O}}{\overset{}{C}}{-}\overset{..}{C}H_2^{\ominus})$$

process. The most efficient and marvelous catalyst systems of all are enzymes, which are able to carry out far more complicated processes than any yet achieved in the laboratory. Enzymatic activity depends not only on the functional-group chemistry of a reaction, but also on the *stereochemistry*, or the arrangement of the functional groups in three-dimensional space, which is discussed in Chapter I-4.

## REVIEW OF NEW TERMS AND CONCEPTS

Define each term and, if possible, give an example of each of the following:

| | |
|---|---|
| Transition state | Intermediate |
| Heterolytic bond cleavage | Homolytic bond cleavage |
| Carbonium ion (carbocation) | Free radical |
| Carbanion | Autoprotolysis |
| Lowry–Brønsted acid–base system | Lewis acid |
| Lewis base | Nucleophile |
| Electrophile | Substitution reaction |
| Addition Reaction | Elimination reaction |
| Oxidation | Reduction |
| Polymerization | Rearrangement |
| Catalyst | |

**1.** Many organic reactions are really simple equilibria:

$$A + B \; \rightleftharpoons \; C + D$$

Reactants          Products

A typical example is the acid-catalyzed reaction between an acid and an alcohol to form an ester. How would you shift the equilibrium one way or the other? For the reaction between butyric acid and methanol, explain how you would maximize the yield of ester. If you had a quantity of pure ester, how would you hydrolyze it back to the acid plus alcohol?

$$CH_3OH + CH_3CH_2CH_2COOH \; \overset{H^+}{\rightleftharpoons} \; CH_3CH_2CH_2\overset{\displaystyle O}{\overset{\displaystyle \|}{C}}OCH_3 + H_2O$$

Alcohol         Acid                 Ester

**2.** In the transition state for a substitution reaction, why is it necessary for bond-making and bond-breaking to be occurring simultaneously? Consider the alternatives and discuss each in detail:

(a) Bond-breaking (leaving group) occurs much faster than bond-making.
(b) Bond-making (entering group) occurs much faster than bond-breaking.

**3.** In a *neutral* solution ROH does not react with NaI to produce RI, but in acidic solution ROH does yield RI as a final isolable product. How would you explain this result?

**4.** One of the reasons tertiary carbonium ions form in preference to 2° or 1° carbonium ions is that they are more stable. A possible explanation for their greater stability is that the C—H bond is really slightly polarized $C^{\delta\ominus}-H^{\delta\oplus}$. Explain how this reasoning would lead you to predict that the *t*-butyl carbonium ion is more stable than $CH_3^+$.

**5.** Would you expect HCl to be a stronger acid in $H_2O$ or in a hydrocarbon solvent? Explain your answer.

**6.** Write autoprotolysis reactions for the following solvents. In each case identify the strongest base that can exist in that solvent.

(a) $CH_3CH_2OH$

(b) $CH_3CH_2\overset{\displaystyle O}{\overset{\displaystyle \|}{C}}OH$

(c) $CH_3SH$

(d) $\langle\!\!\!\bigcirc\!\!\!\rangle\!-\!NH_2$

(e) $H_2SO_4$

(f) $HNO_3$

(g) $CH_3CH_2NHCH_2CH_3$

(h) $\langle\!\!\!\bigcirc\!\!\!\rangle\!-\!SO_3H$

(i) $D_2O$

7. Identify each of the following as either a Lewis acid, a Lewis base, or both:

(a) $H_2O$

(b) $Li^{\oplus}$

(c) $AlCl_3$

(d) $FeCl_4^{\ominus}$

(e) $CH_3CH_2^{\oplus}$

(f) $Cl^{\ominus}$

(g) $BF_3$

(h) $GaCl_3$

(i) $NCl_3$

(j)

(k) $H^{\ominus}$

(l) $CH_3CH_3$

8. Oxidation–reduction reactions in organic chemistry can be balanced in the same manner as inorganic reactions, using a few additional rules: (1) C—C bonds are not counted as part of the electrovalence; (2) C atoms not involved in electron gain or loss are ignored; and (3) H is assumed to have an electrovalence of +1 and O is −2.

(a) Calculate the valence of each individual carbon in the following:

(i)
$$\underset{\uparrow}{CH_3}\overset{OH}{\underset{|}{C}}HCH_3$$

(ii) $CH_3CH_2\overset{\uparrow}{CH}=\overset{\uparrow}{CH_2}$

(iii)
$$CH_3-\overset{O}{\overset{||}{\underset{\uparrow}{C}}}-H$$

(iv)
$$CH_3CH_2\overset{O}{\overset{||}{\underset{\uparrow}{C}}}-OH$$

(b) Balance each of the following in terms of electron gain or loss

(i)
$$\overset{OH}{\underset{|}{CH_3CHCH_3}} + MnO_4^- = \overset{O}{\overset{||}{CH_3CCH_3}} + MnO_2$$
In basic solution

(ii)
$$CH_3CH_2CH=CH_2 + Cr_2O_7^{-2} = \overset{O}{\overset{||}{CH_3CH_2C}}-OH + CO_2 + Cr^{+3}$$
In acidic solution

*Hint:* In order to balance an oxidation–reduction reaction you should follow this procedure:

(a) Balance *e*-gain with *e*-loss;
(b) Balance elements involved on both sides of equation;
(c) Balance charges by adding $H^+$ or $OH^-$ to either side to make the equation electrically neutral;
(d) Balance hydrogens by adding $H_2O$ to appropriate side of the equation;
(e) Check to see if the equation is now balanced by comparing number of oxygens on either side;
(f) Equation should now be balanced!

9. A free-radical initiator may abstract hydrogens from hydrocarbons to form hydro-carbon free radicals. Keeping in mind that hydrogens will be removed so as to form

the *most stable free radical*, predict the products of hydrogen abstraction from each of the following.

(a)

$$CH_3-\underset{\underset{CH_3}{|}}{\overset{\overset{CH_3}{|}}{C}}-H$$

(b) —CH$_3$

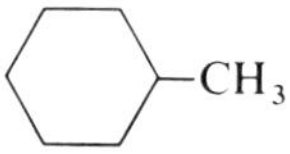

(c) CH$_3$CH$_2$CH$_2$CH$_3$

(d) CH$_3$CH=CH$_2$

(e) 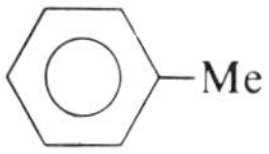 —Me

(f) (CH$_3$)$_2$CHC(CH$_3$)$_3$

10. Arrange the following in the order of increasing base strength and identify your choices:

(a) H$_2$O    (b) NH$_3$    (c) HCl    (d) $\overset{\ominus}{N}H_2$    (e) HO$^{\ominus}$    (f) $\overset{\ominus}{C}H_3$    (g) $\overset{\ominus}{F}$

# 3

# Alkanes:
## The "Saturated" Hydrocarbons

Most studies of organic chemistry begin with the alkanes, the most fundamental and least complicated of the structural units we will investigate. Alkanes are composed of carbon and hydrogen only, hence the designation *hydrocarbon*. They are also referred to as being *saturated*, a term which indicates that all of the carbons are bound to each other and to hydrogen by single ($\sigma$) bonds. Historically, this term also referred to the inability of alkanes to absorb further quantities of hydrogen. In this chapter we will learn how these compounds are constructed and how they may be synthetically prepared. We will also learn that one of their unique properties is their inertness to most chemical reagents other than halogens. *Systematic nomenclature* will be introduced in this chapter and the concept of *structural* or *constitutional isomerism* reviewed. Finally, we will discuss the occurrence of hydrocarbons in nature, and how they may be recovered, purified, and structurally manipulated by modern industrial processes.

## REVIEW OF ALKANE STRUCTURE

Alkanes, or saturated hydrocarbons, are organic molecules composed only of hydrogen and carbon. The carbon backbone is composed of $sp^3$-hybridized carbon atoms linked by $\sigma$ bonds resulting from $sp^3$–$sp^3$ orbital overlap. As previously discussed, the tetrahedral nature of carbon bonding results in carbon chains having zigzag configurations, as shown below for a typical alkane. Although the chains are

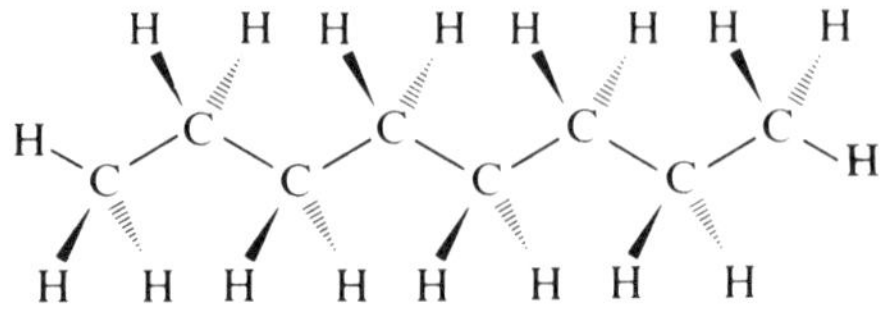

Octane: $C_8H_{18}$

obviously not straight, as can readily be seen in Fig. 3-1, alkanes are often referred to as "straight-chain" hydrocarbons to distinguish them from cyclic structures. Alkane structure corresponds to the general class formula $C_nH_{2n+2}$, thus the hydrogen content of any hydrocarbon may be calculated for any given number of carbons. A close look at the molecular formulas

> For an alkane containing 10 carbons there are $(2 \times 10) + 2 = 22$ hydrogens. Formula is $C_{10}H_{22}$.

of the alkane series indicates that each formula differs from the preceding formula by a constant unit, $CH_2$.

$C_3H_8$
$\phantom{C_3H_8}\rangle CH_2$
$C_4H_{10}$
$\phantom{C_4H_{10}}\rangle CH_2$
$C_5H_{12}$
$\phantom{C_5H_{12}}\rangle CH_2$
$C_6H_{14}$

A homologous series of compounds differs by a constant unit as we proceed from one member to another.

Any structural series whose members differ from one another by such a constant unit is referred to as a *homologous series*, and the individual members of the series are called *homologs*.

## PHYSICAL PROPERTIES OF ALKANES

Alkanes are nonpolar. They are almost totally insoluble in water and other highly polar associated solvents. Ionic or polar compounds have very little solubility in hydrocarbon solvents, but relatively nonpolar compounds such as fats and oils, waxes, and greases are readily soluble. Alkanes have lower boiling points for a given molecular weight than other classes of organic compounds, and even large hydrocarbons such as $C_{20}H_{42}$ have low melting points (36°C). Many waxes isolated from natural sources are essentially mixtures of low-melting hydrocarbon ($C_{20}$–$C_{40}$) chains ending in various functional groups, while the waxy portion of many plant and

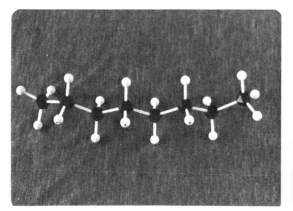

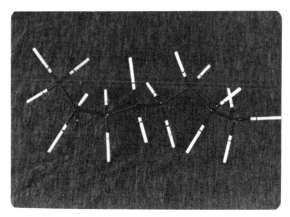

**Figure 3-1.** Molecular models of octane.

fruit cuticles contains large concentrations of $C_{20}$–$C_{40}$ hydrocarbons such as $n$-nonacosane ($C_{29}H_{60}$) and $n$-hentriacontane ($C_{31}H_{64}$). The presence of these constituents explains many of the properties we associate with waxes, such as nonsolubility in water and the relative ease of application on various nonpolar surfaces (for example, wood). Table 3-1 lists some physical properties of a few typical alkanes.

## STRUCTURAL ISOMERISM REVISITED

In Chapter I-1 we discussed the concept of structural isomerism. For any given alkane structure $C_nH_{2n+2}$ where $n$ is greater than 3, we can write more than one formula. There are two butanes, $C_4H_{10}$, and three pentanes, $C_5H_{12}$, and as the carbon content of the formula increases, the number of structural isomers increases

Table 3-1.  Physical Properties of Alkanes

| Name | Formula | M.p. (°C) | B.p. (°C) | Density (20°—liquid) |
|---|---|---|---|---|
| Methane | $CH_4$ | −183 | −162 | gas |
| Ethane | $CH_3CH_3$ | −172 | −88 | gas |
| Propane | $CH_3CH_2CH_3$ | −187 | −45 | gas |
| Butane | $CH_3(CH_2)_2CH_3$ | −138 | 0 | gas |
| Pentane | $CH_3(CH_2)_3CH_3$ | −130 | 36 | 0.626* |
| Hexane | $CH_3(CH_2)_4CH_3$ | −94 | 69 | 0.659 |
| Decane | $CH_3(CH_2)_8CH_3$ | −30 | 174 | 0.730 |
| Hexadecane | $CH_3(CH_2)_{14}CH_3$ | 18 | 280 | 0.775 |
| Isopentane (2-methylbutane) | $CH_3CHCH_2CH_3$ $\vert$ $CH_3$ | −160 | 28 | 0.620 |
| Neopentane (2,2-dimethylpropane) | $(CH_3)_4C$ | −17 | 9.5 | 0.613 |
| 2,3-Dimethylbutane | $(CH_3)_2CHCH(CH_3)_2$ | −129 | 58 | 0.668 |

* Note that all liquid alkanes are lighter than water—they will float on top of a water layer.

dramatically.  In the traditional naming system wherein compounds were named according to their *total* carbon content, this became an insurmountable problem for the larger molecules.  In this chapter we will see how the IUPAC system solves this

$$CH_3CH_2CH_2CH_3 \qquad CH_3\overset{\overset{\displaystyle CH_3}{\vert}}{C}HCH_3$$

Two isomeric butanes

$$CH_3CH_2CH_2CH_2CH_3 \qquad CH_3\overset{}{C}HCH_2CH_3 \qquad CH_3\overset{}{C}CH_3$$

Three isomeric pentanes

problem by naming the compound according to the *longest* carbon chain in the molecule, rather than trying to name the structure as a whole.  Though compounds having the same total number of carbon atoms may have quite different parent names, we can be assured that each compound can be named, and that each name is unique.


## ALKANE NOMENCLATURE

In Chapter I-1 the necessity for a systematic method of naming organic compounds was discussed, and the parent names for hydrocarbon chains were introduced.  The IUPAC system of nomenclature must be mastered before organic chemistry can be

42

discussed intelligently. The rules for naming organic compounds are relatively easy to use, but you must practice their application in order to feel comfortable with them. In the following section we will discuss these rules and demonstrate their application to a wide variety of alkane structures.

*Rule 1:* Identify the *longest* continuous chain of carbon atoms in the molecule and name this chain according to the *number* of carbon atoms in it (see Table I-1). Note that in writing condensed structural formulas it is *not* necessary that the longest chain be written in a straight line. This is a mistake made by many beginning students in organic chemistry.

Example 3-1:

$$CH_3$$
$$\text{---}CH_3CH_2CHCHCH_2CH_3 \rightarrow$$
$$CH_3$$

Longest chain contains 6 carbons
Parent name: hexane

$$C$$
$$\text{---}C\text{---}C\text{---}C\text{---}C\text{---}C$$
$$C\text{---}C\text{---}C\text{---}C \rightarrow$$
$$C$$

Longest chain contains 7 carbons
Parent name: heptane

**Exercise 3-1:**  Identify the longest chain in each of the following compounds:

(a) $CH_2CH_3$
$CH_3CH_2CHCH_2CH_2CHCH_3$
$CH_2CH_2CH_3$

(b) $CH_2CH_3$
$CH_3CH_2CH_2CCH_2CH_3$
$CH_2CH_2CH_3$

*Rule 2:* The groups which exist as branches off the main chain are named as *alkyl groups*, and the parent chain is *numbered* so as to give the substituent alkyl groups the *lowest possible numbers*.

Alkyl group names are obtained by removing the *-e* from the alkane name, and replacing it with *-yl*.

| Alkane name | Formula | Alkyl name | Formula |
|---|---|---|---|
| Methane | $CH_4$ | Methyl | $CH_3-$ |
| Ethane | $CH_3CH_3$ | Ethyl | $CH_3CH_2-$ |

Example 3-2:

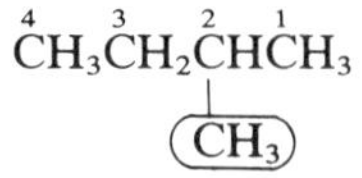

$$\overset{4}{C}H_3\overset{3}{C}H_2\overset{2}{C}H\overset{1}{C}H_3$$     Parent name: butane
                                        Substituent group: methyl

⏐
(CH₃)

IUPAC name: 2-methylbutane

Notice that it is necessary to number the chain from right to left in this example. If we had numbered the chain from left to right, the name would be 3-methylbutane, which is incorrect because the substituent group does not have the lowest possible number. When in doubt about how to number the chain, do it both ways and pick the name with the lowest substituent group numbers.

Example 3-3:

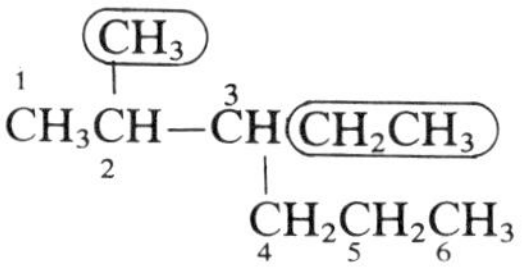

(CH₃)
$$\overset{1}{C}H_3\overset{}{C}H-\overset{3}{C}H(CH_2CH_3)$$     Parent name: hexane
    2
       $$CH_2CH_2CH_3$$       Substituent groups: 2-methyl
        4    5    6                              3-ethyl

IUPAC name: 3-ethyl-2-methylhexane

Note that this name could also be written 2-methyl-3-ethylhexane without violating either Rule 1 or Rule 2. The order in which the substituent groups are named may be either *alphabetical* or in *order of size* (molecular complexity). Which order was followed in the above example?

===

**Exercise 3-2:**   Name the following compounds using the IUPAC system:

(a) 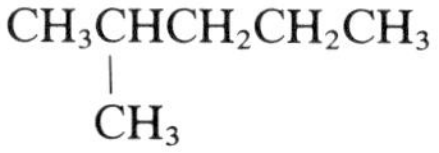  $CH_3CHCH_2CH_2CH_3$                    (b) 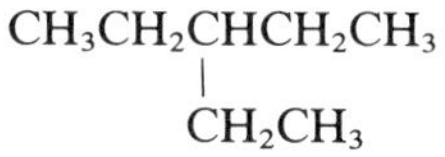  $CH_3CH_2CHCH_2CH_3$
         ⏐                                              ⏐
         $CH_3$                                          $CH_2CH_3$

===

Example 3-4:

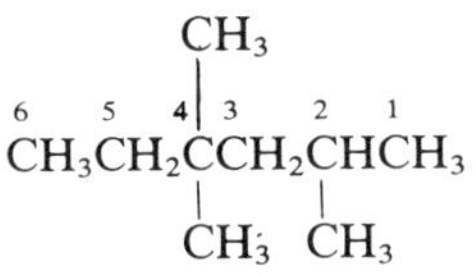

        $CH_3$
         ⏐                        Parent name: hexane
$$\overset{6}{C}H_3\overset{5}{C}H_2\overset{4}{C}\overset{3}{C}H_2\overset{2}{C}H\overset{1}{C}H_3$$     Substituent groups: 2-methyl
         ⏐    ⏐                                          4-methyl
        $CH_3$  $CH_3$                                   4-methyl

IUPAC name: 2,4,4-trimethylhexane

In Example 3-4, several items should be noted. Even though the substituent groups on carbon atom 4 are identical, it is necessary to indicate this by repeating the positional designation (4) for *each* of these methyls. Similarly, the third methyl group is indicated as being on carbon atom 2. Therefore, in the final name, we indicate a numerical position for *each* of the three methyl groups (*tri*methyl). Similarly, two methyls on the same carbon are referred to as "dimethyl," as in "2,2-dimethyl," and four methyl groups in a chain as "tetramethyl," as in "2,2,4,4-tetramethyl." Please note the use of commas and hyphens in the above examples—their correct use is also part of the nomenclature system.

$$\begin{array}{c} CH_3 \\ | \\ -\underset{2}{C}- \\ | \\ CH_3 \end{array} \quad 2,2\text{-Dimethyl} \qquad \begin{array}{ccc} CH_3 & & CH_3 \\ | & 3 & | \\ -\underset{2}{C}-CH_2-\underset{4}{C}- \\ | & & | \\ CH_3 & & CH_3 \end{array} \quad 2,2,4,4\text{-Tetramethyl}$$

**Exercise 3-3:** Name the following compounds using the IUPAC system:

(a)
$$\begin{array}{c} CH_3 \\ | \\ CH_3CCH_2CH(CH_3)_2 \\ | \\ CH_3 \end{array}$$

(b) $(CH_3CH_2)_3CCH_2CH_3$

(c)
$$\begin{array}{c} CH_3 \\ | \\ CH_3CHCHCH_2CH_3 \\ | \\ CH_2CH(CH_3)_2 \end{array}$$

(d) $(CH_3CH_2)_2C(CH_3)_2$

*Hint:* Remember that $(CH_3)_2CH-$ means $CH_3-\overset{\overset{\textstyle CH_3}{\textstyle |}}{CH}-$

## ALKYL GROUPS AND COMMON NAMES

Before the advent of the IUPAC system, alkanes were named rather haphazardly by a nonsystem now referred to as "common" names. Although such usage is discouraged today, the names of simple alkyl groups from $C_1$ to $C_5$ derived from these common names provide us with an easily used and understood nomenclature. In practice, the following common alkyl group nomenclature is still quite useful:

| Hydrocarbon | Alkyl groups | Alkyl group name |
|---|---|---|
| $CH_3CH_2CH_3$ Propane | $CH_3CH_2CH_2-$ $CH_3\overset{|}{CH}CH_3$ | *n*-propyl (Read: normal propyl) *i*-propyl (Read: isopropyl) |

| Hydrocarbon | Alkyl groups | Alkyl group name |
|---|---|---|
| $CH_3CH_2CH_2CH_3$ | $CH_3CH_2CH_2CH_2-$ | *n*-butyl<br>(Read: normal butyl) |
| *n*-Butane | $CH_3\overset{\mid}{C}HCH_2CH_3$ | *s*-butyl<br>(Read: secondary butyl) |
| $CH_3\overset{\overset{\displaystyle CH_3}{\mid}}{C}HCH_3$ | $CH_3\overset{\overset{\displaystyle CH_3}{\mid}}{C}HCH_2-$ | *i*-butyl<br>(Read: isobutyl) |
| Isobutane | $CH_3\overset{\overset{\displaystyle CH_3}{\mid}}{\underset{\mid}{C}}CH_3$ | *t*-butyl or *tert*-butyl<br>(Read: tertiary butyl) |

In the preceding examples, note that the different alkyl groups are obtained by successive removal of *each* different hydrogen in the alkane. Hydrogen may be designated as primary (1°), secondary (2°), or tertiary (3°), depending on the type of carbon to which it is attached.

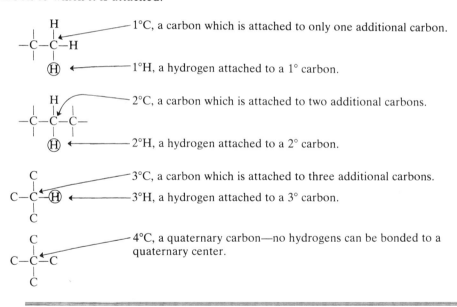

— 1°C, a carbon which is attached to only one additional carbon.

— 1°H, a hydrogen attached to a 1° carbon.

— 2°C, a carbon which is attached to two additional carbons.

— 2°H, a hydrogen attached to a 2° carbon.

— 3°C, a carbon which is attached to three additional carbons.

— 3°H, a hydrogen attached to a 3° carbon.

— 4°C, a quaternary carbon—no hydrogens can be bonded to a quaternary center.

**Exercise 3-4:** In each of the following compounds identify the number of *different* kinds of hydrogens and draw structures for each *different* alkyl group which could be generated by hydrogen removal.

(a) $CH_3CH_2\overset{\overset{\displaystyle }{\mid}}{C}HCH_3$    (b) $CH_3CH_2CH_2CH_2CH_3$    (c) $(CH_3)_4C$
      $\overset{\mid}{C}H_3$

Some additional common names that are still used occasionally are:

$$\begin{matrix} CH_3 \\ | \\ CH_3CCH_2- \\ | \\ CH_3 \\ \text{Neopentyl} \end{matrix} \qquad \begin{matrix} CH_3 \\ | \\ CH_3CCH_2CH_3 \\ | \\ \\ \textit{tert}\text{-amyl} \\ (\textit{tert}\text{-pentyl}) \end{matrix} \qquad \begin{matrix} CH_3CH_2CH_2CH_2CH_2- \\ \\ n\text{-Amyl} \\ (n\text{-Pentyl}) \end{matrix}$$

The *iso-* designation is a curiosity in that it refers to a particular terminal chain structure rather than the type of hydrogen removed. We will see in subsequent

$$\boxed{\begin{matrix} CH_3CH- \\ | \\ CH_3 \end{matrix}} \qquad \text{The \textit{iso}-grouping at the end of a chain}$$

$$\begin{matrix} CH_3CHCH_2- \\ | \\ CH_3 \end{matrix} \qquad \text{Isobutyl}$$

$$\begin{matrix} CH_3CHCH_2CH_2- \\ | \\ CH_3 \end{matrix} \qquad \text{Isopentyl or isoamyl}$$

chapters that the *iso* nomenclature is also used in the common names of aldehydes, carboxylic acids, and the various derivatives of carboxylic acids.

## SYNTHESIS OF ALKANES

Although most alkanes of any commercial importance may be obtained from either natural gas or petroleum deposits, their separation is tedious and totally impractical on anything smaller than an industrial scale. In the laboratory we are more interested in preparing *pure* hydrocarbons than in preparing large quantities, therefore organic chemists have developed *synthetic* methods of producing the desired hydrocarbon from readily available (and, if possible, cheap) starting materials. Alkanes may be prepared from a variety of compounds, but the most important procedures start from an alkyl chloride or bromide, as outlined below. Some of these procedures are associated with the name of their discoverer, and are referred to as *name reactions*.

(1)

$$2\,R-Br + 2\,Na \xrightarrow{\Delta} R-R + 2\,NaBr$$

$R = $ any alkyl group     $\{$ Wurtz reaction $\}$

Examples:

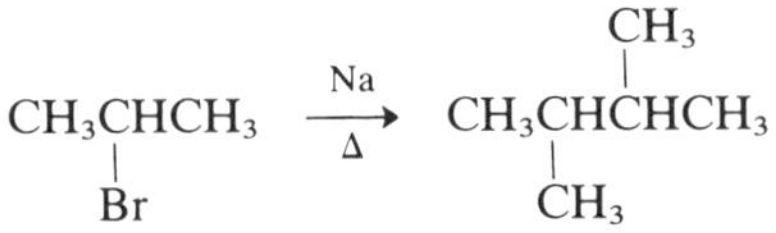

$$CH_3CHCH_3 \xrightarrow[\Delta]{Na} CH_3CHCHCH_3$$
with Br below left and CH$_3$ top/bottom on right.

$$CH_3CH_2CH_2Cl \xrightarrow[\Delta]{Na} CH_3CH_2CH_2CH_2CH_2CH_3$$

In essence, the Wurtz reaction joins two alkyl groups together.  The reaction is known to proceed stepwise

(a)     $RBr + 2\,Na \rightarrow [RNa] + NaBr$

(b)     $RBr + [RNa] \rightarrow R{-}R + NaBr$

through an extremely reactive organosodium intermediate and thus is useful for the preparation of pure hydrocarbons having only *even-numbered* chains.

**Exercise 3-5:**  Why isn't the Wurtz reaction a feasible method of preparing hydrocarbons having odd-numbered chains?  What would be the final product (and product distribution) in the following reaction?

$$CH_3CH_2Br + CH_3CH_2CH_2Br \xrightarrow[\Delta]{Na}$$

(2)

$$R{-}Br \xrightarrow[ether]{Mg} [RMgX] \xrightarrow{H_3O^+} R{-}H$$
Grignard reagent

Examples:

$$CH_3CH_2CH_2CH_2Br \xrightarrow[ether]{Mg} [CH_3CH_2CH_2CH_2MgBr]$$

$\downarrow H_3O^+$

$$CH_3CH_2CH_2CH_3$$
Even-numbered chain

$$CH_3CHCH_2CH_2CH_3 \xrightarrow[ether]{Mg} [CH_3CHCH_2CH_2CH_3]$$
with Br and MgBr substituents

$\downarrow H_3O^+$

$$CH_3CH_2CH_2CH_2CH_3$$
Odd-numbered chain

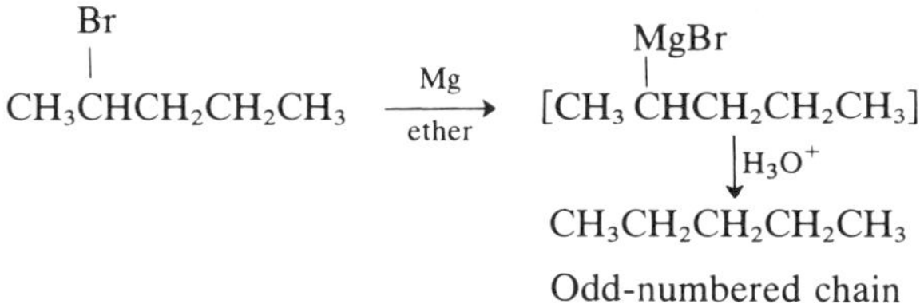

This sequence involves the preparation of an extremely useful organic intermediate known as a Grignard reagent, which will be discussed in more detail in subsequent chapters. Reaction of these reagents with water or any other source of hydrogen produces a pure hydrocarbon. Since we are not adding or subtracting any carbons in this reaction, the product has the *same chain structure* as the starting material; thus it doesn't particularly matter what halide we utilize as long as the carbon chain has the desired structure:

$$
\begin{array}{ccc}
\overset{\displaystyle CH_3}{\underset{\displaystyle CH_3}{\overset{|}{CH_3CMgBr}}}
& \xrightarrow{\ H_3O^+\ }
& \boxed{\overset{\displaystyle CH_3}{\underset{\displaystyle CH_3}{\overset{|}{CH_3CH}}}}
& \xleftarrow{\ H_3O^+\ }
& \overset{\displaystyle CH_2MgBr}{\underset{\displaystyle CH_3}{\overset{|}{CH_3CH}}}
\end{array}
$$

Both reagents produce the same product.

---

**Exercise 3-6:**   What hydrocarbons would be produced from the following halides if they are converted to Grignard reagent followed by reaction with $D_2O$?

(a) $\underset{\displaystyle Br}{\overset{|}{CH_3CHCH_2CH_3}}$

(b) $\underset{\displaystyle CH_3}{\overset{\displaystyle CH_3}{\overset{|}{CH_3CH_2C-Cl}}}$

(c) $(CH_3)_3CCH_2Br$

(d) $\underset{\displaystyle Br}{\overset{|}{CH_3CH_2CHCH_2CH_3}}$

---

(3)

$$
R{-}Br \xrightarrow[H_3O^+]{Zn} R{-}H \quad \text{(Chemical reduction)}
$$

$$
(+\ Zn^{+2})
$$

Many different chemical reducing agents may be utilized for this conversion; the combination of zinc and acetic acid is one widely used method.

Examples:

$$
\underset{\displaystyle Br}{\overset{\displaystyle CH_3}{\overset{|}{CH_3CHCHCH_3}}} \xrightarrow[HOAc]{Zn} \overset{\displaystyle CH_3}{\overset{|}{CH_3CHCH_2CH_3}}
$$

$$
CH_3CH_2CH_2Br \xrightarrow[HOAc]{Zn} CH_3CH_2CH_3
$$

# CYCLOALKANES

So far we have discussed both straight- and branched-chain alkanes. It is possible for the two chain termini to be bonded together to form a *ring*, or *cycloalkane*. Although cycloalkanes are similar to noncyclic (acyclic) hydrocarbons in chemical and physical properties, their preparation from acyclic materials requires special synthetic techniques which we will discuss much later in the text. They are named according to the number of carbon atoms *in the ring*, with the prefix *cyclo*-. Some examples are shown below. Note that these rings can be written as geometric figures with a carbon atom at each apex. If there is only one substituent group attached to the ring, it is not

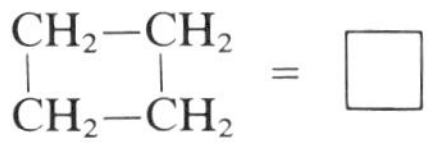

$$
\begin{array}{cc}
CH_2-CH_2 \\
| \quad\quad | \\
CH_2-CH_2
\end{array} = \square
$$

Cyclobutane

$$
\begin{array}{c}
CH_2 \\
H_2C \quad CH_2 \\
| \quad\quad\quad | \\
H_2C \quad CH_2 \\
CH_2
\end{array} =
$$

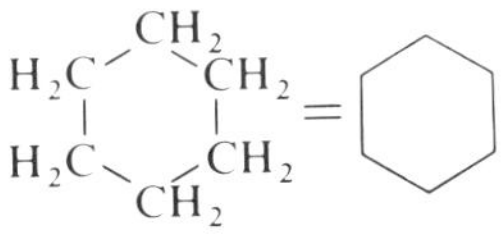

Cyclohexane

necessary to number its position, which is automatically assumed to be C-1. However, if there are two or more substituents it is necessary to number their positions so as to give the lowest possible number.

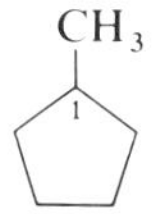

Methylcyclopentane

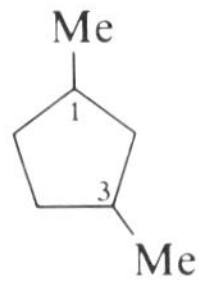

1,3-Dimethylcyclopentane $(Me=CH_3)$

In general, pure cycloalkanes can be prepared from cycloalkyl halides either by chemical reduction or by the Grignard method previously discussed. The smallest member of the series, cyclopropane, may be synthesized by an internal, or

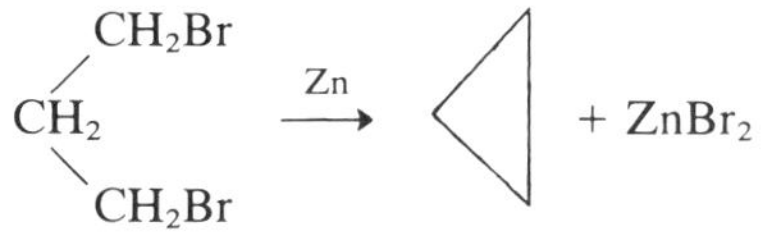

$$
\begin{array}{c}
CH_2Br \\
/ \\
CH_2 \quad \xrightarrow{\;Zn\;} \quad \triangle \quad + ZnBr_2 \\
\backslash \\
CH_2Br
\end{array}
$$

intramolecular, Wurtz reaction from 1,3-dibromopropane and zinc; however, the usefulness of this reaction for creating larger rings decreases sharply as the chain length and distance between the chain termini increases.

# REACTIONS OF ALKANES AND CYCLOALKANES

Alkanes and cycloalkanes are for the most part chemically inert. In fact, the older chemical literature refers to this family as *paraffins* (Latin: *parvum*—little; *affins*—affinity). They do not react with either acids or bases and are one of only two classes of organic compounds (the other being alkyl halides) that are not soluble in concentrated sulfuric acid. They are also inert to most oxidizing and reducing agents. However, they do react with halogens when heated or when exposed to ultraviolet radiation, and of course they burn in the presence of oxygen.

## Halogenation ($Cl_2$ or $Br_2$)

$$CH_4 + Cl_2 \xrightarrow[\text{or } h\nu]{\Delta} CH_3Cl + HCl$$

Large
excess

$$CH_3CH_2CH_3 + Cl_2 \xrightarrow[\text{or } h\nu]{\Delta} CH_3CH_2CH_2Cl + CH_3\overset{\displaystyle Cl}{\underset{|}{C}}HCH_3$$

Large
excess

Methane or propane reacts with $Cl_2$ or $Br_2$ to produce monosubstitution products if they are in large excess. If halogen is in excess, polyhalogenated mixtures of chloromethane ($CH_3Cl$), dichloromethane ($CH_2Cl_2$), trichloromethane ($CHCl_3$),

$$CH_4 + Cl_2 \xrightarrow{\Delta} CH_3Cl + CH_2Cl_2 + CHCl_3 + CCl_4$$

Equimolar                Mixture of products
or in excess

and tetrachloromethane ($CCl_4$) are obtained. These products are quite useful as solvents, particularly for dissolving greases, oils, and other similar materials. Tetrachloromethane (carbon tetrachloride) was once used as a cleaning agent in "dry-cleaning" procedures until its toxicity to the liver over a long period of time was discovered. Dichloromethane is also known as methylene chloride, while $CHCl_3$ is probably more widely referred to as chloroform, one of medicine's early anesthetics.

Alkane halogenation is now known to occur by free-radical reactions, the *initiation step* being the breaking of the halogen bond by either thermal or photochemical energy. The *propagation steps* of the radical chain reaction involve the

$$X_2 \xrightarrow[\text{or } h\nu]{\Delta} 2 :\ddot{\underset{.}{X}}\cdot \quad \text{A halogen free radical}$$

$$X = Cl \text{ or } Br \qquad \text{Initiation step}$$

attack of the halogen free radical on the methane (or alkane) C—H bond, forming HX and an organic free radical R·. The organic radical, in turn, can then attack an undissociated halogen molecule, forming R—X and regenerating another halogen free radical. These two steps can recycle 5,000 to 10,000 times for each initiation step before the reaction is *terminated* by the collision of two free radicals, stopping the chain reaction.

(a)    $CH_4 + X\cdot \longrightarrow CH_3\cdot + HX$
      Recycles

                                      Chain propagation steps

(b)    $CH_3\cdot + X_2 \longrightarrow CH_3X + X\cdot$

$$2X\cdot \longrightarrow X_2$$

$$X\cdot + CH_3\cdot \longrightarrow CH_3X \qquad \text{Chain termination steps}$$

$$2\,CH_3\cdot \longrightarrow CH_3CH_3$$

---

**Exercise 3-7:** Write out the reactions involved in the free-radical bromination of propane. Remember that two products are formed ($CH_3CH_2CH_2Br$ and $CH_3CHBrCH_3$). Write out initiation, propagation, and termination steps for *each product*. If the bromination were completely random, what would the percentages of these two products be? The actual experimental results are shown below. Are they in agreement with your prediction based on random substitution? If not, why not?

$$CH_3CH_2CH_3 \xrightarrow[h\nu\ or\ \Delta]{Br_2} CH_3CH_2CH_2Br + CH_3\overset{\overset{\displaystyle Br}{|}}{C}HCH_3$$

                                            3%              97%

---

## Combustion (Oxidation)

$$CH_4 + 2\,O_2 \longrightarrow CO_2 + 2\,H_2O + \Delta H \quad (213\ kcal/mole)$$

Although the production of halogenated solvents has tremendous industrial importance, it fades into insignificance compared with the use of hydrocarbons as fuel. For better or worse, our society is wedded to the internal combustion engine and the home oil or gas furnace, and it is the release of heat that is the important aspect of hydrocarbon combustion.

| Hydrocarbon | $\Delta H$ combustion (kcal/mole) |
|---|---|
| $CH_4$ | −213 |
| $C_2H_6$ | −368 |
| $C_3H_8$ | −526 |
| $n\text{-}C_4H_8$ | −688 |
| $n\text{-}C_5H_{10}$ | −833 |

Methane is the chief constituent (about 95%) of natural gas. The primary sources of methane are the fossil deposits, and for this reason there is concern that the known resources will not last beyond the year 2000. There is considerable reasearch directed toward the production of synthetic natural gas from coal today, and although this has been accomplished on a small scale, we are still several decades away from the utilization of this technique as a solution to our energy problems. Propane and butane are also utilized as home heating fuels, especially in those rural areas where there are no natural-gas pipelines. These fuels have an advantage over methane in that they are easily liquefied, thus greatly simplifying storage in pressurized tanks. Both these fuels are well known to campers and backpackers as being conveniently transportable fuels for portable stoves and heaters.

The long straight-chain alkanes found in plants, and thus in fossil deposits as well, are derived from fatty acids by a decarboxylation process. Since there is a

$$CH_3(CH_2)_nCOOH \longrightarrow CH_3(CH_2)_{n-1}CH_3 + CO_2\uparrow$$
Fatty acid

preponderance of even-numbered fatty acids in plants, the normal alkanes found in plant residues are odd-numbered. Heptane, for example, is the main constituent of turpentine from Jeffrey pine.

Petroleum, in the form of crude oil, is a complex mixture of thousands of hydrocarbons, both saturated and unsaturated, as well as many more complex sulfur- and nitrogen-containing compounds. Upon distillation, crude oil yields several different fractions which are arbitrarily divided into categories with which we are more familiar.

| Petroleum fraction | Distillation range | Composition |
|---|---|---|
| Gases | Less than room temperature | $C_1$–$C_4$ hydrocarbons |
| Petroleum ether | 20°–60° | $C_5$–$C_6$ |
| Natural gasoline | 40°–180° | $C_5$–$C_{10}$ |
| Kerosine | 180°–315° | $C_{12}$–$C_{18}$ |
| Lubricating oil | Non-volatile liquid | $C_{20}$–$C_{34}$ |
| Asphalt | Non-volatile solid | Polycyclic hydrocarbons |

The familiar paraffin wax and Vaseline are waxy solids obtained by fractional cooling from the higher-melting compounds in the lubricating-oil fraction.

## THE PETROLEUM INDUSTRY AND GASOLINE

To a large extent, the growth and quality of the American way of life in the twentieth century has been linked to the automobile and the availability of inexpensive gasoline. However, even in the best crude oils, the gasoline fraction is rarely more than 25%. The growth of the petroleum industry in this country has reflected the national need for ever-increasing supplies of this valuable fuel.

Not all hydrocarbons burn with equal efficiency. Branched-chain alkanes burn better than straight-chain alkanes, and unsaturated hydrocarbons and aromatics burn even better. The burning characteristics of various hydrocarbon mixtures can be measured by an arbitrary scale known as *octane rating*. A highly branched

$$
\begin{array}{c}
\overset{\displaystyle CH_3}{\underset{\displaystyle CH_3}{\overset{|}{\underset{|}{C}}}} \overset{\displaystyle CH_3}{} \\
CH_3CCH_2CHCH_3 \\
\end{array}
\qquad \text{Octane rating of 100}
$$

Isooctane

$$CH_3CH_2CH_2CH_2CH_2CH_2CH_3 \qquad \text{Octane rating of 0}$$

Heptane

hydrocarbon, isooctane (2,2,4-trimethylpentane), has been assigned an octane rating of 100, and has good burning characteristics. On the other hand, *n*-heptane is a poor-burning fuel that tends to explode, producing engine "knock" and incomplete combustion. It has been assigned an octane rating of 0. An octane rating may be assigned to any mixture of hydrocarbons by comparing it with known mixtures of heptane and isooctane. For example, a fuel with an octane rating of 70 has burning characteristics similar to a mixture of 70% isooctane and 30% heptane. The problem is how to convert a higher percentage of crude oil to mixtures of hydrocarbons having an acceptable octane rating.

To a large extent this problem has been solved by the use of catalytic conversion processes such as *cracking*. When large hydrocarbon chains are subjected to high temperatures in the presence of silica–alumina catalysts, they "crack" into smaller fragments in the $C_5$–$C_{10}$ range. This not only significantly increases the percentage of crude oil that can be utilized as gasoline (as high as 50%), but also produces higher percentages of unsaturation in this fraction, raising the quality (octane) of the mixture.

*Catalytic reforming* is an additional process for converting low-octane hydrocarbons (mostly straight chain) into higher octane fuels. In this process, the straight-chain hydrocarbons are *cyclized* catalytically and then dehydrogenated to yield aromatic compounds such as benzene.

## REVIEW OF NEW TERMS AND CONCEPTS

Define each term and, if possible, give an example of each of the following:

| | |
|---|---|
| Saturated hydrocarbon | 1°, 2°, and 3° hydrogen atoms |
| Homologous series | Grignard hydrocarbon synthesis |
| Alkyl group | Paraffin |

Combustion
Natural gas
Octane rating
Catalytic reforming
Alkane
Parent name
1°, 2°, and 3° carbon atoms

Wurtz reaction
Cycloalkane
Free-radical halogenation
A free radical
Gasoline
Catalytic cracking

## PROBLEMS: CHAPTER I-3

**1.** Name the following compounds according to the IUPAC system:

(a)
$$CH_3$$
$$CH_3CH_2CHCHCH_3$$
$$CH_3$$

(b)
$CH_3$
$CH_3$

(c)
$CH_3$

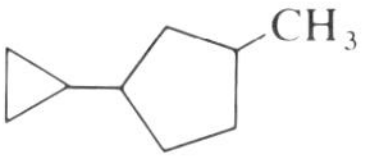

(d)
$$CH_2CH(CH_3)_2$$
$$CH_3CH_2CH_2CH_2CH$$
$$C(CH_3)_3$$

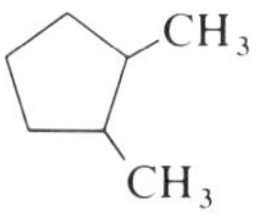

(e) $(CH_3)_3CC(CH_3)_3$

(f) 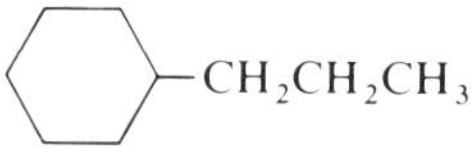
$-CH_2CH_2CH_3$

**2.** What, if anything, is wrong with the following names?

(a) 1-methylbutane
(b) 2,2-diethyl hexane
(c) 2,4-dimethylcyclohexane
(d) 3,5-dimethyl hexane
(e) *tert*-butyl methane
(f) 2-ethyl-5-methyl hexane
(g) 2,2-dimethyl hexane
(h) cyclohexylcyclohexane

**3.** Complete the following reactions showing only organic products:

(a) $CH_3CH_2CHCH_2CH_2MgCl$ $\xrightarrow{H_3O^+}$
$CH_3$

(b)
$Br$
$Me$
$\xrightarrow[H_3O^+]{Zn}$

(c)
$-CH_2CH_2Br$ $\xrightarrow[\Delta]{Na}$

(d)
$CH_3$
$BrCH_2CCH_2Br$ $\xrightarrow[\Delta]{Na}$
$CH_3$

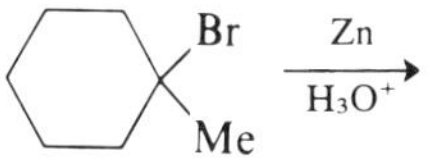

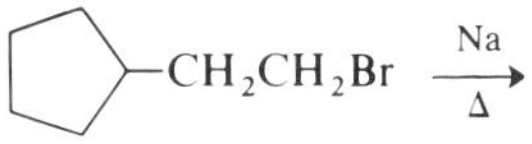

(e) $CH_3CH_2CH_2CH_3$ $\xrightarrow{Cl_2 \atop 500°}$
(Excess)

(f) 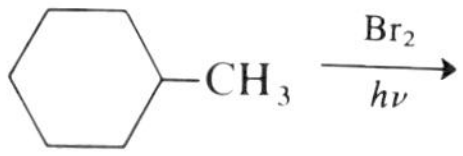
$-CH_3$ $\xrightarrow{Br_2 \atop h\nu}$
(Excess)

**4.** Show how each of the following compounds could be prepared by a Wurtz reaction.

(a) $CH_3(CH_2)_{10}CH_3$    (b)

(c) $(CH_3)_2CHCH_2CH_2CH(CH_3)_2$

**5.** Show how each of the following compounds could be prepared from a Grignard reagent.

(a) ⟨triangle⟩—$CH_3$

(b) $(CH_3)_3CCH_2CH_3$

(c)
$$(CH_3)_2CHCCH_2CH_3$$
with $CH_3$ above and $CH_3$ below the central carbon

(d) ⟨cyclopentane ring⟩—D, with Me below

(e) $(CH_3)_4C$

**6.** A new synthetic gasoline mixture has an octane rating of 110. What does that mean in terms of our discussion of octane rating? Another mixture of straight-chain unbranched hydrocarbons has a rating of −10. What does that mean?

**7.** Many backpackers prefer to carry butane stoves because of their light weight, ease of operation, and cleanliness. However, winter backpackers prefer either propane or pressurized gasoline stoves. Why?

**8.** Why are chlorinated hydrocarbons preferable to ordinary hydrocarbons as dry-cleaning fluids, since they are both nonpolar liquids?

**9.** How many structural isomers are there which correspond to the formula $C_9H_{20}$? Draw them.

**10.** How many different *alkyl* groups could be generated from each of the following?

(a) ⟨cyclohexane ring⟩—Me

(b) $(CH_3)_3CCH_2CH_3$

(c)
$$CH_3CHCH_2CH_2CH_3$$
with $CH_2CH_3$ above the second carbon

**11.** There are three isomeric pentanes, $C_5H_{12}$.

(a) Which isomer has only one monochloronation product?
(b) Which isomer has three monochloronation products?
(c) Which isomer has four monochlorination products?

**12.** A new graduate student proposed the following reaction scheme, but his advisor cautioned him not to try it. Why?

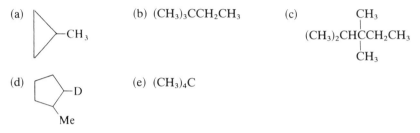

# 4

# Stereochemistry

Stereochemistry is simply the study of the arrangement of the atoms in a molecule in three-dimensional space. Stereoisomers are isomers that differ from one another only by the spatial arrangements of their component atoms. It has always been a difficult task for beginning students of organic chemistry to think of and perceive organic structure in three dimensions, yet this is the key to understanding organic structure. We are perhaps too reliant on two-dimensional study aids, such as chalk and blackboard, pencil and paper, and of course the textbook itself. To some extent we can alleviate this problem by using molecular models, but we still must be able to describe and discuss organic structure by two-dimensional means. In this chapter we will introduce the elementary concepts of stereoisomerism and develop the use of projection formulas and the two-dimensional representation of three-dimensional structure. Familiarity with these concepts will help us to appreciate the overall importance of absolute configuration in organic and biological systems.

## TWO-DIMENSIONAL REPRESENTATION OF THREE-DIMENSIONAL STRUCTURE: FISCHER PROJECTIONS

An appreciation of the importance of the third dimension in organic chemistry really began in 1874 when Van't Hoff and LeBel postulated the tetrahedral nature of carbon bonding in simple organic molecules. This simple concept, that the four valences of carbon are directed toward the corners of a regular tetrahedron, revolutionized orthodox notions of organic structure and laid the foundation for the study of molecular stereochemistry. In Fig. 4-1, the arrangement of atoms or groups about a central carbon atom may be represented in several ways to indicate the tetrahedral arrangement.

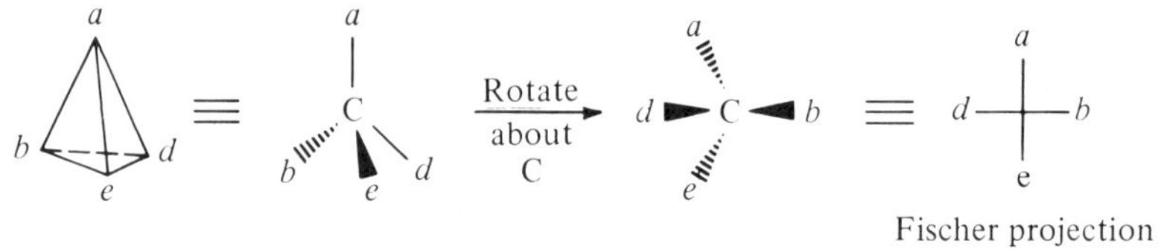

Fischer projection

**Figure 4-1.** Representations of tetrahedral carbon.

The carbon atom itself is inside the tetrahedron, approximately equidistant from each of the groups (corners) $a$, $b$, $d$, and $e$. It might be interesting to note at this point that the original "molecular model" kits were constructed as solid tetrahedrons by Van't Hoff and stored in a large matchbox!* We can replace the cumbersome tetrahedral representations by a graphic figure designed to indicate a three-dimensional model. In this representation, solid lines indicate bonds *in* the plane of the paper, a solid wedge indicates a bond *above* the plane of the paper pointing toward the reader, and the hatched wedge represents a bond *below* the plane of the paper. In Fig. 4-1, rotation about C produces a representation in which C is in the plane of the paper, $b$ and $d$ project above the plane, toward the reader, and $a$ and $e$ are directed away from the reader. If we now "squash" this representation into a two-dimensional projection, we establish a convention for writing three-dimensional formulas in two dimensions. This type of projection formula is widely used in organic chemistry and is known as a *Fischer projection*, after the German chemist who pioneered its use.

---

**Exercise 4-1:** Construct a molecular model with four different groups attached to a central carbon. Orient this model with the drawings in Fig. 4-1 until you feel familiar with drawing and interpreting projections.

---

Some care is necessary in the use of Fischer projections, and we should never forget that we are really dealing with a three-dimensional structure. Three basic operational rules apply to use of these projections:

(1) We *can* rotate projection 180° in either direction *in the plane of the paper* without altering the meaning of the projection.

---

* The originals still exist and may be seen at the Science Museum of the University of Leyden, The Netherlands.

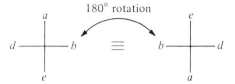

(2) We *cannot* "lift" the projection out of the plane of the paper and flip it over.

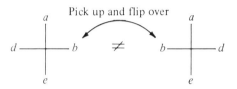

(3) We *cannot* rotate the projection 90° or 270° in either direction.

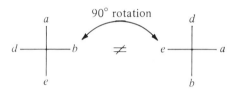

**Exercise 4-2:** Construct a tetrahedral model with four different "groups" attached to the central carbon. Draw the Fischer projection for this model. Carry out operations 1–3 described above to see if your model corresponds to the projections obtained above from each operation. Are you satisfied with the above restrictions?

## SYMMETRY AND ASYMMETRY

When *a*, *b*, *d* and *e* are all different, the molecule is said to be *asymmetric*, or *chiral*. If two or more of the atoms or groups are identical, however, the molecule will possess some type of symmetry. For example, as shown in Fig. 4-2, bromochloromethane

**Figure 4-2.** Symmetric and asymmetric structures.

has a *plane of symmetry* determined by the Cl—C—Br, (the plane of the paper), while bromochlorofluoromethane does not. When a molecule possesses a plane of symmetry, the portions above and below the plane are mirror images. Molecules containing an internal plane of symmetry are superimposable with their mirror images, but asymmetric molecules are not. Thus, in Fig. 4-3 we can see that there are

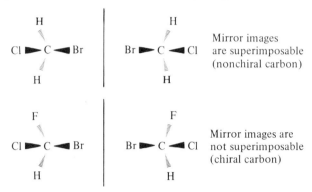

**Figure 4-3.**  Examples of superimposable and nonsuperimposable mirror images.

two forms of bromochlorofluoromethane, differing *only* in the arrangement of the atoms in space. We refer to this type of isomerism as *stereoisomerism*, and to the individual isomers as *enantiomers*. Carbon atoms containing four *different* groups are referred to as *chiral centers*.

**Exercise 4-3:**  Using molecular models, demonstrate that the mirror images of bromochloromethane are superimposable. Also demonstrate that a model of chloromethane is superimposable with its mirror image. Prove to your own satisfaction that bromochlorofluoromethane and its mirror image are *not* superimposable. How many different groups must be bonded to a carbon in order to observe this behavior?

## OPTICAL ACTIVITY—THE INTERACTION BETWEEN ASYMMETRIC MOLECULES AND PLANE-POLARIZED LIGHT

Since enantiomeric stereoisomers differ from one another only by the arrangement of their atoms in space, you might well ask how they can be distinguished from each other. Since the bonding and the charge distribution within each enantiomer is identical, one might expect that their physical properties will also be identical; such is the case. However, enantiomers do differ in one respect—their interaction with plane-polarized light.

Ordinary light is composed of waves vibrating in many different planes. This may be visualized by imagining a beam of light directed at the reader from a source behind the page of this book. (See Fig. 4-4.) Plane-polarized light, however, vibrates

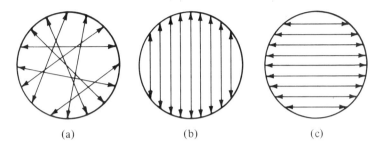

(a)                              (b)                              (c)

**Figure 4-4.**   Ordinary and plane-polarized light: (a) ordinary light vibrating in all planes; (b) and (c) plane-polarized light vibrating in planes 90° apart.

in only one plane as it approaches the reader. Plane-polarized light may be produced from ordinary light by passing it either through a prism (usually crystalling $CaCO_3$, known as *Icelandic spar* or *calcite*) which reflects all but one plane orientation, or through an artificial polarizing medium such as oriented organic crystals in a plastic medium (Polaroid). The two enantiomers will rotate a plane of polarized light equally, but in opposite directions (Fig. 4-5). This rotation of plane-polarized light is known as *optical activity*.

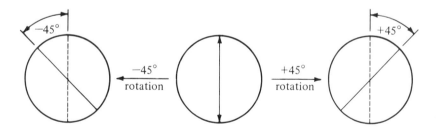

**Figure 4-5.**   Rotation of plane-polarized light in opposite directions by enantiomers.

## POLARIMETRY AND THE MEASUREMENT OF OPTICAL ACTIVITY

The rotation of plane-polarized light by asymmetric molecules may be measured with an instrument known as a *polarimeter*. It is most convenient to measure optical rotations for solutions of enantiomers in inactive solvents. When a solution of a known concentration of an optically active material is placed in the polarimeter, the beam of plane-polarized light is rotated a certain number of degrees, either to the right or to the left. Rotation of the beam to the right is designated as *dextrorotatory* behavior, and may be symbolized by the letter *d* or by a (+) sign. Rotation to the left is symbolized by the letter *l* (*levorotatory*) or by a (−) sign. The magnitude of the rotation, in degrees, is referred to as the *observed rotation*, α. The observed rotations for equal concentrations of each enantiomer are identical, but opposite in sign. A 50–50 mixture of *d* and *l* forms is known as a *racemic mixture*, or *racemate*, and will not rotate a beam of plane-polarized light in either direction. A racemate is designated by the symbol *dl* or by (±). A schematic of a polarimeter in operation is shown in Fig. 4-6.

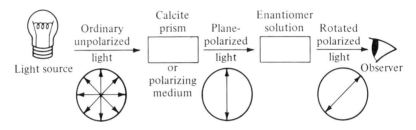

**Figure 4-6.**   Representation of a polarimeter in operation.

## SPECIFIC ROTATION

The greater the concentration of enantiomers, the greater will be the experimentally observed rotation. In order to compare the relative magnitudes of rotation for different compounds, it is necessary to define the rotation for a solution of a certain concentration. Similarly, it is necessary to define a standard wavelength of light, the path length of the cell used in the polarimeter, and the temperature at which the measurement is made. This quantity is then referred to as the *specific rotation*, $[\alpha]_\lambda^T$. If one is measuring the specific rotation of an unknown compound for the first time it is necessary to measure α for at least two different concentrations or path lengths in order to determine the direction of rotation (+ or −) as well as the calculated specific rotation.

$$[\alpha]_\lambda^T = \text{specific rotation} = \frac{[\alpha]\,\text{observed}}{l \times c}$$

where $[\alpha]_{\text{obs.}}$ = experimental rotation, in degrees
$l$ = length of polarimeter tube, in dm
$C$ = concentration of enantiomer, in g/mℓ
$T$ = temperature in °C
$\lambda$ = wavelength of light, usually the sodium D line (589.3 nm).

Example:   A solution of $\alpha$-phenethylchloride (2.5 g/100 m$\ell$) has an observed rotation of $-1.23°$.  A second solution in which the concentration has been doubled has an observed rotation of $-2.46°$. What is the specific rotation (at 25°C using the sodium D line)?

$$[\alpha]_D^{25} = \frac{\alpha}{l \times c} = \frac{[-1.23]}{(1) \times (2.5/100)} = -49.2°$$

Therefore, we have the $(-)$ enantiomer of $\alpha$-phenethylchloride and its specific rotation is $-49.2°$.

---

**Exercise 4-4:**   A solution of sucrose (table sugar) in water (5.0 g/100 ml) in a 1 dm tube has an observed rotation of $+3.3°$. Calculate the specific rotation. What effect would doubling the concentration have on the experimental rotation? On the specific rotation? Similarly, what effect would be observed if we halved the length of the polarimeter tube?

---

# ABSOLUTE CONFIGURATION

*Absolute configuration* is defined as the actual three-dimensional arrangement of the atoms in an asymmetric molecule in space. The assignment of an absolute configuration to any particular enantiomer is one of the more vexing problems in organic chemistry. When we draw projection formulas for a pair of enantiomers, the problem of assigning these structures to the $(+)$ and $(-)$ isomers remains. The direction of observed rotation actually tells us little about the true arrangement of the four groups around the asymmetric carbon atom. There are many examples in which a $(+)$ enantiomer may be chemically converted into a new structure with opposite rotation without disturbing the absolute configuration. Obviously, then, we need to determine absolute configuration by other physical means. By far the most reliable method is X-ray diffraction, but it is an extremely difficult and expensive method, and the assignment of *all* absolute configurations by the process is clearly impractical.

In practice, absolute configurations may be inferred by *comparing* configurations for a series of molecules obtained from one another by chemical reactions that do not disturb the bonds to the asymmetric centers, as shown in Fig. 4-7.

Thus it can be shown that the relative configurations of $(+)$-glyceraldehyde and $(-)$-glyceric acid are identical even though they rotate light in opposite directions. If we know the *absolute* configuration of $(+)$-glyceraldehyde, we can infer that $(-)$-glyceric acid has the *same* absolute configuration. By careful interconversions, a series of relative configurations may be built up, and a large number of absolute configurations assigned from a single X-ray determination.

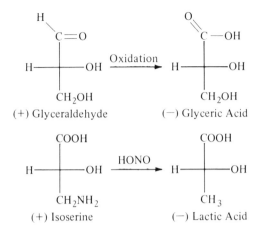

Figure 4-7. Chemical conversions in which absolute configuration is conserved.

## SEQUENCE RULES FOR DESCRIBING
## ABSOLUTE CONFIGURATION: *R* AND *S*

Cahn, Ingold, and Prelog have developed a unique system for describing the arrangement of groups about an asymmetric carbon which allows us to discuss absolute configuration in quite simple terms. This system is illustrated in Figs. 4-8 and 4-9. Priorities are assigned to the groups attached to the asymmetric carbon, say,

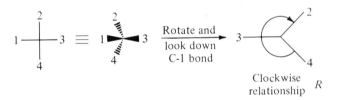

Figure 4-8. Definition of *R* configuration.

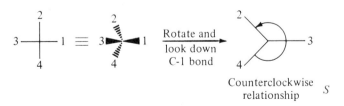

Figure 4-9. Definition of *S* configuration.

64

1–4, with 4 having the highest priority. If one then orients the molecule so that the group of *lowest* priority points directly *away* from the viewer, the remaining groups will assume either a clockwise or a counterclockwise relationship (4 → 3 → 2). The counterclockwise rotation, or rotation to the left, is defined as an *S* (Latin, *sinister*: left) configuration, while the clockwise rotation to the right defines an *R* (Latin, *rectus*: right) configuration. Thus mirror images can be assigned an *R* or an *S* configuration based solely on the internal ordering, or priorities, of the groups attached to the asymmetric center.

The priority rules for groups 1–4 are based primarily on the degree of molecular complexity, and for simple groupings the preference is first assigned on the basis of the atomic number of the atom attached to the asymmetric center, as illustrated in Fig. 4–10. If two groups are attached to the asymmetric center by the

Priority order: Br (35) > Cl (17) > C (6) > H (1)

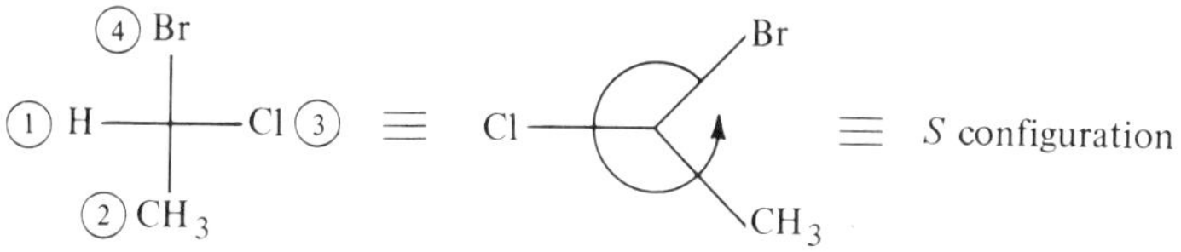

**Figure 4-10.** Assignment of priorities for chiral centers.

same atom, for example, methyl and ethyl, then we consider the atomic numbers of the atoms attached to each of the carbons. Since the sum of the atomic numbers of the groups attached to the methyl carbon is less than the sum for the groups attached to the ethyl carbon, the ethyl group has a higher priority, as shown in Fig. 4–11.

$$-\boxed{C}H_3 \quad \equiv \quad C\,(H,\,H,\,H) \quad \equiv \quad 6\,(1,\,1,\,1)$$

$$-\boxed{C}H_2CH_3 \quad \equiv \quad C\,(H,\,H,\,CH_3) \quad \equiv \quad 6\,(1,\,1,\,6)$$

**Figure 4-11.** Assignment of priorities in alkyl groups.

**Exercise 4-5:** Work out the priority order for the following alkyl groups: methyl, ethyl, *n*-propyl, *i*-propyl, *n*-butyl, *s*-butyl, *i*-butyl, and *tert*-butyl.

For groups containing multiply-bonded atoms, the sequence rules count doubly-bonded atoms as being the equivalent of two singly-bonded atoms, and a triply-bonded atom as the equivalent of three singly-bonded atoms. Figure 4–12

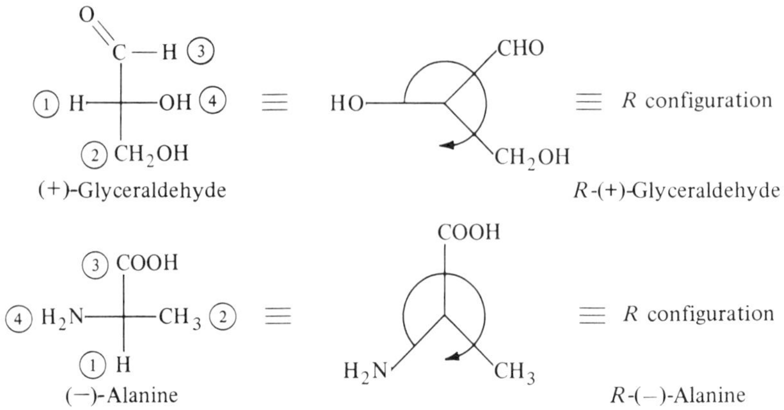

Figure 4-12. Configurations of molecules containing complex groups.

demonstrates the application of these rules for (+)-glyceraldehyde and (−)-alanine.

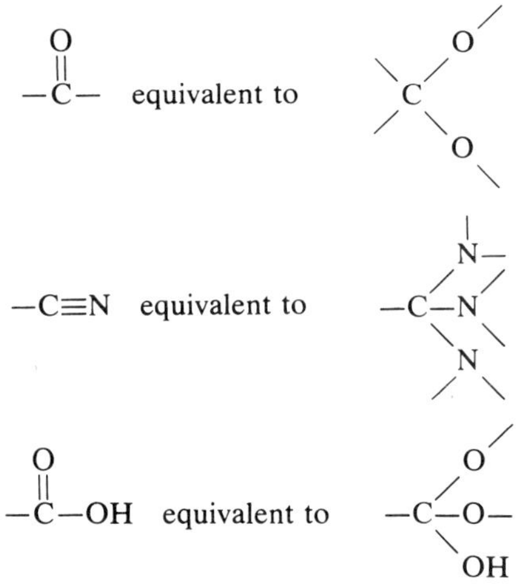

---

**Exercise 4-6:** Prove to your satisfaction that mirror images of aymmetric molecules have opposite absolute (*R* and *S*) configurations by drawing several mirror-image pairs and working out their individual configurations.

---

## MOLECULES CONTAINING MORE THAN ONE ASYMMETRIC CENTER

When molecules contain more than one asymmetric center, the number of possible isomers increases rapidly. If there are no internal symmetry elements (plane of symmetry, point of symmetry, etc.), then the total number of possible stereoisomers can be predicted accurately. Thus

$$\text{Total number of sterioisomers} = \boxed{2^n}$$

where $n$ = number of different asymmetric centers. A molecule with two asymmetric centers has a total of four stereoisomers, while a molecule containing four centers has a total of sixteen possible stereoisomers. Let us consider the relationship between the various stereoisomers of a simple compound containing two asymmetric centers, the monomethyl ester of tartaric acid (Fig. 4-13). Structure I has a 2-$R$, 3-$R$ configuration, while II is 2-$S$, 3-$S$. These structures are obviously mirror images.

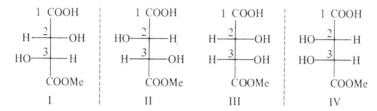

**Figure 4-13.**   Examples of molecules containing two asymmetric centers.

Similarly III and IV are also enantiomers, having 2-$R$, 3-$S$ and 2-$S$, 3-$R$ configurations, respectively. However, one might ask what relationship I bears to either III or IV. They are *diastereomeric*, in that they differ in configuration about only one of the two asymmetric centers. *Molecules that are stereoisomers but are not enantiomers are diastereomers.* Diastereomers can differ markedly in their physical properties; in fact, diastereomers generally have quite different boiling points, melting points, specific rotations, and solubilities. This latter property often enables diastereomers to be separated quite readily. In molecules containing an internal symmetry element, the total number of stereoisomers is less than predicted by the $2^n$ formula. For example, there are only three forms of tartaric acid itself. Tartaric acid exists as an enantiomeric pair and a *meso* form which is optically inactive (Fig. 4-14). The *meso* isomer has an internal mirror plane where one half of the molecule is a reflection of the other half. A *meso* compound, therefore, is a molecule which is superimposable with its mirror image, even though it contains chiral centers.

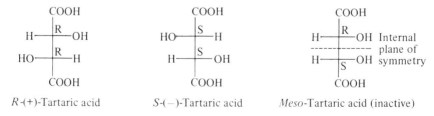

**Figure 4-14.**   Stereoisomers of tartaric acid.

Working out the absolute configurations of the two asymmetric centers in the meso acid reveals that one is $R$ and the other $S$. One can envision that for each $(+)$ rotation of the plane-polarized light by one of these atoms, an equal and opposite rotation by the other center occurs. Thus the *net* rotation is zero. A *meso* form, then, may be termed an *internal racemate*.

---

**Exercise 4-7:**   Draw structures for all possible stereoisomers of

$$HOCH_2-\underset{\underset{OH}{|}}{CH}-\underset{\underset{OH}{|}}{CH}-\underset{\underset{OH}{|}}{CH}-\underset{\underset{OH}{|}}{CH}-CH_2OH.$$

How many enantiomeric pairs are there?
How many *meso-* forms? Draw two isomers that have a diastereomeric relationship.

---

## RESOLUTION OF ENANTIOMERIC PAIRS (RACEMATES)

Enantiomers, as discussed earlier, have identical physical properties except for the rotation of plane-polarized light, thus physical methods of separation are precluded. However, we can make use of the fact that most diastereomers *do* have different physical properties—in particular, different solubilities. If we could convert our racemic mixture into a pair of diastereomers, then physical separation would become practical. In order to accomplish this, we must react the racemate with an optically active reagent, and the reaction *must not* disturb any of the four bonds to the asymmetric center. After physical separation of the diastereomeric pair obtained in this manner, we must also be able to regenerate our original enantiomer, again without disturbing any of the bonds to the chiral center. The formation of salts of either acids or bases (amines) is a typical example of resolution which satisfies these criteria, as shown in Fig. 4-15. Naturally occurring optically active alkaloids and amines such as brucine, strychnine, ephedrine, and menthylamine are normally utilized as resolving agents for racemic acids, while racemic amines may be resolved by such naturally occurring acids as tartaric acid, malic acid, and camphoric acid.

Brucine          Menthylamine          Ephedrine

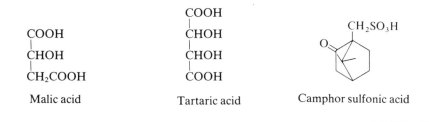

COOH
|
CHOH
|
CH₂COOH

Malic acid

COOH
|
CHOH
|
CHOH
|
COOH

Tartaric acid

Camphor sulfonic acid

---

**Exercise 4-8:**   Show how (−)-malic acid or (+)-tartaric acid might be used to resolve a racemic amine such as α-phenethylamine, $C_6H_5CHCH_3$, by utilizing the preceding
$$\overset{|}{NH_2}$$
scheme.

---

A second resolution method utilizes the high degree of stereospecificity in the chemical reactions of enzymes. Enzymatic resolution depends, to a large extent, on the "lock-and-key" or "hand-in-glove" nature of enzymatic reactions. Enzymes are usually quite large molecules composed of many different functional groups and possessing a large number of asymmetric centers. In order to react on an enzyme surface, the molecular chirality must match the chirality of the enzyme binding sites.

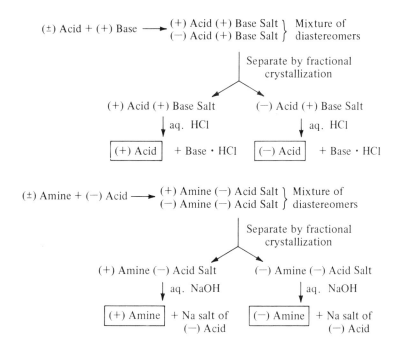

**Figure 4-15.**   Resolution of racemic acids and amines.

Thus enzymes will normally react with only one enantiomer of a racemic pair. In many respects, trying to fit an enantiomer of opposite chirality to that of the enzyme binding site is like trying to force your right hand into your left-hand glove—it just doesn't fit.

Enzymatic resolution can be of two different types. In the first, or destructive, resolution, the enzyme catalyzes destruction of one enantiomer of a racemic pair. *Penicillium glaucum* enzymes selectively destroy (+)-tartaric acid, enabling pure (−)-tartaric acid to be obtained from the racemate. However, other enzymes will form derivatives of one enantiomer in preference to the other, allowing a separation of one pure enantiomer. The second enantiomer can then be recovered by nonenzymatic means. The nondestructive resolution of (±)-alanine with fresh hog kidney acylase is a classic example of the latter technique, as is the use of $\alpha$-chymotrypsin's specificity in the hydrolysis of $S$-amino acid methyl esters.

to "protect" amino group

$$R'-\underset{\underset{NH_2}{|}}{CH}COOH \xrightarrow[\substack{(Acetic \\ anhydride)}]{Ac_2O} R'-\underset{\underset{\underset{\underset{CH_3}{|}}{C=O}}{\underset{|}{NH}}}{CH}-COOH \xrightarrow[H^+]{MeOH} R'-\underset{\underset{\underset{\underset{CH_3}{|}}{C=O}}{\underset{|}{NH}}}{CH}-\overset{\overset{O}{\|}}{C}-OCH_3$$

$R,S$-amino acid
(racemic mixture)

Acetyl-$R,S$-acid     Acetyl-$R,S$-methyl ester

$$CH_3\overset{\overset{O}{\|}}{C}NH-\underset{\underset{R'}{|}}{\overset{\overset{H}{|}}{C}}-COO^-Na^+ \quad\xleftarrow{\substack{\alpha\text{-Chymotripsin} \\ MeOH-H_2O \\ pH\,7}}$$

Acetyl-$S$-amino acid salt

+

$$CH_3\overset{\overset{O}{\|}}{C}NH-\underset{\underset{H}{|}}{\overset{\overset{R'}{|}}{C}}-COOMe$$

Acetyl-$R$-methyl ester

The two products are now chemically separable. This method is readily adaptable to the resolution of any racemic $\alpha$-amino acid.

## THE IMPORTANCE OF STEREOCHEMISTRY

In this chapter we have developed methods by which stereochemistry may be discussed and studied. As we proceed with further study of functional-group chemistry, the importance of stereochemistry will continually be emphasized, particularly in molecules of biochemical interest. Familiarity and confidence come only with practice, however, and the techniques learned in this chapter should be reviewed whenever new stereochemical ideas are introduced throughout the text.

But why is stereochemistry so important? Nature seems to favor asymmetry in general, and specific enantiomers in particular. For example, cholesterol is a steroid produced in the human body, and is linked to arteriosclerosis. There are eight asymmetric centers in cholesterol, and yet the body produces only this *one isomer* out of a possible 256. Many asymmetric molecules such as the many *S*-amino acids that

$2^8 = 256$ possible isomers for this molecular structure.

Cholesterol

are essential in human nutrition, cannot be utilized by the human system unless they have the correct stereochemistry. The *R*-acids cannot be metabolized directly. *R*-sugars are important food and energy sources, but some *S*-sugars are actually poisonous to the human system. Many physiologically active molecules apparently derive their activity from their absolute configurations. LSD (lysergic acid diethylamide), an active, illicit hallucinogen, is dependent on the absolute configuration at the amide center—one enantiomer produces "trips," while the other has no effect. Similarly, any long-distance trucker can tell you that dexedrine (the pure (+) isomer of amphetamine) keeps the sandman away much longer than benzedrine (the racemic mixture).

Hallucinogenic properties depend on the configuration of this carbon.

LSD

We are only beginning to comprehend the full implications of stereochemistry in connection with life processes. The importance of stereochemistry in nutrition and physiological activity is well known, and scientists are even now formulating additional stereochemical theories of taste and odor. Although it is impossible to predict *what* advances will be made in the near future in the field of stereochemistry, it is safe to predict that they will be both important and exciting.

## REVIEW OF NEW CONCEPTS AND TERMS

Define each term and, if possible, give an example of each:

Chirality                                          Asymmetric (chiral) carbon atom
Molecular plane of symmetry          Enantiomers
Diastereomers                                  *Meso* form
Racemic mixture (racemate)           Resolution of racemate
*R* and *S*                                       Specific rotation
Absolute configuration


## PROBLEMS: CHAPTER I-4

**1.** Determine the number of asymmetric centers for each of the following compounds and predict the number of possible stereoisomers (*dl* pairs and *meso* forms). When in doubt, build models!

(a) $CH_3CHCH_2CH_3$
        |
        Br

(b) 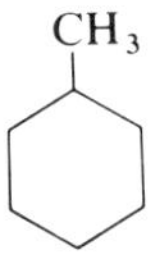 $CH_3$

(c) 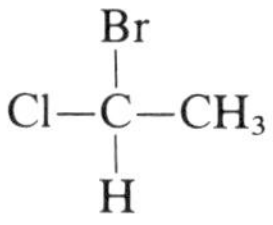 Br
        |
    Cl—C—CH₃
        |
        H

(d) $(CH_3)_2CHCHCOOH$
                 |
                 $NH_2$

(e) HOOC—CH—CHCOOH
             |        |
             Cl     OH

(f) 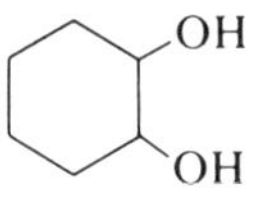
OH
OH

(g) 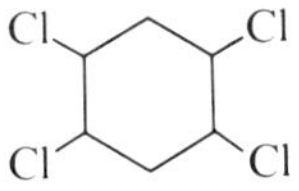 Cl        Cl
Cl        Cl

(h) 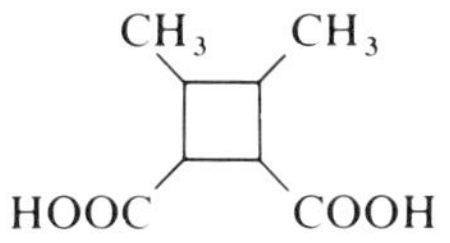 $CH_3$     $CH_3$
HOOC        COOH

(i) CHO
    |
    CHOH
    |
    CHOH
    |
    CHOH
    |
    $CH_2OH$

**2.** Consider the following projection formulas of *R*-(+)-glyceraldehyde.

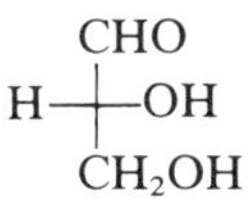

$$CHO$$
$$H—\!\!\!\!\!—OH$$
$$CH_2OH$$

With the aid of molecular models show whether structures (a)–(f) are identical to the *R* form or to the enantiomeric *S*-(−)-glyceraldehyde.

(a) 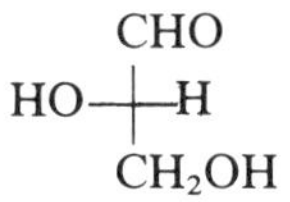        CHO
     HO—\!\!—H
         $CH_2OH$

(b) 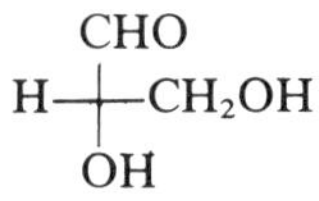        CHO
     H—\!\!—$CH_2OH$
         OH

(c) 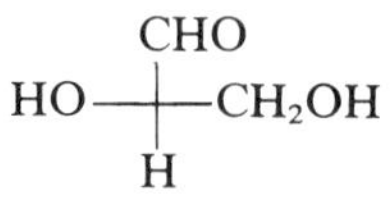        CHO
     HO—\!\!—$CH_2OH$
         H

(d) 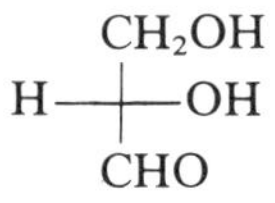    CH$_2$OH

H——OH

CHO

(e) 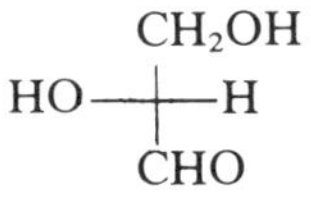    CH$_2$OH

HO——H

CHO

(f) 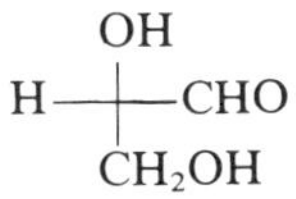    OH

H——CHO

CH$_2$OH

What conclusions can be drawn regarding the exchanging of groups in a projection formula?

3. (a) The experimentally observed rotation for a newly discovered alkaloid is found to be +60°. How would you determine if this was really a *dextro* stereoisomer or a *levo* stereoisomer with a rotation of −120°?

   (b) What would be the specific rotation of the above *dextro* isomer if the concentration was 0.20 g/m$\ell$ and the measurement was carried out in a 2 dm tube at 25° (Na D line)?

4. Arrange the following groups in Cahn–Ingold–Prelog priority order and then draw projection formulas for the *R* isomers.

   (a) —COOH, —H, —NH$_2$, —CH$_3$        (b) —CH$_2$OH, —H, —OH, —CH$_3$
   (c) —Cl, —Br, —CH$_2$CH$_3$, —CH(CH$_3$)$_2$   (d) —H, —D, —CH$_3$,
                                                     —CH$_2$CH$_2$CH$_2$CH$_3$

   (e)

   , —CH=CH$_2$, —Cl, —CH$_3$

5. Assign absolute (*R* or *S*) configurations to each of the following structures. If the structure contains more than one chiral center, assign configurations to each center.

   (a) 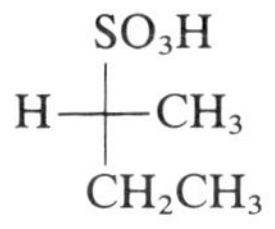    SO$_3$H

   H——CH$_3$

   CH$_2$CH$_3$

   (b) 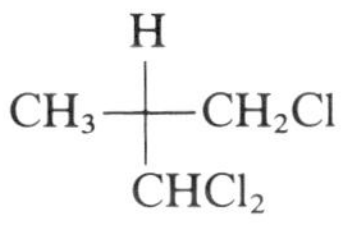    H

   CH$_3$——CH$_2$Cl

   CHCl$_2$

   (c) 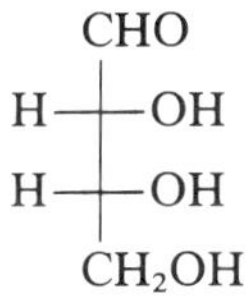    CHO

   H——OH

   H——OH

   CH$_2$OH

   (d) 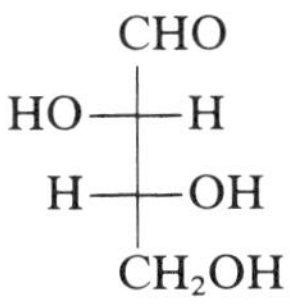    CHO

   HO——H

   H——OH

   CH$_2$OH

   (e) 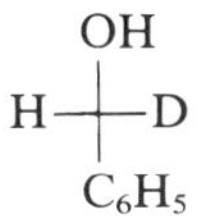    OH

   H——D

   C$_6$H$_5$

   (f) 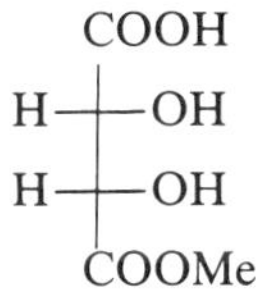    COOH

   H——OH

   H——OH

   COOMe

6. The addition of HCl to 1-butene, shown below, yields 2-chlorobutane. A chiral center has been created by chemical reaction. Would you expect the product to rotate plane-polarized light? Explain your answer.

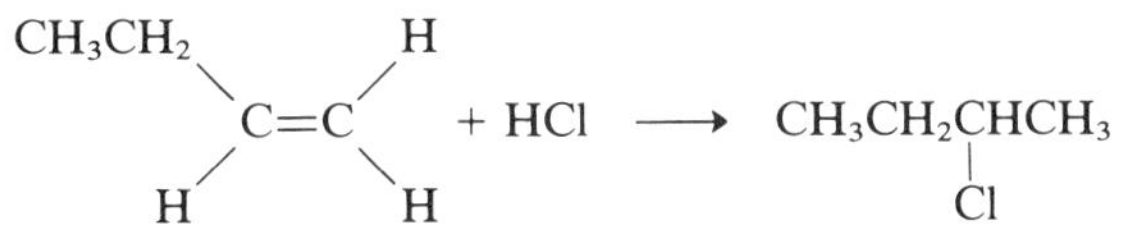

CH$_3$CH$_2$ \      / H
           C=C        + HCl  ⟶  CH$_3$CH$_2$CHCH$_3$
       H  /      \ H                      |
                                          Cl

7. Draw all *structural* isomers corresponding to the formula C$_6$H$_{13}$Br. For each compound, calculate the total number of *stereoisomers* possible.

**8.** Some molecules can be chiral without containing an asymmetric carbon. However, like the chiral molecules we have studied in this chapter, they have nonsuperimposable mirror images. Consider the two molecules below; one is optically active, the other is not. Build appropriate models and determine which is which.

(a) $CH_3CH=C=CHCH_3$    2,3-pentadiene

(b) $CH_3CH=C=C=CHCH_3$   2,3,4-hexatriene

# 5

# Alkenes, Alkynes, and Dienes:
## The "Unsaturated" Hydrocarbons

Even though alkanes may be considered as the "parent" structures in organic chemistry, their chemistry is relatively uninteresting compared with that of the unsaturated hydrocarbons. Alkenes and alkynes are extremely reactive compounds whose chemistry is controlled by the $\pi$ bond. In this chapter we will study both the structure and the reactions peculiar to this fundamental group. Two major reaction classes, *addition* and *elimination*, are intimately involved in both the reaction chemistry and the formation of $\pi$ systems, and we will learn both the similarities and the distinctions between C=C and C≡C reactions. Restricted rotation about the carbon–carbon double bond also gives rise to *geometric isomerism*, and we will see that configurational assignments can be made by priority rules similar to those we have previously learned for asymmetric molecules. Finally, we will learn that molecules with extended $\pi$ systems, such as dienes, have an interesting chemistry all their own, which results from *delocalization* of $\pi$-electron density.

## THE $\pi$ BOND IN ALKENES AND ALKYNES

In Chapter I-2 we learned that the carbon–carbon single bond is formed by the overlap of two carbon $sp^3$ hybrid orbitals resulting in the now familiar tetrahedral arrangement in saturated hydrocarbons. At that time we deferred discussion of

75

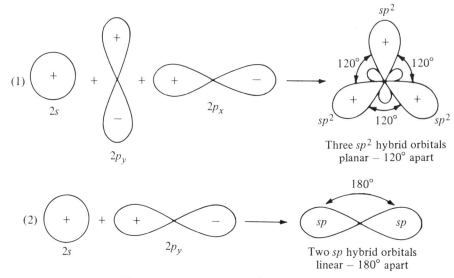

**Figure 5-1.** Formation of $sp^2$ and $sp$ hybrids.

other possible carbon hybridizations. Consider again the ground state electron configuration of the carbon atom:

$$C(He2s^2 2p_x^1 2p_y^1 2p_z^0) \equiv \frac{\uparrow\downarrow}{2s}\ \frac{\uparrow}{2p_x}\ \frac{\uparrow}{2p_y}\ \frac{}{2p_z}$$

If we promote a $2s$ electron to the vacant $2p_z$ orbital, it is not mathematically necessary to hybridize all four atomic orbitals. Two additional mixes are permissible: (1) the blending of the $2s$, $2p_x$, and $2p_y$ orbitals to form three $sp^2$ hybrid orbitals; and (2) the blending of the $2s$ with only one $2p$ orbital to form two equivalent $sp$ hybrids. In forming these new hybrids, recall that they should assume spatial orientations allowing for *maximum* separation, which results in the configurations in Fig. 5-1. The unused $2p$ orbitals remain orthogonal (90° apart) to the new hybrids, and in the case of the $sp$ hybridization, to each other. This leads to a

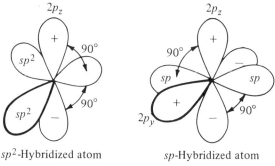

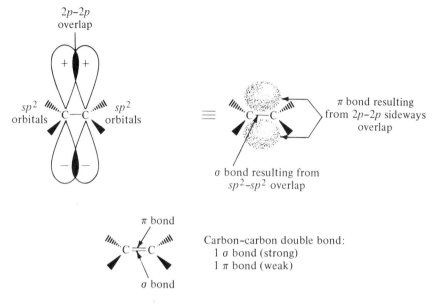

**Figure 5-2.** Formation of $\pi$ bond in alkenes.

unique situation when we allow C—C $\sigma$ bonds to form between either $sp^2$ hybrids or $sp$ hybrids. The $2p$ orbitals on the adjacent carbon atoms overlap one another—but *from the side*, as shown in Fig. 5-2. This $2p$–$2p$ overlap is not efficient, nor does it produce a strong bond. However, the sum total of the $\sigma$ and $\pi$ overlap *does* result in a stronger bond than the $sp^3$–$sp^3$ single bond in alkanes.

Overlap of two $sp$-hybrids produces another type of $\sigma$ bond (Fig. 5-3), and we now have *two* sets of $2p$ orbitals that have sideways overlap resulting in the formation

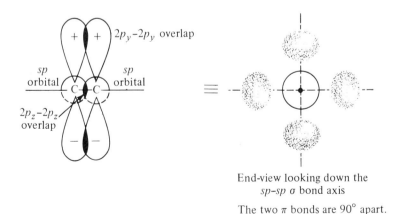

End-view looking down the
$sp$–$sp$ $\sigma$ bond axis

The two $\pi$ bonds are 90° apart.

**Figure 5-3.** Formation of $\pi$ bonds in alkynes.

of *two* $\pi$ bonds. Thus the carbon–carbon *triple* bond is in reality one $\sigma$ + two $\pi$

Carbon–carbon triple bond:
1 $\sigma$ bond (strong)
2 $\pi$ bonds (weak)

$$-C{\equiv}C-$$

bonds. The $\sigma$ bonds in both the C=C and C≡C are stronger than a $\sigma$ bond in C—C because of the greater percentage of *s* character. Thus the combination of stronger $\sigma$ bonds and the addition of $\pi$-bond overlap makes the double and triple bonds shorter and stronger than the carbon–carbon single bond, as shown in Table 5-1. It has been estimated that the $\sigma$ bond in ethylene, resulting from $sp^2$–$sp^2$ overlap, has a bond strength of 95 kcal as opposed to 88 kcal for an $sp^3$–$sp^3$ $\sigma$ bond in ethane. This means that a $\pi$ bond has an approximate strength of 68 kcal/mole, well below the value of known $\sigma$ carbon–carbon bonds.

Table 5-1.   Comparisons of Carbon–Carbon Bonds

| Bond type | Bond length | Bond strength (kcal/mole) |
| --- | --- | --- |
| C—C | 1.54 Å | 88 (in $CH_3{-}CH_3$) |
| C=C | 1.34 Å | 163 (in $CH_2{=}CH_2$) |
| C≡C | 1.21 Å | 198 (in HC≡CH) |

## THE GEOMETRY OF UNSATURATION: ALKENES AND ALKYNES

You will recall that $\sigma$ bonds have cylindrical symmetry about the bond axis. Rotation of one carbon atom with respect to the other does not alter this symmetry. However, the $\pi$ bond in a carbon–carbon double bond has planar symmetry, and

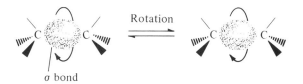

Rotation about $\sigma$ bond does not alter bond

rotation about this bond would cause the overlap between the adjacent *p* orbitals to decrease. A 90° rotation would reduce this overlap to zero, as seen in Fig. 5-4, in essence breaking the bond. Thus, while rotation about a $\sigma$ bond requires little energy (5–10 kcal—see Chapter II-5), rotation about a $\pi$ bond would require 60–70 kcal/mole. To put this in perspective, rotations about single bonds occur even

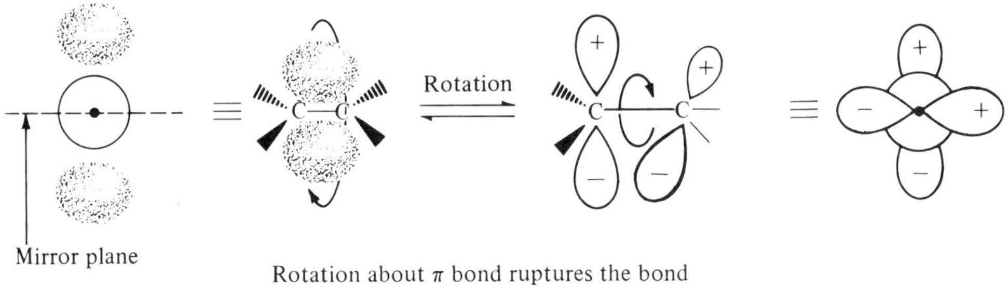

Rotation about π bond ruptures the bond

**Figure 5-4.** Consequences of rotation about σ and π bonds.

at 0°C, while double bonds remain comparatively rigid even at 300°C. One can also see that there should be little rotation about the triple-bond axis in alkynes, where we would have two π bonds breaking upon rotation.

## THE CONSEQUENCES OF STEREOCHEMICAL RIGIDITY: GEOMETRIC ISOMERISM

Consider the consequences of substitution in both alkanes and alkenes. In alkanes, rotation about the σ bond axis only *momentarily* produces molecules having different spatial arrangements of the constituent atoms (Fig. 5-5). This is an example of a dynamic equilibrium in which the *more stable* spatial arrangement will predominate at any particular instant. However, if we have a rigid bond, then different spatial arrangements (Fig. 5-6) will produce different isomers. The different

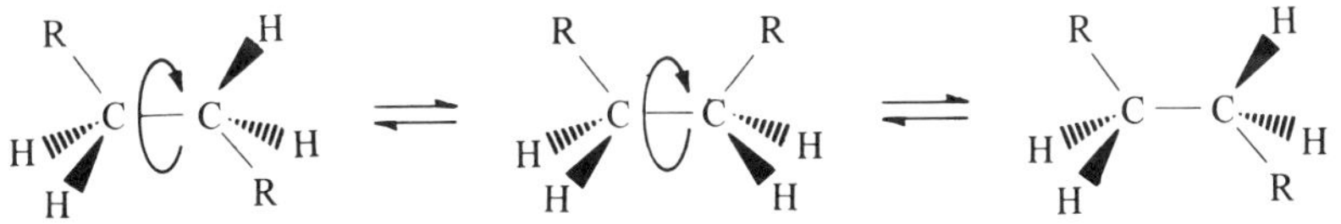

**Figure 5-5.** Different stereochemical arrangements during rotation about a σ bond.

|  | Isomer | M.p., °C | B.p., °C | Density |
|---|---|---|---|---|
| *cis:* | 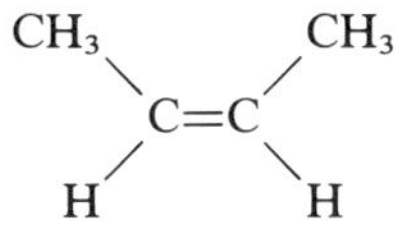 | −139 | 3.73 | 0.621 |
| *trans:* | 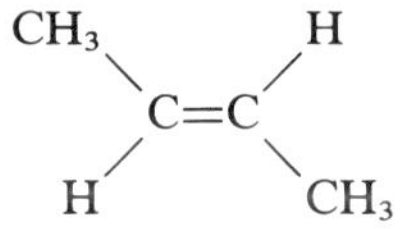 | −106 | 0.96 | 0.604 |

79

configurations which result from the rigidity of the double bond are referred to as being *cis* (Latin: on the same side) or *trans* (Latin: across from), and are referred to as *geometric isomers.* These isomers have different boiling points, melting points, solubilities, etc., as illustrated for the isomeric 2-butenes.

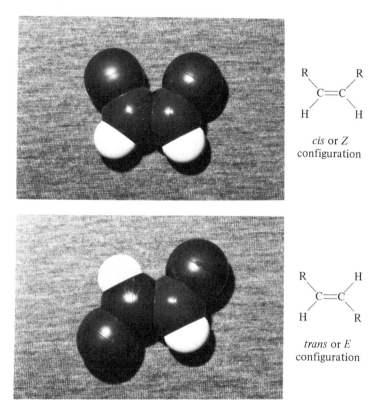

*cis* or Z
configuration

*trans* or E
configuration

**Figure 5-6.** Different stereochemical (geometric) isomers as a consequence of *nonrotation* about π bonds.

**Exercise 5-1:** For each of the following formulas, determine if *cis* and *trans* isomers exist. You may utilize the principle of superimposition learned in Chapter I-3 and also build models to aid you in your decision.

(a)

$CH_2=C$ 〈 $CH_3$ / $CH_2CH_3$

(b) $CH_3CH=CHCH_2CH_3$

(c)

(d) $CH_3$ \ $C=C$ / $CH_2CH_3$  $CH_3$ / $C=C$ \ $CH_2CH_3$

(e) $CH_3$ \ $C=CHCl$  $Cl$ /

(f) $(Cl)(F)C=C(Br)(I)$

## ABSOLUTE GEOMETRIC CONFIGURATION

In Chapter I-4 we saw that we needed a systematic method for determining *absolute* configurations of stereochemical isomers. This same problem is encountered again with geometric isomers. The labels *cis* and *trans* are easy to apply when the groups bonded to the $C=C$ are similar in nature. Confusion abounds, however, when no clear distinction between groups exists, as shown in the following example—*which* is *cis* and *which* is *trans*?

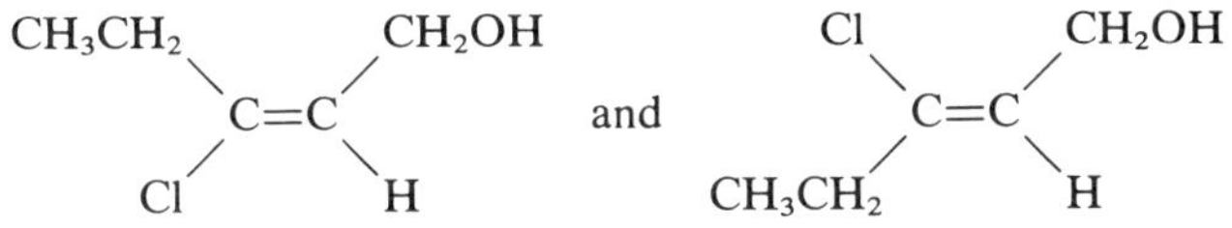

The problem may be solved easily enough if we apply the same sequence (priority) rules we previously learned for the assignment of $R$ and $S$ configurations. Consider the following:

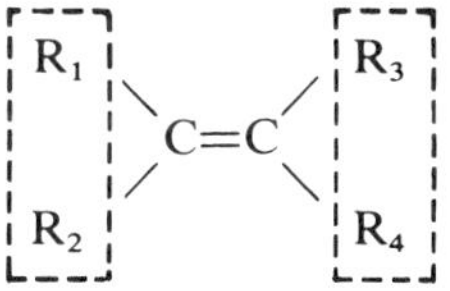

If $R_1 > R_2$, $R_3 > R_4$, then configuration is $Z$ (German: *zusammen*, together), and if $R_1 > R_2$, $R_3 < R_4$, then configuration is $E$ (German: *entgegen*, opposite).

According to the sequence rules:

$$Cl—(at. \# 17) > CH_3CH_2—(at. \# 6)$$
$$and \quad HOCH_2—(at. \# 6) > H—(at. \# 1)$$

and thus we can assign absolute geometric configurations.

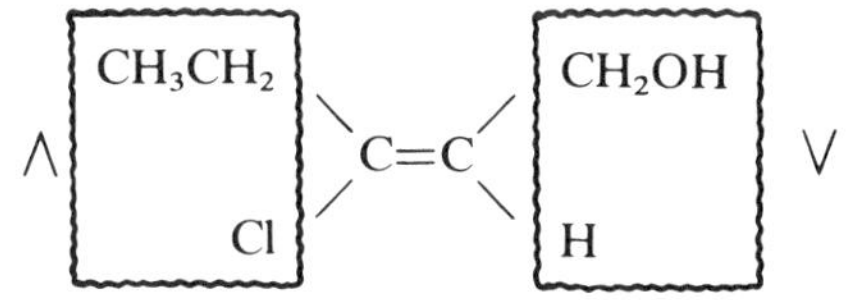

$Cl > CH_2CH_3$
$CH_2OH > H$
$E$ configuration

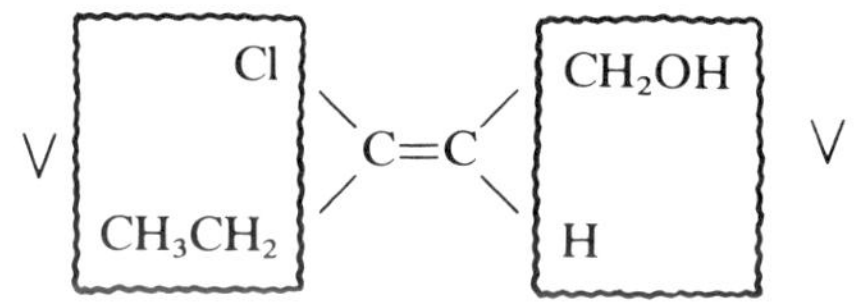

$Z$ configuration

As we will see in the next section, the $E$ and $Z$ designations actually are included as part of the IUPAC name.

**Exercise 5-2:**   Assign absolute configurations ($E$ or $Z$) to each of the following:

(a)

$$CH_3 \diagdown \atop H \diagup C=C \diagup CH_2CH_3 \atop \diagdown H$$

(b)

$$Cl \diagdown \atop Br \diagup C=C \diagup H \atop \diagdown CH_3$$

(c)

$$Cl \diagdown \atop F \diagup C=C \diagup Br \atop \diagdown H$$

(d)

$$CH_3 \diagdown \atop H \diagup C=C \diagup H \atop \diagdown CH_2OH$$

(e)

(f)

## ALKENE AND ALKYNE NOMENCLATURE

Parent names for alkenes ($C_nH_{2n}$) and alkynes ($C_nH_{2n-2}$) are determined in the same manner as for alkanes. The longest continuous carbon chain *containing* the C=C or C≡C is identified, and the -*ane* ending is replaced with either -*ene* (alkenes) or -*yne* (alkynes). The *position* of the C=C or C≡C in the chain is determined by numbering the chain so as to give the functional group the lowest possible number. Although the double or triple bond must exist between two different carbons, it is necessary only to identify the atom of *lowest* number, as seen in the following examples.

(1)    $\overset{1}{C}H_3\overset{2}{C}H=\overset{3}{C}H\overset{4}{C}H_2\overset{5}{C}H\overset{6}{C}H_3$
$\qquad\qquad\qquad\qquad \underset{CH_3}{|}$

(1) identify the longest chain containing the C=C
   parent: hexene
(2) assign C=C lowest-numbered position in chain: 2-hexene
(3) name and number substituent groups

IUPAC name: 5-Methyl-2-hexene

(2)    $\overset{1}{C}H_3\overset{2}{C}H\overset{3}{C}\equiv\overset{4}{C}\overset{5}{C}H\overset{6}{C}H_2\overset{7}{C}H_3$
$\qquad\quad \underset{CH_3}{|} \qquad \underset{CH_3}{|}$

(1) identify longest chain containing C≡C
   parent: heptyne
(2) assign C≡C lowest-numbered position in chain: 3-heptyne
(3) name and number substituent groups

IUPAC name: 2,5-Dimethyl-3-heptyne

In the above examples, we have not assigned $E$ or $Z$ configurations; as the structures are written, we cannot tell which configuration to assign to the alkene. If we wish to determine the $E$ or $Z$ designation, we must write the structure in more detail:

$$CH_3 \diagdown \atop H \diagup C=C \diagup H \atop \diagdown \underset{\underset{CH_3}{|}}{CH_2CHCH_3}$$

is $E$-5-Methyl-2-hexene

Alkynes, R—C≡C—R, are linear, and do not have $E$ and $Z$ isomers.

Common names for both alkenes and alkynes are still used for some of the more common structures, and some of the more familiar usages are listed in Table 5-2. It is also interesting to note that alkenes are still often referred to as "olefins," meaning *oil-forming*.

Table 5-2.  Some Common Names for Alkenes and Alkynes

| Compound | Common name | IUPAC name |
|---|---|---|
| $CH_2$=$CH_2$ | Ethylene | Ethene |
| $CH_3CH$=$CH_2$ | Propylene | Propene |
| $CH_3CH_2CH$=$CH_2$ | Butylene | 1-Butene |
| $(CH_3)_2C$=$CH_2$ | Isobutylene | 2-Methylpropene |
| HC≡CH | Acetylene | Ethyne |
| $CH_3C$≡CH | Methylacetylene | Propyne |
| $CH_3C$≡$CCH_2CH_3$ | Methylethylacetylene | 2-Pentyne |

Cycloalkenes are common intermediates in many organic reactions and are named in a similar manner to cycloalkanes, with *-ene* replacing *-ane*. It is not necessary to number the double-bond position; the C=C is understood to exist between $C_1$ and $C_2$

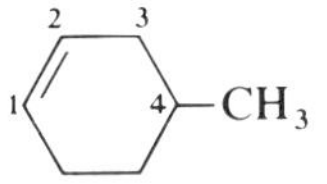

4-Methylcyclohexene

---

**Exercise 5-3:**  Name the following by the IUPAC system:

(a)

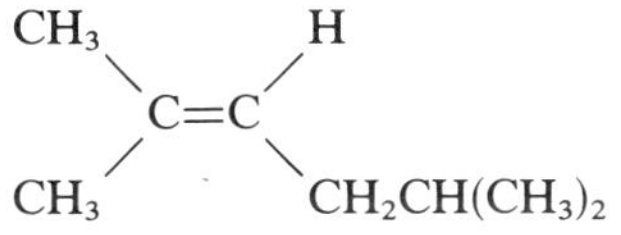

(b)

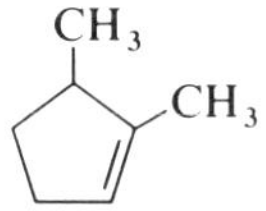

(c) $(CH_3)_3CC$≡$CCH(CH_3)_2$

---

# SYNTHESIS OF ALKENES

Carbon–carbon double bonds are normally introduced into carbon chains by elimination reactions, particularly acid-catalyzed dehydration of alcohols or dehydrogenation of alkyl halides.

(1)

Examples:

$$CH_3\overset{\overset{\displaystyle OH}{|}}{CH}CH_2CH_3 \xrightarrow[100°]{Aq.\ H_2SO_4} CH_3CH{=}CHCH_3 \quad \text{Major product}$$

$$+$$

$$CH_2{=}CHCH_2CH_3 \quad \text{Minor product}$$

$$\xrightarrow[350°]{Al_2O_3}$$

(2)

$$RCH_2\overset{\overset{\displaystyle X}{|}}{C}HR' \xrightarrow[\Delta]{\text{Alcoholic KOH}} \{RCH{=}CHR'\} + K^+X^- + H_2O$$

X = Cl, Br, I

Examples:

$$CH_3\overset{\overset{\displaystyle Br}{|}}{CH}CH_2CH_3 \xrightarrow[\Delta]{KOH} CH_3CH{=}CHCH_3 \quad \text{Major product}$$

$$+$$

$$CH_2{=}CHCH_2CH_3 \quad \text{Minor product}$$

$$CH_3\overset{\overset{\displaystyle CH_3}{|}}{\underset{\underset{\displaystyle CH_3}{|}}{C}}Cl \xrightarrow[\Delta]{\text{Alcoholic KOH}} \overset{CH_3}{\underset{CH_3}{>}}C{=}CH_2$$

In general, if two products are possible, as in the first example above, the *more highly substituted product* (the alkene with the larger number of alkyl groups) is favored. This phenomenon is referred to as the *Saytzeff orientation rule*. Increasing the number of alkyl groups attached directly to a double bond increases compound stability, thus in elimination paths which follow the Saytzeff rule, the *more stable* olefin is formed.

Ease of formation in elimination reactions

Increasing stability

In the second example, $Al_2O_3$ functions as a *Lewis acid*. Dehydrohalogenation also produces the more highly substituted alkene as the major product.

## SYNTHESIS OF ALKYNES

Carbon–carbon triple bonds may be introduced into a carbon chain by a double dehydrohalogenation reaction.

$$
\underset{\substack{\text{H} \ \text{H}}}{\overset{\substack{\text{Br} \ \text{Br}}}{\text{R}-\text{C}-\text{C}-\text{R}}} \quad \xrightarrow[\Delta]{\text{Alcoholic KOH}} \quad \{RC\equiv CR\} \quad \xleftarrow[\Delta]{\text{Alcoholic KOH}} \quad \underset{\substack{\text{H} \ \text{Br}}}{\overset{\substack{\text{H} \ \text{Br}}}{\text{R}-\text{C}-\text{C}-\text{R}}}
$$

Example:   $\underset{\substack{\text{Br Br}}}{\text{CH}_3\text{CH}_2\text{CHCHCH}_3} \quad \xrightarrow[\Delta]{\text{Alcoholic KOH}} \quad CH_3CH_2C\equiv CCH_3$

Acetylene itself is one of the few common organic compounds that can be prepared from common inorganic chemicals on a large scale, and thus is an extremely important industrial intermediate. Limestone ($CaCO_3$) and coke (C) produce calcium carbide when heated strongly in an electric furnace (2000°C). Hydrolysis of calcium carbide produces acetylene. At the turn of the century, simple lamps

(1)   $CaCO_3 \xrightarrow{\Delta} CO_2\uparrow + CaO$

(2)   $CaO + C \longrightarrow CO\uparrow + CaC_2$   (Calcium carbide)

(3)   $CaC_2 \xrightarrow{H_2O} \{HC\equiv CH\} + Ca(OH)_2$

(bicycle, automobile, and miners' lamps) in which acetylene was burned as a source of illumination utilized calcium carbide on a wide scale. We will discuss the synthesis of substituted alkynes from the parent hydrocarbon later in this chapter.

## REACTIONS OF THE $\pi$ BOND

Both alkenes and alkynes undergo a wide variety of *addition reactions*. These additions normally involve the attack of an electrophile on the $\pi$-electron cloud of the $\pi$ bond, thus producing an intermediary carbonium ion. The carbonium ion then reacts with an available nucleophile to yield the addition product. Examples of the variety of reagents that can behave in this fashion with both alkenes and alkynes are summarized below.
   General reaction:

(1)

Intermediate

(2)   $-C\equiv C- + 2X^{\oplus}Y^{\ominus} \longrightarrow$

$$
\begin{array}{c}
\text{X} \quad \text{Y}\\
-\overset{|}{C}-\overset{|}{C}-\\
\text{X} \quad \text{Y}
\end{array}
$$

$\downarrow X^+$

$$
\overset{\oplus}{C}=C\overset{X}{\diagdown} \xrightarrow{Y^-} \quad Y\diagdown C=C\diagup^{X} \xrightarrow{X^+} \quad \overset{Y}{\diagdown}\overset{+}{C}-\overset{X}{\underset{X}{C}}- \quad Y^-
$$

The addition to a triple bond occurs stepwise, and under appropriate conditions, the intermediary $-\underset{\underset{X}{|}}{C}=\underset{\underset{Y}{|}}{C}-$ can be isolated.

(1) Addition of halogen:

$$
\diagup^{\diagdown}C=C\diagdown^{\diagup} + X_2 \longrightarrow \quad -\overset{X}{\underset{|}{\underset{X}{C}}}-\overset{|}{\underset{|}{C}}- \quad X = Cl \text{ or } Br
$$

$$
-C\equiv C- + 2X_2 \longrightarrow \quad -\overset{X}{\underset{X}{\overset{|}{C}}}-\overset{X}{\underset{X}{\overset{|}{C}}}-
$$

Examples:

$$
CH_3CH=CHCH_3 \xrightarrow{Br_2} CH_3\overset{Br}{\underset{Br}{\overset{|}{C}H}}CHCH_3
$$

$$
CH_3CH_2C\equiv CCH_3 \xrightarrow{2\,Cl_2} [CH_3CH_2\overset{Cl}{\underset{Cl}{\overset{|}{C}}}=CCH_3] \longrightarrow CH_3CH_2\overset{Cl}{\underset{Cl}{\overset{|}{C}}}-\overset{Cl}{\underset{Cl}{\overset{|}{C}}}CH_3
$$

Both these additions occur in essentially quantitative yield and can be carried out either by adding the halogen directly to the unsaturated compound, or in solution ($CHCl_3$ or $CCl_4$). The absorption of bromine, for example, occurs so rapidly that it can be utilized as a *test* for the presence of *unsaturation*, the red bromine color being discharged during the addition.

$$\text{\Large$\diagdown$}C=C\text{\Large$\diagup$} + Br_2 \xrightarrow{\text{fast}} \begin{array}{c} \quad\ \ Br \\ | \quad | \\ -C-C- \\ | \quad | \\ Br \end{array}$$

(Colorless)    (Red)          (Colorless)

(2) Addition of hydrogen halide:

$$\text{\Large$\diagdown$}C=C\text{\Large$\diagup$} + HX \longrightarrow \begin{array}{c} H \\ | \quad | \\ -C-C- \\ | \quad | \\ \quad\ X \end{array} \quad X = Cl, Br, I$$

$$-C{\equiv}C- + 2\,HX \longrightarrow \begin{array}{c} H\quad X \\ | \quad | \\ -C-C- \\ | \quad | \\ H\quad X \end{array}$$

Examples:

$$CH_3CH_2CH{=}CH_2 \xrightarrow{HBr} \begin{array}{c} \quad\quad\quad Br \\ | \\ CH_3CH_2CHCH_3 \end{array}$$

$$CH_3C{\equiv}CCH_3 \xrightarrow{HCl} \begin{array}{c} \quad\quad\quad Cl \\ | \\ CH_3CH_2CCH_3 \\ | \\ Cl \end{array}$$

The addition of HX may be considered as a model for all additions of unsymmetrical reagents. The positive electrophile (in this case H⁺) always adds to form the *most stable* carbonium ion (3° > 2° > 1°) and thus adds to the carbon containing the greatest number of hydrogens (*Markownikoff's rule*). Thus we can always predict which product will form from the addition of an unsymmetrical reagent to an unsymmetrical alkene or alkyne. Carbonium-ion stability is related to the *number* of alkyl groups attached to the positive carbon center. Thus a 3° carbonium ion, with three attached alkyl groups, is more stable than a 1° carbonium ion (one group). The rationale behind these stability differences is that the alkyl groups may be polarized by the positive charge, resulting in overall charge dispersion. The greater the number of these groups, the more polarization and charge dispersal, and thus greater stability. In Markownikoff addition, then, the *more stable* carbonium ion is normally formed and we can predict the direction of addition.

$$\underset{R}{\overset{R}{R-\overset{|}{\underset{|}{C}}\oplus}} > \underset{H}{\overset{R}{R-\overset{|}{\underset{|}{C}}\oplus}} > \underset{H}{\overset{H}{R-\overset{|}{\underset{|}{C}}\oplus}} > \underset{H}{\overset{H}{H-\overset{|}{\underset{|}{C}}\oplus}}$$

Carbonium ion stability

Alkyl groups help disperse charge by polarization.

$$H-\overset{\underset{|}{H}}{\underset{|}{C}}\oplus$$

Hydrogen cannot disperse charge by polarization (too small).

(3) Addition of $H_2O$ (hydration):

A "vinyl" alcohol            A carbonyl compound

Examples:

$$CH_3CH_2CH=CH_2 \xrightarrow{H_3O^+} CH_3CH_2\overset{\overset{\displaystyle OH}{|}}{CH}CH_3$$

$$CH_3C\equiv CH \xrightarrow[(Hg^{+2})]{H_3O^+} \left[ CH_3\overset{\overset{\displaystyle OH}{|}}{C}=CH_2 \right] \xrightarrow[shift]{H} CH_3\overset{\overset{\displaystyle O}{||}}{C}CH_3$$

Cannot be isolated            Acetone (a ketone)

The addition of water to either an alkyne or an alkene requires an acid catalyst, whose purpose is to convert the reactant to the more reactive carbonium ion. The addition of $H_2O$ to an alkyne is more interesting in that the intermediary vinyl alcohol is unstable with respect to rearrangement to a carbonyl compound. This type of rearrangement is really a dynamic equilibrium called *tautomerism*, in which one form is greatly favored over another. For example, in acetone the "keto" form is greatly favored over the "enol" form.

$$CH_2-\overset{\overset{\textstyle O}{\|}}{C}-CH_3 \quad \rightleftharpoons \quad CH_2=\overset{\overset{\textstyle O}{\mid}}{C}-CH_3$$

$$H \quad >99.9\% \qquad\qquad <0.1\%$$

Keto form $\rightleftharpoons$ Enol form

Tautomerism in Acetone

(4) Catalytic hydrogenation:

$$\overset{\diagdown}{\underset{\diagup}{}}C=C\overset{\diagup}{\underset{\diagdown}{}} \quad \xrightarrow[\text{Pt, Pd, or Ni}]{H_2} \quad -\overset{\overset{\textstyle H}{\mid}}{C}-\overset{\overset{\textstyle H}{\mid}}{C}-$$

$$-C\equiv C- \quad \xrightarrow[\text{Catalyst}]{H_2} \quad \overset{\diagdown}{\underset{H}{}}C=C\overset{\diagup}{\underset{H}{}} \quad \xrightarrow{H_2} \quad -\overset{\overset{\textstyle H}{\mid}}{\underset{\underset{\textstyle H}{\mid}}{C}}-\overset{\overset{\textstyle H}{\mid}}{\underset{\underset{\textstyle H}{\mid}}{C}}-$$

Examples:

$$CH_3CH_2CH=CH_2 \quad \xrightarrow[\text{Pd}]{H_2} \quad CH_3CH_2CH_2CH_3$$

$$CH_3C\equiv CCH_3 \quad \xrightarrow[\text{Pd}]{H_2} \quad \left[\overset{CH_3}{\underset{H}{}}C=C\overset{CH_3}{\underset{H}{}}\right] \quad \xrightarrow[\text{Pd}]{H_2} \quad CH_3CH_2CH_2CH_3$$

In the absence of a heavy metal catalyst (Pd, Pt, Ni, etc.) hydrogen does not react with either alkenes or alkynes. However, hydrogen is adsorbed on certain metal surfaces and presumably converted to adsorbed atomic hydrogen. Molecules containing $\pi$ bonds may also be adsorbed on the catalyst surface, and if the proper orientation is attained adjacent to the adsorbed hydrogen, addition takes place. Addition to a triple bond occurs more quickly than to a double bond and thus addition takes place in discrete steps. "Poisoning" of the catalyst surface by the addition of small quantities of other heavy-metal ions can accentuate this difference. For example, a well-known selective catalyst, known as *Lindlar catalyst*, utilizes $Pb^{+2}$ ion as a poison. After the addition of one mole of hydrogen, the $Pb^{+2}$ inhibits the addition of a second mole.

$$R-C\equiv C-R \quad \xrightarrow[\substack{(Pb^{+2}) \\ \text{Lindlar} \\ \text{catalyst}}]{Pd/BaCO_3} \quad \overset{R}{\underset{H}{}}C=C\overset{R}{\underset{H}{}}$$

$$\text{99\% } Z \text{ alkene}$$

**(5) Oxidation:**

Alkenes may be oxidized either by conventional chemical oxidizing agents, such as potassium permanganate or potassium dichromate, or by a special procedure known as *ozonolysis* (breaking of bonds by ozone).

(a) Ordinary oxidation:

$$=CH_2 \longrightarrow CO_2\uparrow$$

$$=CHR \longrightarrow RCOOH \quad \text{A carboxylic acid}$$

$$=CR_2 \longrightarrow \underset{\underset{O}{\|}}{RCR} \quad \text{A ketone}$$

Example:

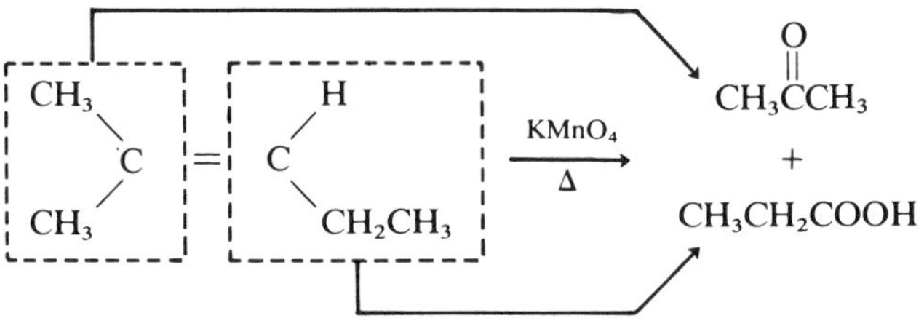

(b) Ozonolysis (under reducing conditions):

$$=CH_2 \longrightarrow H-\underset{\underset{O}{\|}}{C}-H \quad \text{Formaldehyde}$$

$$=CHR \longrightarrow R-\underset{\underset{O}{\|}}{C}-H \quad \text{An aldehyde}$$

$$=CR_2 \longrightarrow R-\underset{\underset{O}{\|}}{C}-R \quad \text{A ketone}$$

Example:

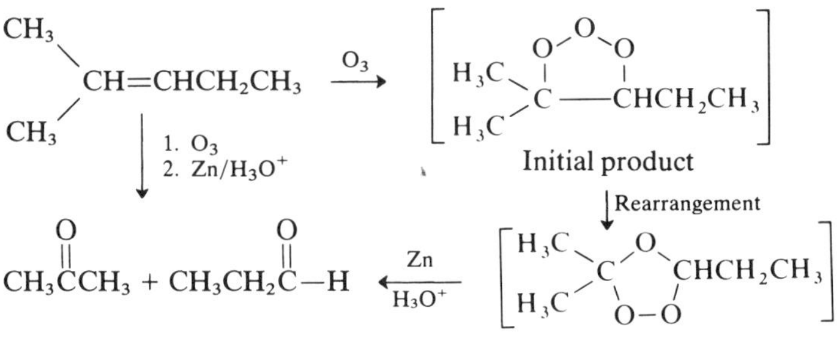

An ozonide

**Exercise 5-4:** (a) An unknown hydrocarbon (A) yielded 2-butanone and propanal when subjected to ozonolysis. What was the structure of A?

$$A \xrightarrow[\text{H}_3\text{O}^+]{\text{O}_3 \quad \text{Zn}} \underset{\text{O}}{\overset{\text{O}}{\text{CH}_3\text{CCH}_2\text{CH}_3}} + \underset{\text{O}}{\overset{\text{O}}{\text{CH}_3\text{CH}_2\text{C}-\text{H}}}$$

(b) Another unknown (B) was found to yield two different carboxylic acids when oxidized with potassium permanganate. Suggest a possible structure for B.

$$B \quad (C_5H_{10}) \xrightarrow[\Delta]{\text{KMnO}_4} \text{Two carboxylic acids}$$

Both these procedures have their own peculiar synthetic advantages, and both may be utilized to identify unknown alkenes. Several problems, in this and subsequent chapters, will demonstrate how such information aids structural analysis.

(c) Glycol formation:

Example:

$$\text{CH}_2{=}\text{CH}_2 \xrightarrow[\text{(cold)}]{\text{KMnO}_4} \underset{\text{OH} \quad \text{OH}}{\text{CH}_2-\text{CH}_2} \quad \begin{array}{l}\text{Ethylene glycol}\\ \text{1,2-ethanediol}\\ \text{(antifreeze)}\end{array}$$

If permanganate oxidation of alkenes is carried out under mild conditions in alkaline solution, fair yields of glycols (1,2-diols) may be isolated. This reaction is often used as a test for the presence of an *easily oxidized functional group*. A positive test results when purple permanganate solution is converted to a brown precipitate ($\text{MnO}_2$).

(d) Allylic halogenation:

When propene is treated with bromine at 25° in solution, an addition reaction takes place, forming the dibromide. However, at 400° in the gas phase, a substitution reaction occurs, forming allyl bromide.

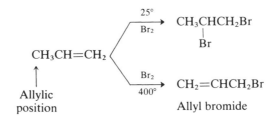

The reaction follows a free-radical chain mechanism, as previously proposed for alkane halogenation.

$$Br_2 \xrightarrow{\Delta} 2\ Br\cdot$$

$$Br\cdot + CH_3CH=CH_2 \longrightarrow HBr + \dot{C}H_2CH=CH_2$$

recycle

$$CH_2=CHCH_2\cdot + Br_2 \longrightarrow Br\cdot + CH_2=CHCH_2Br$$

The positions immediately adjacent to the double bond are extremely reactive, and substitution occurs exclusively in these *allylic positions*.

Example:

Allylic positions

$$\xrightarrow[400°]{Br_2}$$

Only product formed

Br

The intermediate in this reaction is unique in that the free electron is not localized on one carbon as are the alkyl free radicals. By electron redistribution we can draw two equivalent structures for this radical.

$$\dot{C}H_2=CH-\dot{C}H_2 \longleftrightarrow \dot{C}H_2-CH=CH_2$$

Allyl free radical:    $\underset{\cdot}{CH_2 \cdots CH \cdots CH_2}$    A resonance hybrid

Reaction may occur at either end!

In reality neither exists, and our best representation of the true formula for the allyl free radical is the delocalized *resonance hybrid* shown above. Delocalization of electron density in any intermediate or ground state structure leads to greater stability. We use the double-headed arrow (⟷) to signify the resonance phenomenon, while we write opposing arrows (⇌) for equilibria. *Do not* confuse them!

---

**Exercise 5-5:** Predict the allylic bromination products for each of the following hydrocarbons:

(a)    $CH_3$

(b) $CH_3CH_2CH=CH_2$

(c) $\overset{*}{C}H_3CH=CH_2$

$* = C^{14}$

---

# SYNTHESIS OF ALKYNES FROM ACETYLENE

Neither alkane nor alkene C—H bonds can be broken by strong base. An acetylene proton (≡C—H), however, is sufficiently acidic to be removed by a strong base such

$$\left.\begin{array}{c} -\overset{|}{\underset{|}{C}}-H \\[2em] =C\overset{/}{\underset{\backslash}{\phantom{}}}H \end{array}\right\} \xrightarrow[\text{base}]{\text{Strong}} \text{No reaction}$$

as the amide ion, or to react with sodium metal. This weak acidity also allows the identification of terminal alkynes, which form precipitates with silver and copper(I)

$$-C{\equiv}C-H \xrightarrow[\text{Liquid NH}_3]{\text{NaNH}_2} -C{\equiv}C{:}^{\ominus}, \text{Na}^{\oplus}$$

A sodium acetylide

salts. These metal acetylides, when dry, are shock sensitive and may detonate without warning.

$$-C{\equiv}C-H \begin{cases} \xrightarrow{\text{Ag(NH}_3)_2^+} -C{\equiv}C{:}^{\ominus}\text{Ag}^{\oplus}\downarrow \quad \text{(white ppt.)} \\ \xrightarrow[\text{Cu(NH}_3)_2^+]{} -C{\equiv}C{:}^{\ominus}\text{Cu}^{\oplus}\downarrow \quad \text{(red ppt.)} \end{cases}$$

$$RC{\equiv}CR \xrightarrow[\substack{\text{or} \\ \text{Cu(NH}_3)_2^+}]{\text{Ag(NH}_3)_2^+} \text{No reaction}$$

The acetylide ions are powerful nucleophiles and will displace halide ion from alkyl halides, thus producing substituted alkynes. This synthetic route is perhaps the best method of preparing alkylated acetylenes, and may be utilized to prepare either mono- or disubstituted products as shown below:

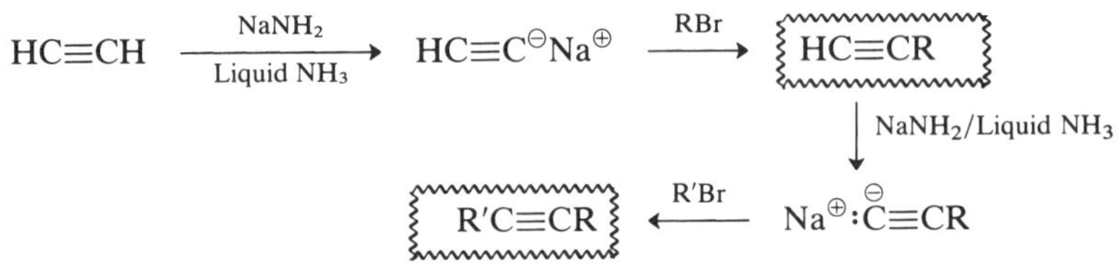

$$HC\equiv CH \xrightarrow[\text{Liquid NH}_3]{\text{NaNH}_2} HC\equiv C^{\ominus}Na^{\oplus} \xrightarrow{\text{RBr}} HC\equiv CR$$

$$\downarrow \text{NaNH}_2/\text{Liquid NH}_3$$

$$R'C\equiv CR \xleftarrow{\text{R'Br}} Na^{\oplus}:\overset{\ominus}{C}\equiv CR$$

Example:

$$HC\equiv C^{\ominus}Na^{\oplus} \xrightarrow{\text{CH}_3\text{CH}_2\text{Br}} HC\equiv CCH_2CH_3$$

$$\downarrow \text{NaNH}_2/\text{liquid NH}_3$$

$$\underset{CH_3}{\overset{CH_3}{\diagdown}} CHC\equiv CCH_2CH_3 \xleftarrow{(\text{CH}_3)_2\text{CHBr}} Na^{\oplus}:\overset{\ominus}{C}\equiv CCH_2CH_3$$

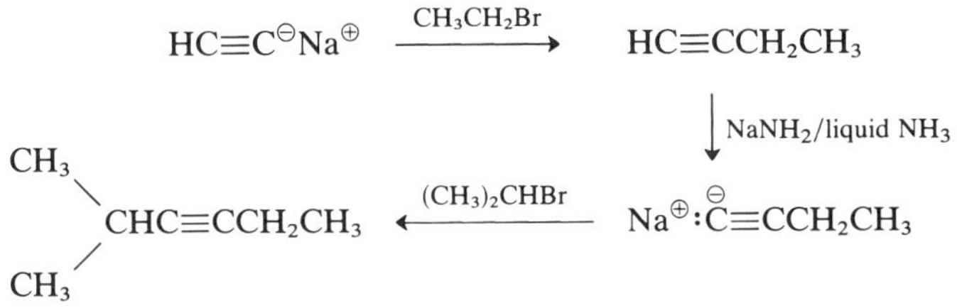

## THE CHEMISTRY OF DIENES

Dienes are unsaturated molecules containing two double bonds. They fall into three distinct categories which depend upon the relative positions of the double bonds.

(1)   $CH_2=C=CHCH_3$   A *cumulative* system
          1,2-Butadiene
          (an allene)

(2)   $CH_2=CH-CH=CH_2$   A *conjugated* system
          1,3-Butadiene

(3)   $CH_2=CHCH_2CH_2CH=CH_2$   An *isolated* system
          1,5-Hexadiene

Isolated dienes, having at least one $-CH_2-$ unit separating the two $\pi$ bonds, react similarly to mono-enes, or, in other words, as if the two $\pi$ bonds were not aware of each other's presence. Allenic dienes are most interesting because the two $\pi$ bonds are mutually orthogonal, giving rise to a special class of optically active molecules whose asymmetry is due to the total molecular structure rather than to the presence of an asymmetric center (see Fig. 5-7).

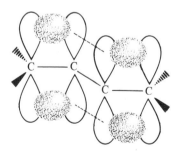

**Figure 5-7.** Nonsuperimposable enantiomers of 2,3-pentadiene.

The most interesting dienes, from a chemical viewpoint, are the conjugated dienes in which the $\pi$ bonds are separated from each other by a single $\sigma$ bond. It is possible to align both $\pi$ bonds parallel to one another. In this orientation they interact to delocalize electron density. This interaction lowers the total ground state

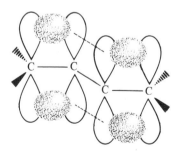

energy of the molecule compared with the isolated dienes. This energy difference can be approximated by comparing the *heats of hydrogenation* for the two systems.

|  | $\Delta H_{hydrogenation}$ (kcal/mole) |
|---|---|
| $CH_2{=}CHCH{=}CHCH_3$<br>1,3-Pentadiene | 54.1 |
| $CH_2{=}CHCH_2CH{=}CH_2$<br>1,4-Pentadiene | 60.8 |

As can readily be seen from the above values, the conjugated diene releases 6.7 kcal *less* energy than the nonconjugated diene, thus approximating the greater ground state stability that the conjugated diene achieves from electron delocalization.

Electron delocalization also plays a dominant role in the reactions of conjugated dienes. When 1,3-butadiene is allowed to react with HBr, two products are formed:

$$CH_2{=}CHCH{=}CH_2 \xrightarrow[\text{HBr}]{\text{1 equiv.}} \begin{array}{c} \overset{\displaystyle Br}{\underset{\displaystyle |}{CH_2{=}CHCHCH_3}} \\ + \\ CH_3CH{=}CHCH_2Br \end{array}$$

If we assume that the mechanism of this reaction involves a stepwise addition of $H^+Br^-$ similar to that previously described for mono olefins, then we see that the carbonium ion generated by electrophilic addition of $H^+$ results in an *allylic* system which can be stabilized by electron delocalization (resonance):

$$CH_2{=}CHCH{=}CH_2 \xrightarrow{H^+} \left\{ \begin{array}{c} CH_3\overset{\oplus}{C}H{-}\overset{\frown}{CH}{=}CH_2 \\ \updownarrow \\ CH_3CH{=}\overset{\frown}{CH}{-}\overset{\oplus}{CH}_2 \end{array} \right\}$$

The observed reaction products may then be derived by attack at either of the positive centers.

$$CH_3\overset{\oplus}{\overbrace{CH{\cdots}CH{\cdots}CH_2}} \underset{\underset{b}{\searrow} Br^{\ominus} \overset{a}{\swarrow}}{} \quad \begin{array}{l} \xrightarrow[\text{at } a]{\text{Attack}} \overset{\displaystyle Br}{\underset{\displaystyle |}{CH_3CHCH{=}CH_2}} \\ \\ \xrightarrow[\text{at } b]{\text{Attack}} CH_3CH{=}CHCH_2Br \end{array}$$

Reactions of conjugated molecules invariably involve resonance-stabilized intermediates (carbonium ions, free radicals, or carbanions). In the following chapter, we will learn how the concept of electron delocalization and aromaticity are related, and the extent to which these concepts can dominate the chemistry of a class of compounds.

## REVIEW OF NEW CONCEPTS AND TERMS

Define each term and, if possible, give an example of each:

| | |
|---|---|
| Alkene | Allylic halogenation |
| $\pi$ bond | Resonance hybrid |
| *sp* orbital | Test for a terminal alkyne |
| *E* and *Z* configurations | A nonconjugated diene |
| Dehydration reaction | Allylic carbonium ion |
| Test for unsaturation | Alkyne |
| Ozonolysis | *sp$^2$* orbital |

Geometric isomers
Cycloalkene
Dehydrohalogenation reaction
Tautomerization
Test for easily oxidized functional groups

Allylic free radical
Electron delocalization
A conjugated diene
Heat of hydrogenation

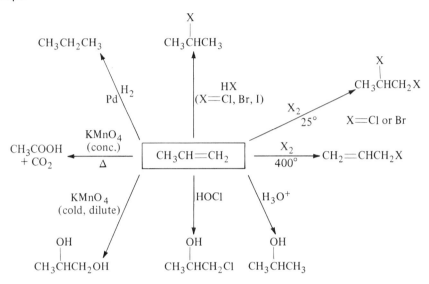

Review of reactions of
a typical alkene

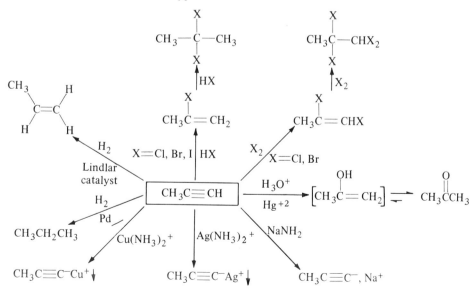

Review of reactions of
a typical alkyne

**1.** Name each of the following compounds by the IUPAC system:

(a) 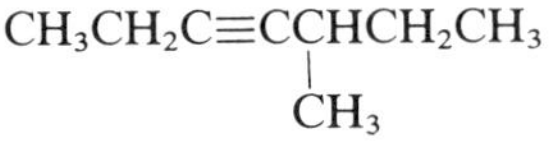 $CH_3CH_2C{\equiv}CCHCH_2CH_3$
$\quad\quad\quad\quad\quad\quad\quad\ |$
$\quad\quad\quad\quad\quad\quad\quad CH_3$

(b) 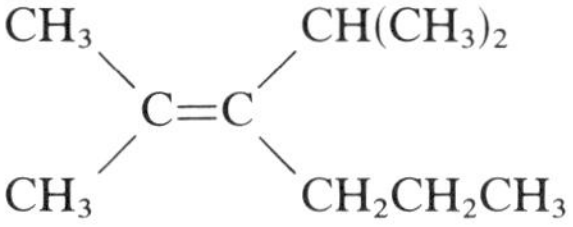 $CH_3$ $\quad\quad\quad\quad CH(CH_3)_2$
$\quad\quad\quad\quad\diagdown\quad\quad\quad\diagup$
$\quad\quad\quad\quad\quad C{=}C$
$\quad\quad\quad\diagup\quad\quad\quad\diagdown$
$\quad\quad CH_3\quad\quad\quad CH_2CH_2CH_3$

(c)

(d) $CH_2{=}CHCH{=}CHCH_2CH_2CH_3$

(e) 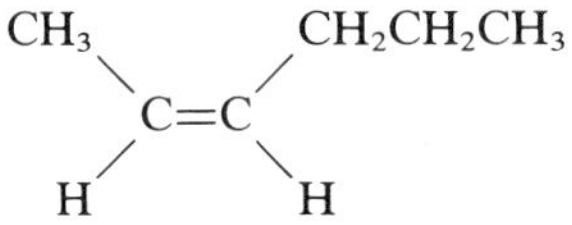 $CH_3$ $\quad\quad\quad\quad CH_2CH_2CH_3$
$\quad\quad\diagdown\quad\quad\quad\diagup$
$\quad\quad\quad C{=}C$
$\quad\quad\diagup\quad\quad\quad\diagdown$
$\quad\ H\quad\quad\quad\quad H$

(f) 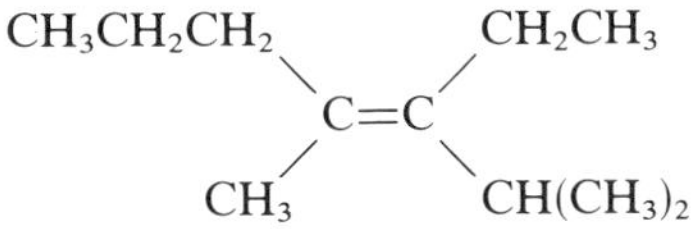 $CH_3CH_2CH_2$ $\quad\quad\quad CH_2CH_3$
$\quad\quad\quad\quad\diagdown\quad\quad\quad\diagup$
$\quad\quad\quad\quad\quad C{=}C$
$\quad\quad\quad\quad\diagup\quad\quad\quad\diagdown$
$\quad\quad\quad CH_3\quad\quad\quad CH(CH_3)_2$

(g)  $-C{\equiv}CCH_2CH_3$

(h) $CH_2{=}CHC{\equiv}CCH_2CH_3$

**2.** Draw all possible structures corresponding to the formula $C_5H_8$ and name all compounds by the IUPAC system.

**3.** Write an example of each *different* class of compounds having the following general formulas:

(a) $C_nH_{2n+2}$     (b) $C_nH_{2n}$     (c) $C_nH_{2n-2}$

(d) $C_nH_{2n-4}$     (e) $C_nH_n$

List the different *types* of unsaturation based upon your answers to this question.

**4.** Build models of 1,2- and 1,3-dimethylcyclopentane. Can you build models corresponding to the definitions of *cis* and *trans* isomers you learned for alkenes? Can a ring dictate the same type of restraint on rotation as a double bond?

**5.** (a) What is the *smallest* ring that you can build with your models that incorporates a *trans* or *E* double bond without severe strain?

(b) What is the *smallest* ring that you can build that has a triple bond incorporated in the ring without severe strain?

**6.** Complete the following reactions showing only the organic products:

(a) $(CH_3)_2C{=}CHCH_3 \xrightarrow{\ HBr\ }$

(b) $CH_3CH_2CH{=}CHCH_3 \xrightarrow[25°]{Br_2}$

(c) $(CH_3)_3CCH{=}CHCH_3 \xrightarrow[400°]{Cl_2}$

(d) $(CH_3)_3CCH=CHCH_3$  $\xrightarrow[25°]{\text{dil. } KMnO_4}$

(e) $(CH_3)_2C=CHCH_3$  $\xrightarrow[\Delta]{\text{conc. } KMnO_4}$

(f) $CH_3C\equiv CCH_3$  $\xrightarrow[Pd]{H_2}$

(g) $CH_3C\equiv CCH_3$  $\xrightarrow[\text{Lindlar cat.}]{H_2}$

(h) $CH_3CH_2CH_2CH=$⬠  $\xrightarrow{O_3}$  $\xrightarrow[H_3O^+]{Zn}$

(i) $CH_2=CHCH=CH_2$  $\xrightarrow[25°]{1 \text{ eq. } Br_2}$

(j) $(CH_3)_2C=CHCH_3$  $\xrightarrow{Cl^{\oplus}OH^{\ominus}}$

7. Alkynes can often be prepared from olefins by a two-step reaction sequence. Fill in the appropriate reagents and intermediate in the following synthesis.

$$CH_3CH=CHCH_2CH_3 \longrightarrow A \longrightarrow CH_3C\equiv CCH_2CH_3$$

8. Give a simple *chemical* test that would distinguish between the following pairs of compounds. Use a test that gives a *visual* result.

(a) $CH_3CH_2CH_2CH_2CH_2CH_3$  and  $CH_3CH=CHCH_2CH_2CH_3$

(b) $CH_3C\equiv CCH_2CH_2CH_3$  and  $CH_3CH_2CH_2CH_2C\equiv CH$

(c) $CH_3CH=CHCH_2CH_2CH_3$  and  $CH_3CH_2CH_2CH_2C\equiv CH$

(d) $CH_3CHCH_2CH_3$  and  $CH_3CH=CHCH_2CH_3$
       |
       OH

9. (a) An unknown alkene (A) $(C_6H_{12})$ yields only one product upon oxidation with $KMnO_4$, a carboxylic acid containing three carbon atoms. What is the structure of A?

(b) An isomeric compound (B) yields two straight-chain acids upon oxidation with $KMnO_4$. Propose a structure for B.

(c) Another isomeric compound (C) is totally inert to oxidation with $KMnO_4$. Suggest a possible structure for C. (*Hint:* There are several possible answers.)

10. γ-terpinene yields two products upon ozonalysis: $CH_3\overset{\overset{\displaystyle O}{||}}{C}CH_2\overset{\overset{\displaystyle O}{||}}{C}H$ and $H-\overset{\overset{\displaystyle O}{||}}{C}CH_2\overset{\overset{\displaystyle O}{||}}{C}CH(CH_3)_2$. Upon hydrogenation it forms a saturated hydrocarbon 4-methyl-1-isopropylcyclohexane. Suggest a structure for γ-terpinene.

**11.** Carry out the following conversions utilizing any inorganic and organic reagents you may deem necessary. None of these conversions can be carried out in one step with your present knowledge of organic chemistry. If you cannot immediately see a method for converting the starting material to the product, try working the problem in reverse by asking yourself how many different ways the product can be prepared. In carrying out your conversions, do not worry about minor products—use only the major product as you go from step to step. Good luck!

(a)

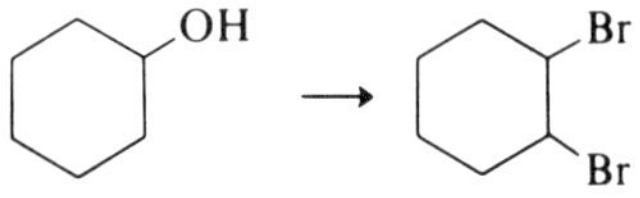

(b)  $CH_3C{\equiv}CH \longrightarrow CH_3C{\equiv}CCH_2CH_3$

(c)

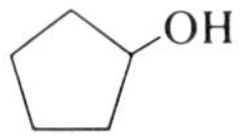

$\longrightarrow HOOC(CH_2)_3COOH$

(d)  $CH_3CH_2CH{=}CHCH_2CH_3 \longrightarrow CH_3CH_2C{\equiv}CCH_2CH_3$

(e)

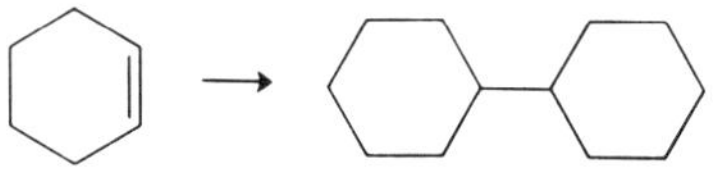

(f)

**12.** Which hydrocarbon would you expect to be more stable: 1,3-cyclohexadiene or 1,4-cyclohexadiene? How would you test your answer?

# 6

# Benzene and the Concept of Aromaticity

In the early days of organic chemistry, the question of benzene's true structure plagued the best minds that science had to offer. From the discovery of benzene by Michael Faraday in 1825 to the development of resonance theory following the early theories of quantum mechanics, the reasons for the seeming unreactivity of the benzene ring was the biggest stumbling block to an understanding of aromatic molecules. In this chapter we will explore the apparent discrepancy between the structure of benzene and its chemistry. We will learn how extensive electron delocalization both stabilizes an aromatic system and also dictates that *substitution* will occur rather than addition. Finally, we will discuss the nature of *aromaticity*, and the criteria that must be satisfied before a particular structure may be considered to be aromatic.

## THE STRUCTURE OF BENZENE— THE NATURE OF THE PROBLEM

The molecular formula of benzene, $C_6H_6$, indicates that it should be highly unsaturated. However, it was recognized very early that benzene does not undergo any of the standard addition reactions characteristic of carbon–carbon double bonds or triple bonds. Despite this seeming anomaly, many early investigators suggested

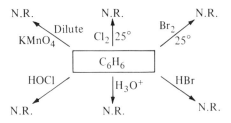

unusual structures for benzene, such as prismane, fulvene, Dewar benzene, benz-valene, and various linear structures, some of which are shown in Fig. 6-1. Although the correct structural assignment for benzene was determined before most of the above structures were even known, they have now all been synthesized, and shown to be quite different in both chemical and physical properties from authentic benzene.

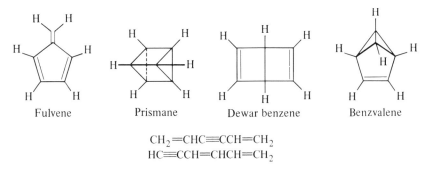

Fulvene       Prismane       Dewar benzene       Benzvalene

$$CH_2=CHC\equiv CCH=CH_2$$
$$HC\equiv CCH=CHCH=CH_2$$

**Figure 6-1.** Various suggested structures for benzene.

## SUBSTITUTION REACTIONS OF BENZENE

The puzzle began to unravel when it was discovered that although benzene did not undergo simple electrophilic addition, it did react with strong electrophiles to yield *substitution* products. Several reactions of benzene were soon discovered, all of which were catalyzed by the presence of a *strong Lewis acid* (Fig. 6-2). Another important discovery was that benzene yielded cyclohexane under high-pressure

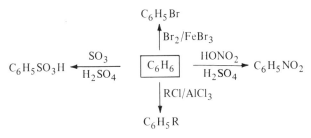

**Figure 6-2.** Some substitution reactions of benzene.

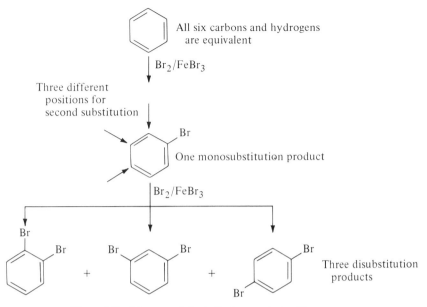

All six carbons and hydrogens
are equivalent

Br₂/FeBr₃

Three different
positions for
second substitution

One monosubstitution product

Br₂/FeBr₃

Three disubstitution
products

**Figure 6-3.** Mono- and disubstitution products of benzene.

hydrogenation with a Ni catalyst, which suggested a cyclic structure.

$$C_6H_6 \xrightarrow[\substack{\text{Ni catalyst} \\ \Delta,\ \text{pressure}}]{3H_2}$$

Benzene absorbs three moles of hydrogen

All the monosubstitution products of benzene are monoisomeric—only one isomer exists. This suggests that all the carbons (and hydrogens) in benzene are equivalent. If we carry this line of reasoning one step further, we find that when a monosubstituted benzene derivative is submitted to a second substitution reaction, three isomers are formed (Fig. 6-3). On the basis of this data, Kekulé proposed in 1865 that benzene was in reality 1,3,5-cyclohexatriene. Kekulé also suggested that the double bonds in cyclohexatriene were in motion and that this dynamic equilibrium somehow affected benzene's ability to undergo addition reactions (Fig. 6-4).

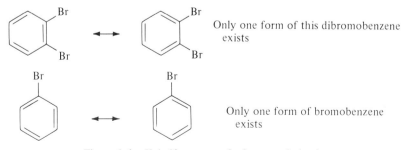

Only one form of this dibromobenzene
exists

Only one form of bromobenzene
exists

**Figure 6-4.** Kekulé structures for benzene derivatives.

**Exercise 6-1:** Examine the alternate $C_6H_6$ formulas in Fig. 6-1, which at one time or another were thought to represent the structure of benzene.
(a) How many monobrominated substitution products would each yield (benzene has only one)?
(b) How many dibrominated products would each yield (benzene has three)?

## THE STRUCTURE OF BENZENE— A MODERN VIEW

Kekulé's proposal of rapidly alternating structures is actually remarkably close to our description of resonance behavior—with one *important* difference. While Kekulé imagined that both extreme forms (A and B) were real entities in equilibrium with

one another, resonance theory would favor a delocalized structure, C (a hybrid), as the best possible description of benzene. In other words, benzene's $\pi$-electron density is delocalized and classical double bonds do not really exist in the benzene ring. Let us see how this delocalization might arise naturally from our description of bonding from hybrid orbitals.

If we form the benzene skeleton from six $sp^2$-hybridized carbons, a planar ring

Carbon–hydrogen skeleton

is formed, with the extra $sp^2$ orbitals overlapping six hydrogen $1s$ orbitals. The $2p_z$ orbitals of each of the six carbons will be perpendicular to the plane of this ring, and will *overlap each adjacent $p_z$ orbital to form a continuous $\pi$ network* above and below the carbon plane. Thus the $\pi$-electron cloud is *totally delocalized* by virtue of the

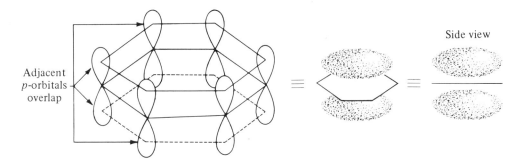

circular track described by the six parallel $p$ orbitals. We have previously seen that such delocalization lowers the ground state energy of $\pi$ systems. Benzene does not undergo classical double bond additions because the stability gained from this circular $\pi$ track may be estimated from a consideration of the *heats of hydrogenation* of cyclohexane, cyclohexadiene, and benzene. Benzene has 36 kcal *less energy* than

$$\bigcirc \xrightarrow[\text{Cat.}]{\text{H}_2} \bigcirc \qquad \Delta H_h = -28.6\,\text{kcal (observed)}$$

$$\bigcirc \rightarrow \bigcirc \qquad \begin{aligned}\Delta H_h &= -55.4\,\text{kcal (observed)}\\ (2 \times -28.6) &= -57.2\,\text{(expected)}\end{aligned}$$

$$\bigcirc \rightarrow \bigcirc \qquad \begin{aligned}\Delta H_h &= -49.8\,\text{kcal (observed)}\\ (3 \times -28.6) &= -85.8\,\text{(expected)}\end{aligned}$$

would be *predicted* for the hypothetical 1,3,5-cyclohexatriene structure. Thus one can say that this energy difference is a measure of the degree of stabilization achieved by delocalization of six $\pi$ electrons in a circular molecular orbital. When we have this bonding condition, molecules are said to be *aromatic*.* Later in this chapter we will discuss the concept of *aromaticity* in more generalized terms, and we will see how we can predict whether a given molecular structure is aromatic.

* The original derivation of the term *aromatic* for benzene type molecules arose from the fact that many of these molecules have pleasant odors.

# NOMENCLATURE OF BENZENE DERIVATIVES

The parent compound in the aromatic series is named *benzene* in the IUPAC system, and simple monosubstituted compounds are named as derivatives of the parent. However, many aromatic compounds bear common names which themselves give

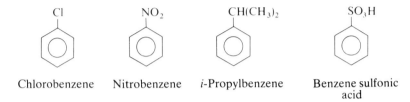

| Chlorobenzene | Nitrobenzene | *i*-Propylbenzene | Benzene sulfonic acid |

rise to parent systems. In naming derivatives of these parent systems, the parent group is assumed to occupy the 1-position and the derivative groups the lowest numbers consistent with this assignment.

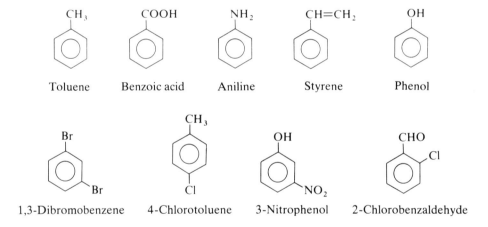

Toluene     Benzoic acid     Aniline     Styrene     Phenol

1,3-Dibromobenzene     4-Chlorotoluene     3-Nitrophenol     2-Chlorobenzaldehyde

Before the widespread usage of the IUPAC system, an alternate nomenclature was utilized which identified ring positions as *ortho*, *meta* and *para* with respect to the parent substituent.

Parent substituent

*Ortho* position

*Meta* position

*Para* position

Examples:

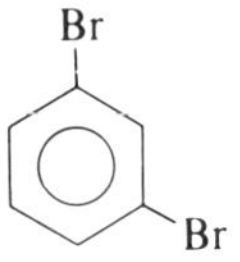

m-Dibromobenzene 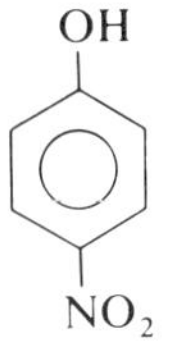    p-Nitrophenol 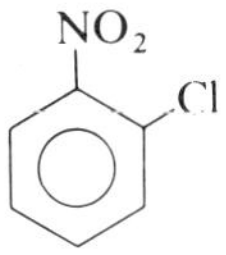    o-Chloronitrobenzene

The $C_6H_5$-group itself may be regarded as a substituent group when part of a larger molecule, or when used as a common name.

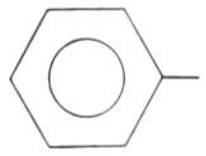

Phenyl group (or subsituent)
(also abbreviated as φ, or Ph)

Example:

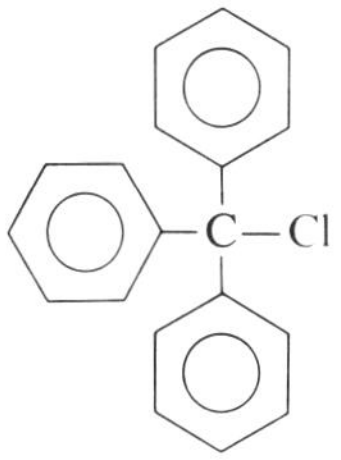

$CH_3CHCH_2CH_2CH_3$    2-Phenylpentane
        |
        φ

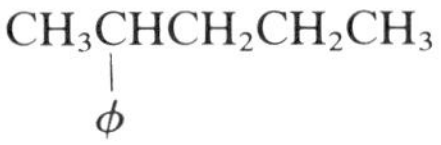

---

**Exercise 6-2:**  Name the following compounds utilizing both IUPAC and common (*ortho, meta, para*) systems.

---

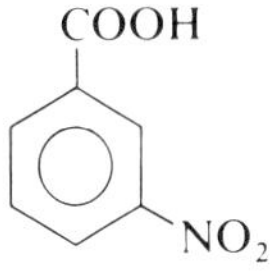

## ELECTROPHILIC AROMATIC SUBSTITUTION

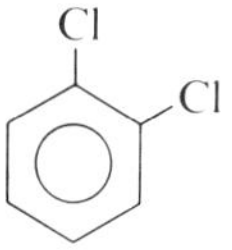

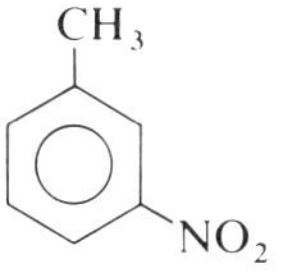

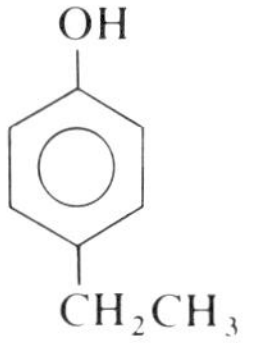

Benzene, like olefins, has an easily accessible π-electron cloud, and it should not be surprising that benzene is susceptible to electrophilic attack.  The major difference occurs after σ bond formation with the incoming electrophile is complete.  In olefin

addition, the carbonium ion is neutralized by any available nucleophile, with resultant formation of an addition compound. In aromatic chemistry, the driving force after initial electrophilic attack is the re-establishment of the aromatic delocalized electron sextet. Therefore the nucleophile present in the reaction system abstracts a proton from the intermediary $\sigma$ complex to reform the benzene nucleus. Some simple electrophilic substitution reactions are outlined below in terms of the general mechanism (Fig. 6-5).

$\pi$ complex          $\sigma$ complex

**Figure 6-5.**   General mechanism of electrophilic substitution.

(a)  Chlorination:

$$\xrightarrow[\text{FeCl}_3]{\text{Cl}_2}$$  + HCl

Attacking electrophile: $\overset{\delta^+}{\text{Cl}}\cdots\overset{\delta^-}{\text{Cl}}\cdots\text{FeCl}_3 \rightleftharpoons \text{Cl}^{\oplus}, \text{FeCl}_4^{\ominus}$
Base which removes proton: $\text{FeCl}_4^{\ominus}$

(b)  Bromination:

$$\xrightarrow[\text{FeBr}_3]{\text{Br}_2}$$  + HBr

Electrophile: $\overset{\delta^+}{\text{Br}}\cdots\overset{\delta^-}{\text{Br}}\cdots\text{FeBr}_3 \rightleftharpoons \text{Br}^{\oplus}, \text{FeBr}_4^{\ominus}$
Base: $\text{FeBr}_4^{\ominus}$

(c)  Nitration:

$$\xrightarrow{\text{NO}_2^{\oplus}\text{BF}_4^{\ominus}}$$  + HF + BF$_3$

Electrophile: $\text{NO}_2^{\oplus}$
Base: $\text{BF}_4^{\ominus}$

(d)  Sulfonation:

$$\xrightarrow[\text{H}_2\text{SO}_4]{\text{SO}_3}$$  SO$_3$H

Electrophile: SO$_3$ (neutral)
Base (internal):

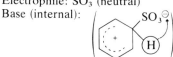

(e) Alkylation (Friedel–Crafts Reaction):

Electrophile: $\overset{\delta^+}{R}\cdots\overset{\delta^-}{Cl}\cdots AlCl_3 \rightleftharpoons R^{\oplus}, AlCl_4^{\ominus}$
Base: $AlCl_4^{\ominus}$

---

**Exercise 6-3:**  Predict the product and write a detailed mechanism for the following reaction (refer to general mechanism if you have trouble):

$$\bigcirc \xrightarrow[AlCl_3]{BrCl} \text{Substitution product}$$

---

In the preceding reactions, the $\sigma$ complex is made up of three contributing resonance forms. At this stage of your knowledge of organic chemistry, you are advised to practice writing and interconverting these forms. The secret is quite

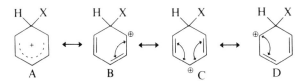

simple—just shift an electron pair from an adjacent position to the positively charged carbon. The delocalized ion is a composite of B, C, and D, showing that the positive charge is distributed over a five-atom system.

---

**Exercise 6-4:**  Write all resonance forms and the composite delocalized ion for each of the following:

(a)  $\overset{\oplus}{C}H_2CH=CHCH=CH_2$

(b)  (c)

---

## ALKYL BENZENES AND REACTIONS OF THE SIDE CHAIN

Alkyl benzenes are normally prepared by a Friedel–Crafts reaction whenever possible. There are limitations to the process, however, because of carbonium-ion rearrangements. For example, n-propylbenzene is *not* the major product of the reaction between *n*-propyl chloride and benzene with a Lewis acid catalyst ($AlCl_3$). Apparently the Lewis acid catalyzes a carbonium-ion rearrangement of the *n*-propyl chain to the isopropyl chain *via* a 1,2-hydride ($H:^-$) shift, thus producing a more

stable intermediate.  Other rearrangements are equally well known, such as the rearrangement of neopentyl carboniun ions to the *tert*-amyl structure which involves a 1,2-*alkide shift* (migration of $CH_3:^-$)

Unrearranged straight side chains may be introduced by a two-step procedure involving a *Friedel–Crafts acylation*.  An acyl halide, RCOCl, is used in place of an alkyl halide, thus eliminating any possibility of rearrangement.  The product of this reaction, a ketone, may then be chemically reduced by zinc and concentrated HCl.

Example:

Aromatic side chains may be halogenated easily in the alpha position by either photochemical or thermal means.

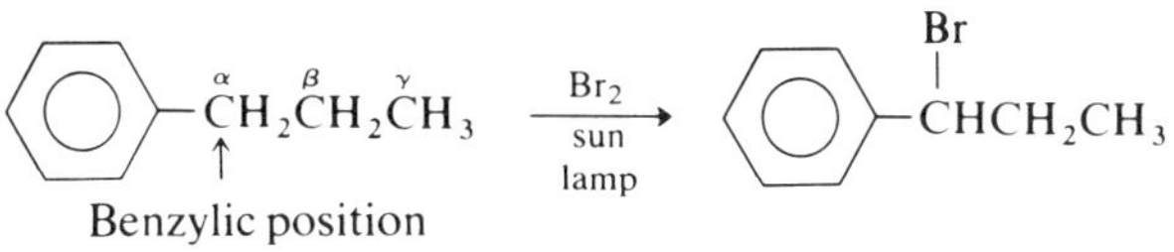

Benzylic position

The reaction occurs by a free-radical chain process, and is limited to $\alpha$-substitution by the stability of the *benzyl free radical*, which is delocalized over the entire ring.

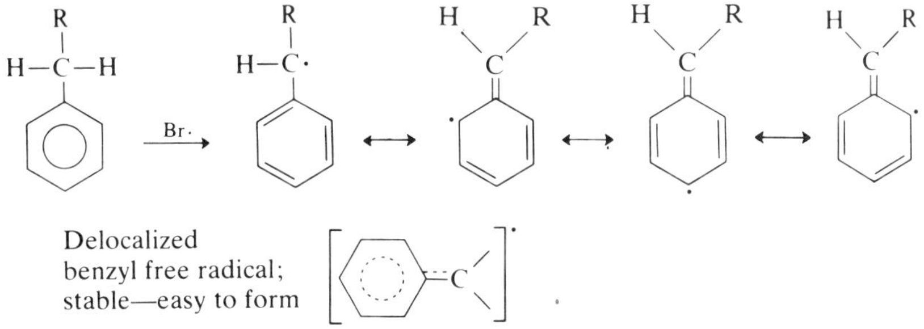

Delocalized
benzyl free radical;
stable—easy to form

**Exercise 6-5:**   Would you expect the benzyl carbonium ion and the benzyl carbanion to be particularly stable?  If so, explain—if not, why not?

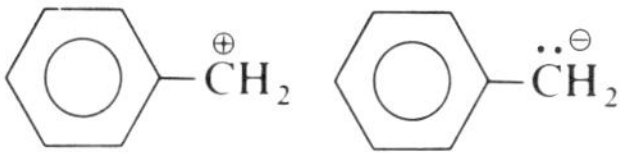

## OXIDATION OF SIDE CHAINS

Side chains may be oxidized by strong oxidizing agents such as potassium permanganate as long as there is at least one $C_\alpha$—H present.  All side chains that meet this condition are oxidized to —COOH, regardless of their length or chain structure.  However, if no $C_\alpha$—H's are present, the chain is inert to oxidation.  This procedure may be utilized in the analysis of unknown aromatic compounds to determine the number of attached alkyl groups.

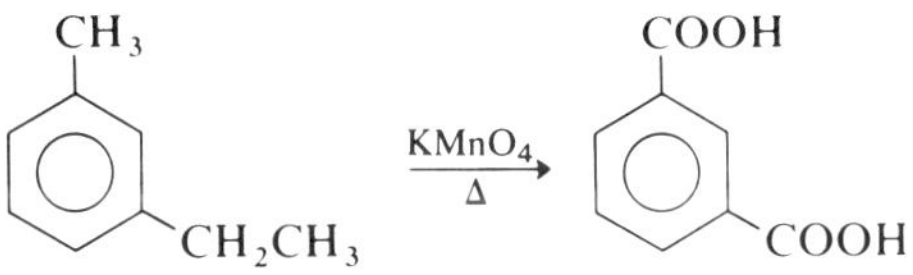

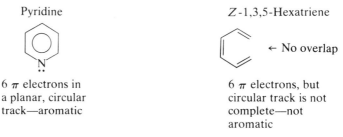

$$No\ \alpha\ hydrogens \qquad \xrightarrow[\Delta]{KMnO_4}$$

---

**Exercise 6-6:**   Two unknown aromatic hydrocarbons, A and B, have the same molecular formula, $C_9H_{12}$. Upon oxidation, the following results were obtained:

A $\xrightarrow{[O]}$ (structure: benzene ring with COOH, HOOC, and COOH)   ;  B $\xrightarrow{[O]}$ (structure: benzene ring with COOH and COOH)

Suggest structures for A and B.

---

## AROMATICITY AND THE 4n + 2 HÜCKEL RULE

Is benzene unique, or can the phenomenon of aromaticity be displayed by other structures? Let us review those structural features inherent in the benzene molecule which might allow us to predict aromatic behavior in nonbenzenoid compounds:

1. Six $\pi$ electrons are delocalized (aromatic sextet).
2. The electrons are delocalized over a planar, circular track.

Using only these two criteria, we can determine whether a large number of molecules should be aromatic. Thus pyridine ($C_6H_5N$) would be expected to be aromatic, while 1,3,5-hexatriene would not.

Pyridine

6 $\pi$ electrons in
a planar, circular
track—aromatic

$Z$-1,3,5-Hexatriene

← No overlap

6 $\pi$ electrons, but
circular track is not
complete—not
aromatic

Is a six-membered ring necessary? Let us see if we could satisfy the aromaticity conditions present in benzene with both larger and smaller rings. In cyclopentadiene, There are four $\pi$ electrons. If we could remove a proton, an additional electron pair would become available for delocalization—6 electrons in a planar 5-membered ring. The cyclopentadiene ring is aromatic.

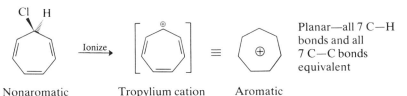

Nonaromatic                                    Aromatic

Planar—all 5 C—H bonds equivalent; all 5 C—C bonds equivalent

In a seven-membered ring, such as 1,3,5-cycloheptatriene (in which we already have six $\pi$ electrons), we require an empty orbital to complete the necessary circular track. This can be achieved by carbonium-ion formation (tropilium ion)

Nonaromatic          Tropylium cation          Aromatic

Planar—all 7 C—H bonds and all 7 C—C bonds equivalent

The German chemist Hückel has been able to show by quantum mechanical arguments that a further restriction on aromaticity is that the molecule must have 4n + 2 $\pi$ electrons in a delocalized *planar* track. Thus we can predict whether a molecule will display aromatic properties by applying this condition.

---

**Exercise 6-7:**    Determine whether the following molecules are aromatic by applying the Hückel condition. If you predict nonaromatic behavior, state which of the aromaticity conditions are not met.

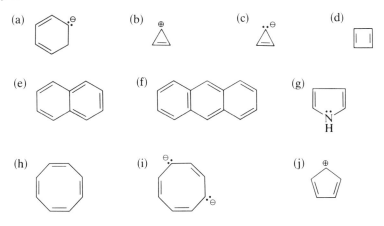

# REVIEW OF NEW TERMS AND CONCEPTS

Define each term and, if possible, give an example of each:

Electrophilic aromatic substitution

Benzene resonance hybrid

Carbonium-ion rearrangement in Freidel–Crafts alkylation

$\sigma$ complex

Benzylic free radical

Hückel rule for aromaticity

Aromatic sextet

Resonance energy

1,2-hydride shift

Clemmensen reduction

Aromaticity

## PROBLEMS: CHAPTER I-6

**1.** Benzene may be nitrated in a mixture of $HNO_3$ in concentrated $H_2SO_4$. We have already discussed the fact that aromatic nitration involves attack of $NO_2{}^+$ on the benzene ring. Explain how this electrophile may be generated in the medium.

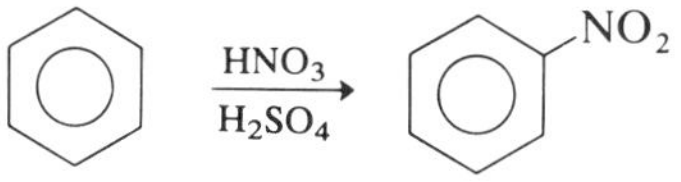

**2.** Name the following compounds by the IUPAC system:

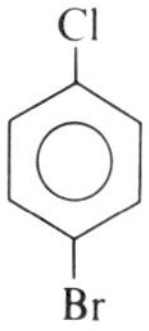

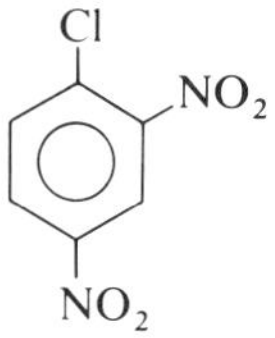

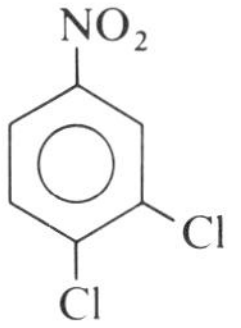

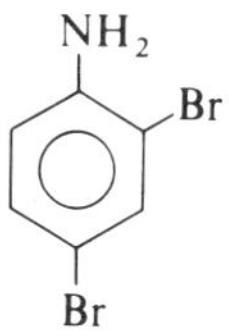

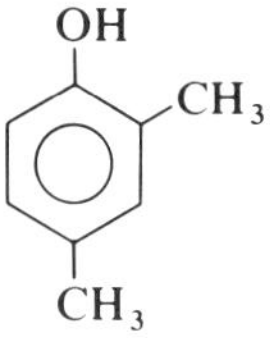

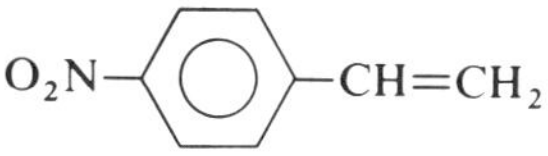

**3.** Complete the following reactions by supplying the missing reagents:

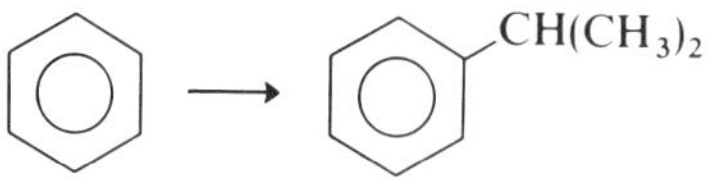

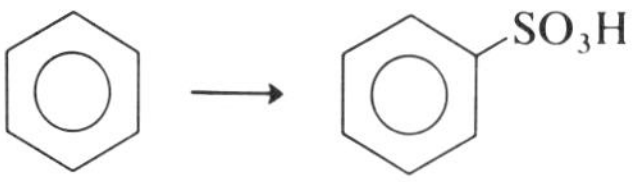

(c)

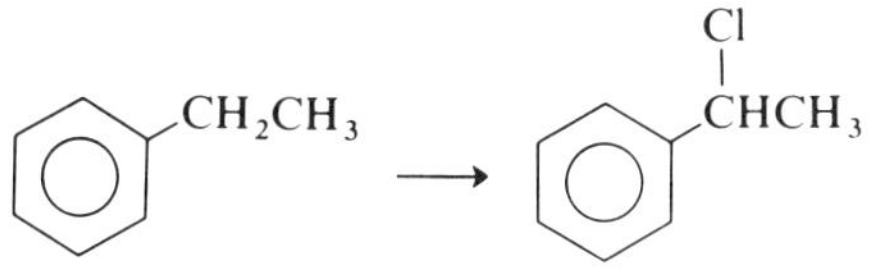

(d)

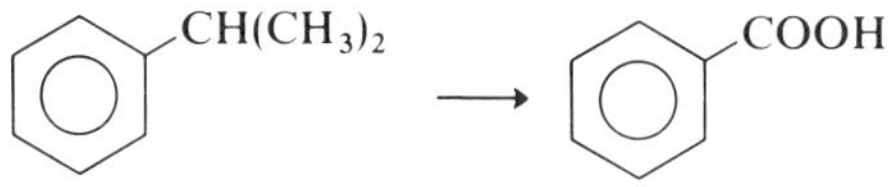

(e)

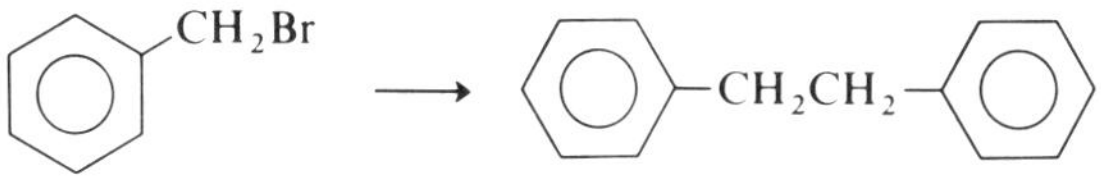

**4.** Predict the products from the following reactions. To what reaction that we have discussed in this chapter are these reactions related?

(a)

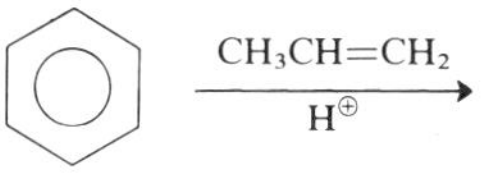

(b)

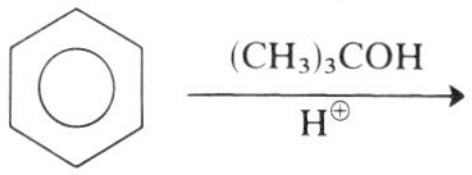

**5.** Prepare each of the following compounds from benzene and any other reagents you might need.

(a)  (b) 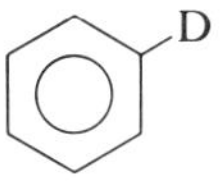  (c)

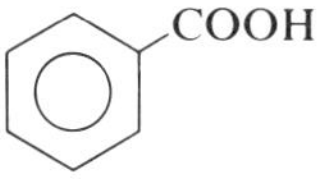

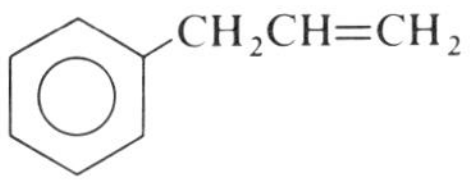

(d) 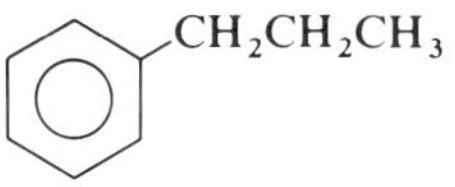  (e)

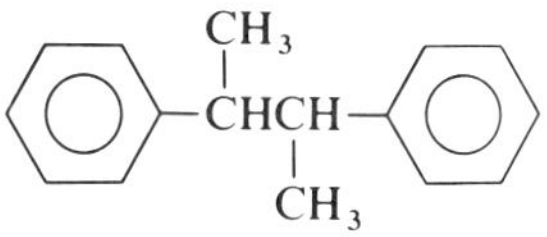

**6.** Consider the following reaction. Would you expect A, B, or a mixture of A and B as a result of acid-catalyzed dehydration?

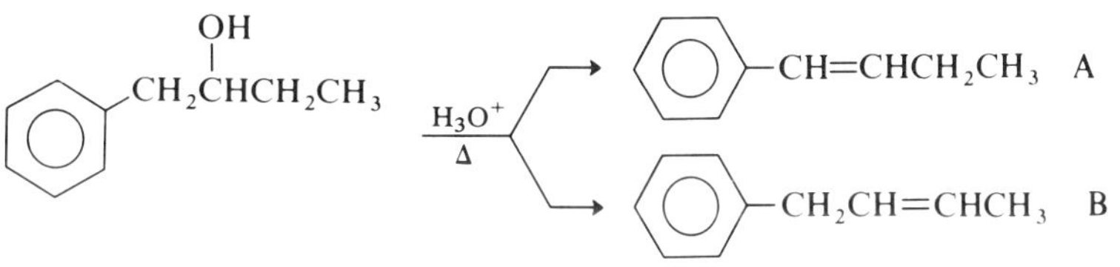

**7.** Cyclodecapentaene (A) is a ten $\pi$ electron system which should be aromatic according to the Hückel rule; however, it is not. Build a model of this molecule and determine why it cannot be aromatic.

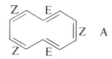

**8.** Suggest a few simple chemical tests that you might perform in order to distinguish between aromatic and nonaromatic hydrocarbons, say, between benzene and cyclooctatetraene.

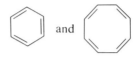

and

**9.** Predict how tropylium bromide and cycloheptyl bromide might differ in such physical properties as
(a) melting point
(b) solubility in $H_2O$
(c) solubility in hydrocarbon solvents
(d) ease of AgBr formation when reacted with $AgNO_3$ solution. Why is there a difference, if any?

Tropylium bromide              Cycloheptyl bromide

**10.** Explain how the three xylenes ($o$, $m$, and $p$) may be distinguished from one another by consideration of the number of mono- and dibromination products they can form.

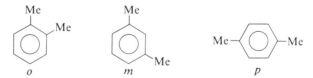

# 7

# Halogen-Containing Compounds:
## Alkyl and Aryl Halides

Organic halides are extremely valuable intermediates in organic synthesis. They are generally easy to prepare and undergo a variety of interesting reactions. However, organic halides are rarely found in living systems and are often poisonous to both plant and animal life. In this chapter we will examine both the preparations and reactions of organic halides, and how they are related to functional-group transformations. We will also explore the uses of these halides as insecticides, cleansing agents, aerosol propellants (in spray cans), and refrigerants, as well as the environmental problems that arise from their abuse.

## PHYSICAL PROPERTIES AND NOMENCLATURE

Most inorganic halides are ionic and fairly soluble in polar solvents such as water. Organic halides, on the other hand, are surprisingly covalent in nature. In fact organic halides are the only group of organic compounds other than saturated hydrocarbons that are insoluble in concentrated sulfuric acid. Polyhalides, such as carbon tetrachloride ($CCl_4$), are quite useful solvents for waxes, fats, and oils, but not for ionic or highly polar compounds. And yet it is the inherent polarity of the C—X bond that is responsible for all the *chemistry* of organic halides. It is this strange dichotomy between hydrocarbon-like molecular behavior and chemical reaction by ionic or polar intermediates that makes the study of organic halides so intriguing.

In the IUPAC system, alkyl and aryl halides are named as derivatives of the parent hydrocarbon, with the halogen atoms being named as substituents.

$$\overset{1}{C}H_3\overset{2}{C}H_2\overset{3}{C}HCH_2CH_3$$
$$\qquad\qquad\quad |$$
$$\qquad\qquad\quad Cl$$

Parent hydrocarbon: pentane
Name: 3-chloropentane

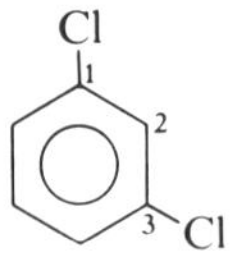

Parent hydrocarbon: benzene
Name: 1,3-dichlorobenzene

—F   fluoro-
—Cl  chloro-
—Br  bromo-   } Halo substituents
—I   iodo-

However, an alternative system, referred to as *common names*, has developed and persisted over the past century. This system calls attention to the fact that these compounds *are*, indeed, halides. The alkyl or aryl group is named followed by the appropriate halide, similar to the manner in which inorganic halides are named. Although this "system" is not systematic at all, its usefulness for naming simple halides has guaranteed its survival. Several examples of both methods of naming are shown below:

| Compound | Common | IUPAC |
|---|---|---|
| $CH_3CH_2Br$ | Ethyl bromide | Bromoethane |
| $(CH_3)_2CHCl$ | *i*-Propyl chloride | 2-Chloropropane |
| $CH_3CH_2C(CH_3)_2Cl$ | *t*-Amyl chloride | 2-Chloro-2-methyl butane |
| 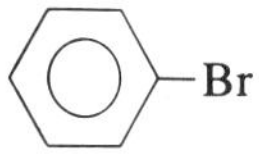 ⟨◯⟩—Br | Phenyl bromide | Bromobenzene |
|  ⟨◯⟩—$CH_2Cl$ | Benzyl chloride | Chlorophenylmethane |
| $CH_2{=}CHCl$ | Vinyl chloride | Chloroethene |
| $CH_2{=}CHCH_2Br$ | Allyl bromide | 3-Bromopropene |

## PREPARATION OF ALKYL AND ARYL HALIDES

Alkyl halides can be prepared from a number of different functional groups, but by far the most useful starting materials are alkenes and alcohols. Direct halogenation of hydrocarbons has been discussed previously (Chapter I-3) but is only synthetically useful when a particularly stable free radical can be formed as an intermediate.

Aromatic halides are best prepared by direct electrophilic aromatic substitution (Chapter I-6) or from diazonium salts. Vinylic, allylic, and benzylic halides are all useful intermediates and can be prepared by specific methods.

Examples of each of these conversions are outlined below, as well as any specific limitations associated with the method.

1. Halides from alkenes—addition to $\underset{\diagdown}{\overset{\diagup}{C}}=\underset{\diagup}{\overset{\diagdown}{C}}$ :

$$RCH{=}CH_2 \xrightarrow{\text{HX}} \underset{\underset{X}{|}}{RCHCH_3} \quad X = Cl, Br, I$$

Examples:

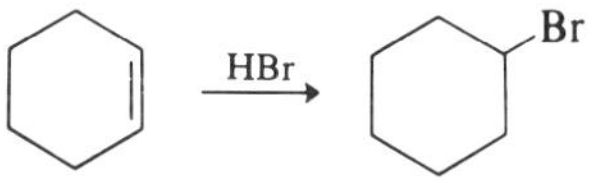

Bromocyclohexane

$$(CH_3)_2C{=}CHCH_3 \xrightarrow{\text{HCl}} (CH_3)_2\overset{\overset{\text{Cl}}{|}}{C}CH_2CH_3$$

2-Chloro-2-methylbutane

The addition occurs stepwise, with the proton adding to give the *most* stable carbonium ion (see Chapter I-5), followed by addition of $:\ddot{X}:^{\ominus}$.

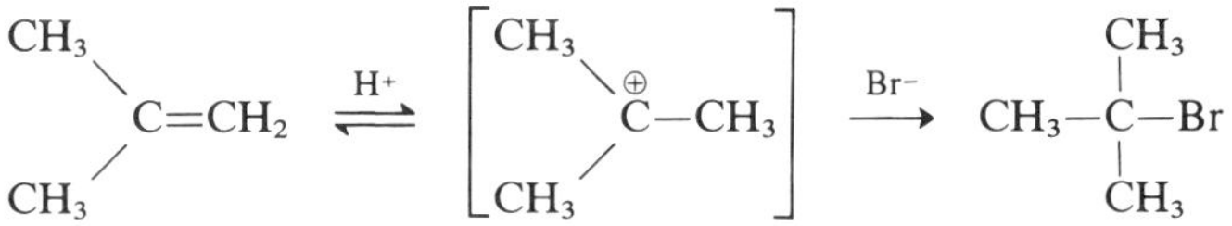

Dihalides may also be prepared from alkenes by addition of either bromine or chlorine in an inert solvent such as carbon tetrachloride.

$$CH_3CH{=}CHCH_3 \xrightarrow[\text{CCl}_4]{X_2} \underset{\underset{X}{|}}{CH_3\overset{\overset{X}{|}}{C}HCHCH_3} \quad X = Cl \text{ or } Br$$

2. Halides from alcohols
   (a)

$$ROH + HX \rightleftharpoons RX + H_2O$$
$$X = Cl, Br, I$$

$3° > 2° > 1° > CH_3OH$
Ease of ROH reaction

This reaction works best for tertiary alcohols. Since it is essentially an equilibrium, excess acid drives the reaction to the right.  Hydrogen halide may also be generated *in situ* by utilizing NaX and $H_2SO_4$.

Examples:

$$CH_3CH_2\overset{\overset{\displaystyle CH_3}{|}}{\underset{\underset{\displaystyle CH_3}{|}}{C}}-OH \xrightarrow{\text{Conc. HCl}} CH_3CH_2\overset{\overset{\displaystyle CH_3}{|}}{\underset{\underset{\displaystyle CH_3}{|}}{C}}-Cl$$

2-Chloro-2-methylbutane

$$CH_3CH_2CH_2CH_2CH_2OH \xrightarrow[H_2SO_4]{\text{NaBr}} CH_3CH_2CH_2CH_2CH_2Br$$

1-Bromopentane

(b)

$$ROH + PX_3 \longrightarrow RX + P(OH)_3$$
$$X = Cl \text{ or } Br$$

Examples:

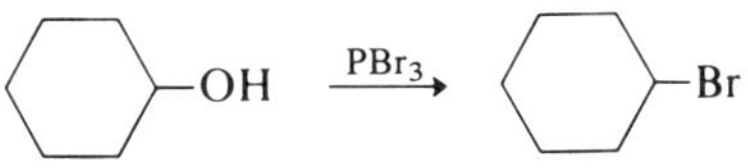

—OH $\xrightarrow{PBr_3}$ —Br  Bromocyclohexane

$$CH_2{=}CHCH_2OH \xrightarrow{PCl_3} CH_2{=}CHCH_2Cl$$

3-Chloropropene

(c)

$$ROH + SOCl_2 \longrightarrow RCl + SO_2{\uparrow} + HCl{\uparrow}$$

This reaction is particularly clean because the inorganic products are gaseous and easily removed from the reaction mixture.

Example:

$$CH_3CH_2\underset{\underset{\displaystyle OH}{|}}{CH}CH_2CH_3 \xrightarrow{SOCl_2} CH_3CH_2\underset{\underset{\displaystyle Cl}{|}}{CH}CH_2CH_3$$

3-Chloropentane

3. Halides by direct substitution of alkanes or alkenes:

$$RX + X_2 \xrightarrow[\text{or } h\nu]{\Delta \cdot} R{-}X + HX$$

Direct halogenation is a free-radical process and thus nonspecific, unless all hydrogens in the molecule are equivalent, or if a particularly stable free radical can be formed:

Hydrogen reactivity:    Allylic    $> 3^{\circ} > 2^{\circ} > 1^{\circ} > C{=}C$    Vinylic

Benzylic

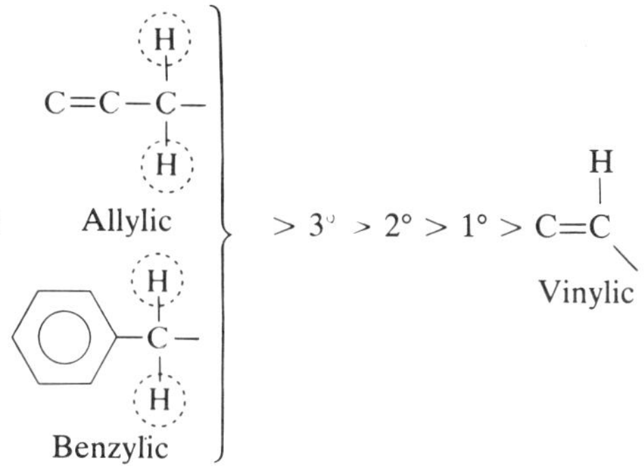

Examples:

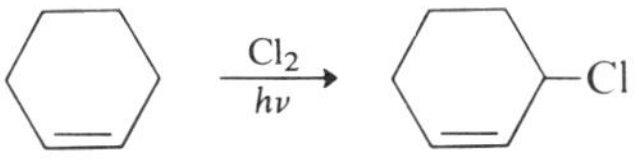

3-Chlorocyclohexene

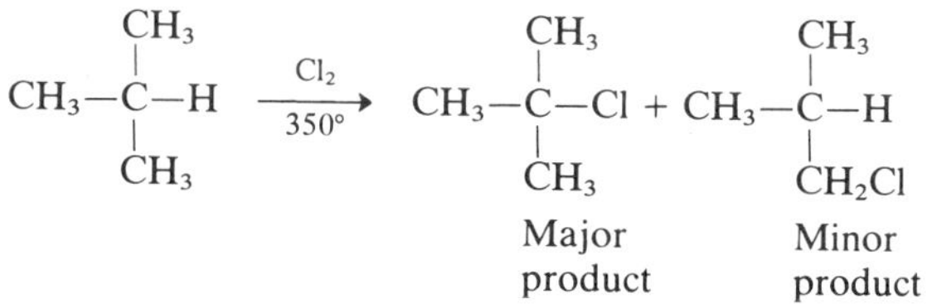

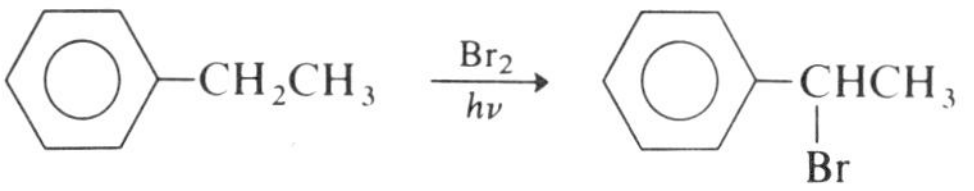

$$CH_3{-}\underset{\underset{CH_3}{|}}{\overset{\overset{CH_3}{|}}{C}}{-}H \xrightarrow[350^{\circ}]{Cl_2} CH_3{-}\underset{\underset{CH_3}{|}}{\overset{\overset{CH_3}{|}}{C}}{-}Cl + CH_3{-}\underset{\underset{CH_2Cl}{|}}{\overset{\overset{CH_3}{|}}{C}}{-}H$$

Major          Minor
product        product

$-CH_2CH_3 \xrightarrow[h\nu]{Br_2}$ $-\underset{\underset{Br}{|}}{CHCH_3}$    1-Bromo-1-phenylethane

4. Vinylic halides from alkynes:

$$RC{\equiv}CH \xrightarrow[\text{1 equiv.}]{HX} R{-}\underset{\underset{X}{|}}{C}{=}CH_2 \quad X = Cl, Br, I$$

Example:

$$HC{\equiv}CH \xrightarrow[HCl]{\text{1 equiv.}} CH_2{=}CHCl$$

Vinyl chloride

5. Aromatic halides:
   (a) Electrophilic substitution:

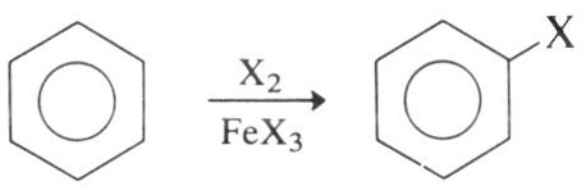

$$\xrightarrow[FeX_3]{X_2} \quad X \qquad X = Cl, Br$$

(b) Nucleophilic substitution of diazonium salts:

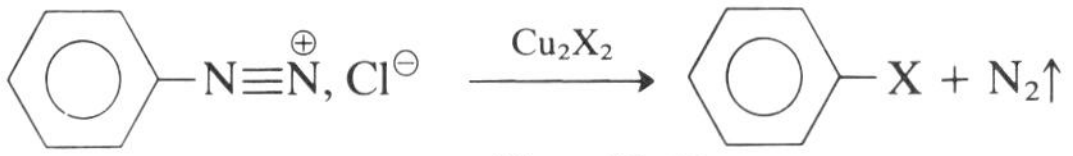

$$\langle\bigcirc\rangle-N\equiv\overset{\oplus}{N}, Cl^{\ominus} \xrightarrow{\text{Cu}_2\text{X}_2} \langle\bigcirc\rangle-X + N_2\uparrow$$

$$X = Cl, Br$$

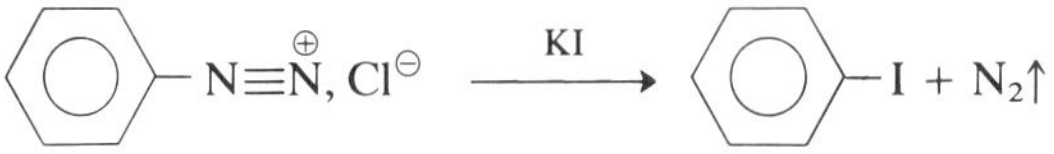

$$\langle\bigcirc\rangle-N\equiv\overset{\oplus}{N}, Cl^{\ominus} \xrightarrow{\text{KI}} \langle\bigcirc\rangle-I + N_2\uparrow$$

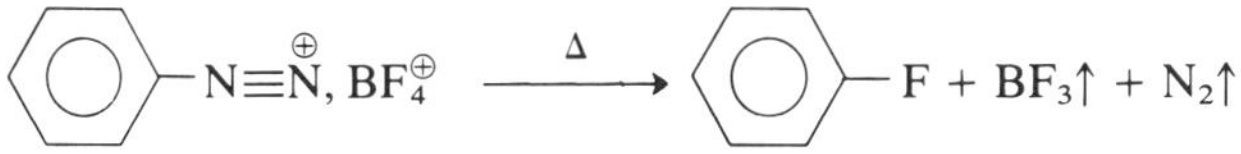

$$\langle\bigcirc\rangle-N\equiv\overset{\oplus}{N}, BF_4^{\oplus} \xrightarrow{\Delta} \langle\bigcirc\rangle-F + BF_3\uparrow + N_2\uparrow$$

Diazonium salts are usually prepared from aromatic hydrocarbons as follows:

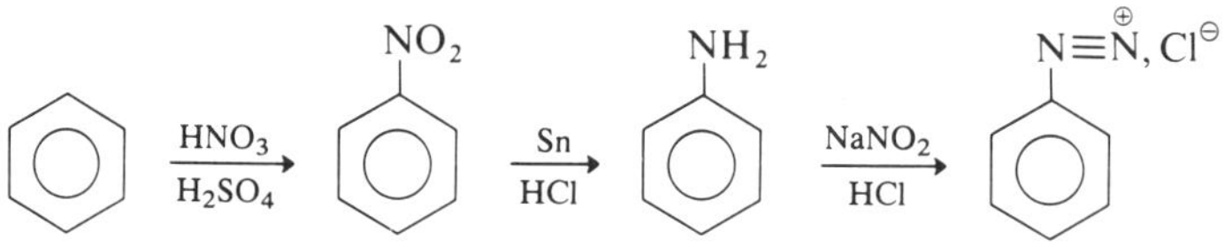

$$\bigcirc \xrightarrow[\text{H}_2\text{SO}_4]{\text{HNO}_3} \overset{NO_2}{\bigcirc} \xrightarrow[\text{HCl}]{\text{Sn}} \overset{NH_2}{\bigcirc} \xrightarrow[\text{HCl}]{\text{NaNO}_2} \overset{N\equiv\overset{\oplus}{N},\,Cl^{\ominus}}{\bigcirc}$$

Benzene
diazonium
chloride

Other reactions of these valuable synthetic intermediates will be discussed in Chapter I-13.

**Exercise 7-1:** Give at least *two* different methods of preparing each of the following halides from appropriate starting materials:

(a) 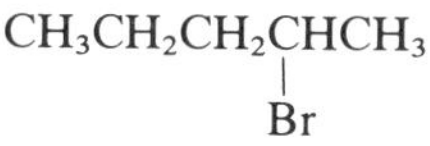 $CH_3CH_2CH_2\underset{\underset{Br}{|}}{C}HCH_3$

(b) 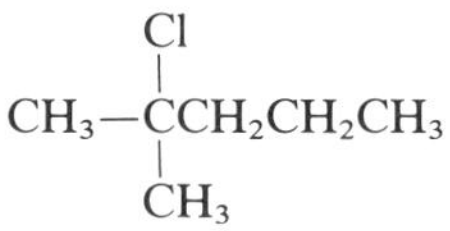 $CH_3-\underset{\underset{CH_3}{|}}{\overset{\overset{Cl}{|}}{C}}CH_2CH_2CH_3$

(c)

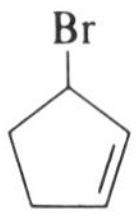

(d)

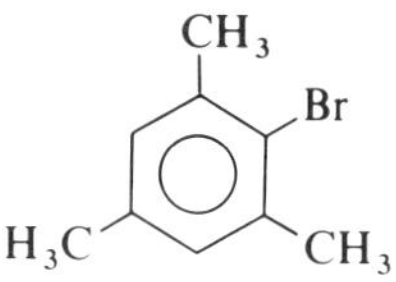

Example:

$$
\underset{\underset{CH_3}{|}}{\overset{\overset{Cl}{|}}{CH_3CHCHCH_3}} \qquad \text{May be prepared from an alcohol or alkene}
$$

$$
\underset{\underset{CH_3}{|}}{\overset{\overset{OH}{|}}{CH_3CHCHCH_3}} \xrightarrow{\text{SOCl}_2} \underset{\underset{CH_3}{|}}{\overset{\overset{Cl}{|}}{CH_3CHCHCH_3}}
$$

$$
\underset{\underset{CH_3}{|}}{CH_3CHCH{=}CH_2} \xrightarrow{\text{HCl}} \underset{\underset{CH_3}{|}}{\overset{\overset{Cl}{|}}{CH_3CHCHCH_3}}
$$

## REACTIONS OF ALKYL HALIDES

Alkyl halides undergo a number of different types of reactions, almost all of which have tremendous synthetic importance. Basically, these reactions fall naturally into one of two different categories: (1) replacement of the halogen by some other reactive group; or (2) elimination of the halide and some other atom to form an alkene or alkyne. In the latter case, HX eliminations are by far the most important example and are referred to as *dehydrohalogenations.* In the following sections we will discuss both kinds of reaction and show examples of their synthetic utility.

(1)
$$
\overset{\overset{X}{|}}{RCHR} \xrightarrow{Z:} \overset{\overset{Z}{|}}{R{-}CH{-}R} + :\ddot{X}:^{\ominus} \quad \text{Replacement of halide by Z}
$$

(2)
$$
\overset{\overset{X}{|}}{RCHCH_2R} \xrightarrow{-HX} RCH{=}CHR \quad \text{Elimination of HX}
$$

### Replacement Reactions—
### Nucleophilic Substitution

Alkyl halides react with a large number of different nucleophiles to produce substitution products. Such reactions may be referred to in a shorthand notation as $S_N$ reactions (*S*ubstitution, *N*ucleophilic). The nucleophile, either a negative ion or a molecule containing an unshared electron pair, displaces the halide as a negative ion, as illustrated below in the formation of alcohols:

$$R\ddot{\ddot{X}}: + :\ddot{O}H^{\ominus} \longrightarrow R\ddot{O}H + :\ddot{\ddot{X}}:^{\ominus}$$

$$R\ddot{\ddot{X}}: + \underset{H}{HÖ:} \longrightarrow R\ddot{O}H + H^{\oplus} + :\ddot{\ddot{X}}:^{\ominus}$$

The substitution process may be regarded as a nucleophilic attack at a carbon, with subsequent displacement of the halide ion (leaving group). A more complete discussion of the mechanism of this process may be found in Chapter II-2. Some typical nucleophilic substitution reactions are shown for $n$-propyl halide in Table 7-1.

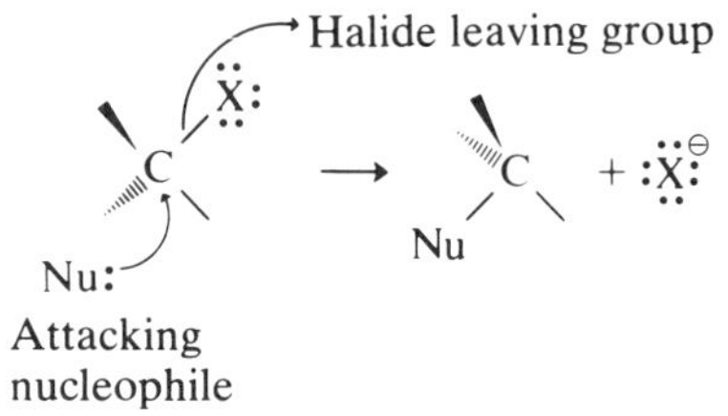

Halide leaving group

Nu:
Attacking
nucleophile

Table 7-1.  Typical Nucleophilic Substitution Reactions
of Alkyl Halides

| Halide | Nucleophile* | Substitution product |
|---|---|---|
| $CH_3CH_2CH_2Cl$ | $Na^+OH^-$ | $CH_3CH_2CH_2\underline{OH}$ <br> $n$-Propyl <u>alcohol</u> |
| $CH_3CH_2CH_2Br$ | $Na^+OCH_3^-$ | $CH_3CH_2CH_2\underline{OCH_3}$ <br> $n$-Propyl methyl <u>ether</u> |
| $CH_3CH_2CH_2Br$ | $Na^+\overset{-}{C}\equiv CH$ | $CH_3CH_2CH_2\underline{C\equiv CH}$ <br> 1-pentyne |
| $CH_3CH_2CH_2I$ | $K^+CN^-$ | $CH_3CH_2\underline{CH_2C\equiv N}$ <br> $n$-Propyl <u>cyanide</u> |
| $CH_3CH_2CH_2Br$ | $Na^+\overset{-}{O}COCH_3$ <br> (NaOAc) | $CH_3CH_2CH_2O-\underset{\underset{O}{\|\|}}{C}-CH_3$ <br> $n$-Propyl <u>acetate</u> |
| $CH_3CH_2CH_2Br$ | $Na^+\overset{-}{S}H$ | $CH_3CH_2CH_2\underline{SH}$ <br> $n$-Propyl <u>mercaptan</u> |

* Reactions normally run at room temperature in polar solvents.

As can readily be seen from Table 7-1, nucleophilic substitution reactions yield an amazing variety of compounds by a relatively simple reaction process. We shall soon see that the addition of this "weapon" to our "arsenal" of synthetic procedures greatly extends out ability to interconvert and synthesize organic molecules.

**Exercise 7-2:** Postulate a nucleophilic substitution reaction that would produce each of the following products from an alkyl halide:

(a) $CH_2=CHCH_2OCH_2CH_3$

(b) 

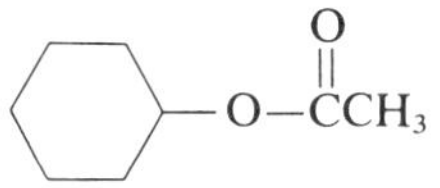

(c) $(CH_3)_2CHCH_2CH_2C\equiv N$

(d) 

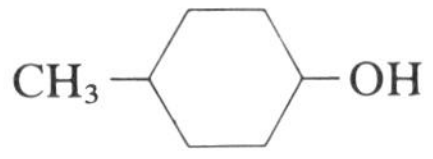

## Elimination Reactions—Dehydrohalogenation

The elements of $H^+X^-$ may be eliminated from an organic halide by strong base, producing a double bond. Triple bonds may also be produced from dihalides by eliminating two HX equivalents. Eliminations are usually carried out in a polar solvent at elevated temperatures, as indicated in the following examples.

(a)

$$CH_3CH_2CHCH_3 \xrightarrow[\text{Reflux}]{\text{KOH} \atop \text{EtOH}} CH_3CH=CHCH_3$$
$$\underset{Cl}{|}$$

Major product

\+

$$CH_3CH_2CH=CH_2$$

Minor product

(b)

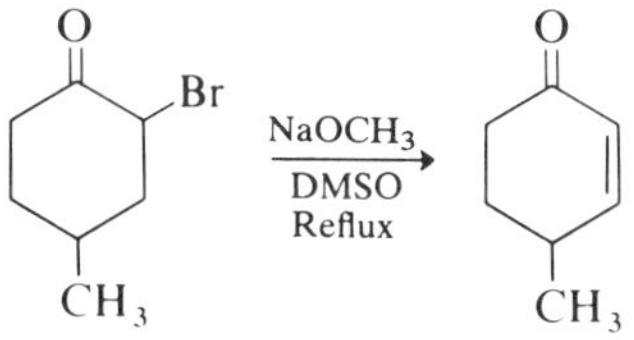

(c)

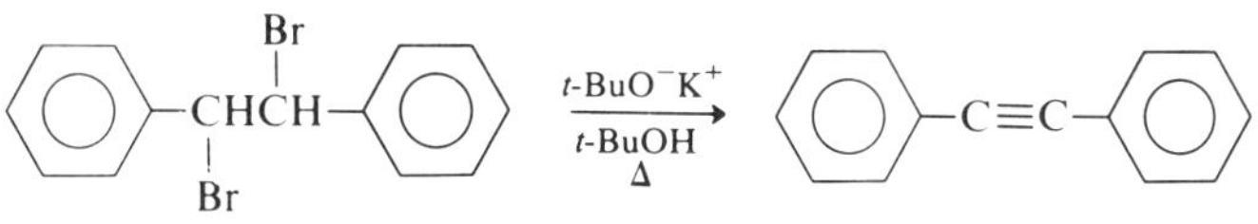

In all the preceding cases prolonged heating, usually at the solvent reflux temperature, is required. In example (a) two products are possible and although both are formed, the major product is the more highly substituted olefin. This observation has been generalized as the *Saytzeff rule*, previously mentioned in Chapter I-5:

*In an ionic elimination reaction in which more than one olefin may be formed, the more highly substituted product is favored.*

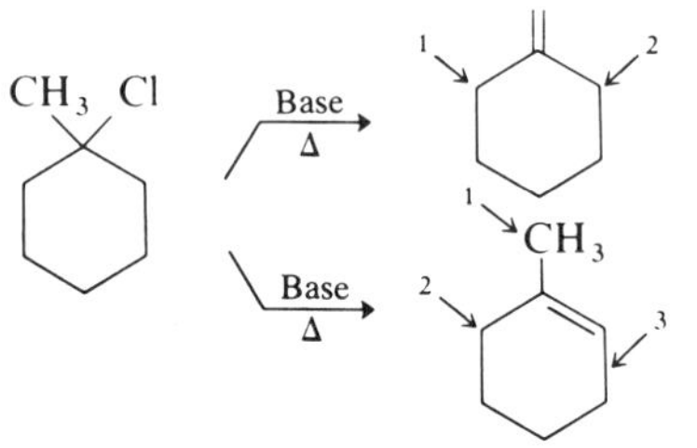

Two alkyl groups—
minor product

Three alkyl groups—
major product

In Saytzeff elimination the more highly substituted olefin predominates.

Other bases that are often utilized for elimination reactions include sodium acetate ($NaOCOCH_3$), sodamide ($NaNH_2$), and any tertiary aromatic amine ($PhNMe_2$ or $PhNEt_2$).

**Exercise 7-3:** Predict the products of each of the following elimination reactions:

(a)

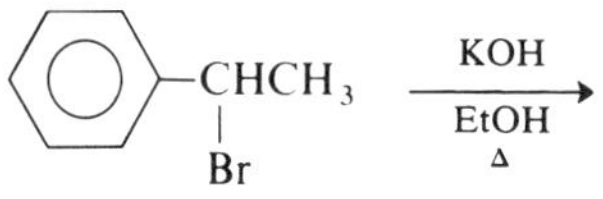

(b)

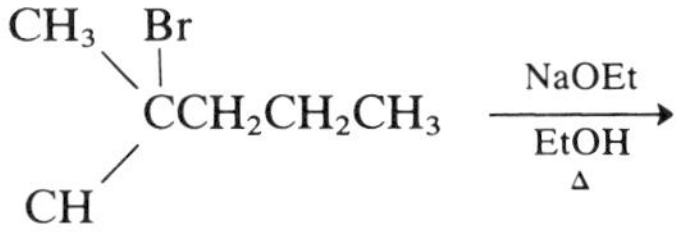

(c)

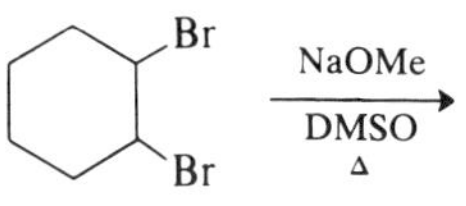

# COMPETITION BETWEEN SUBSTITUTION AND ELIMINATION

Bases are nucleophiles. It should not be surprising, then, that substitution and elimination pathways are related and compete with one another for reaction dominance. In any reaction in which one pathway dominates, the other is always present and relegated to a role as a side reaction. In general, substitution is favored under mild conditions (dilute solutions and moderate to low temperatures), while elimination is favored under stringent conditions (concentrated base and high temperature).

$$RCH_2CHCH_3 \overset{Br}{\underset{}{|}} \begin{cases} \xrightarrow[\text{Room temp.}]{\text{Dil. NaOH}} & RCH_2\overset{OH}{\underset{}{|}}CHCH_3 \quad S_N \\ \xrightarrow[100°]{\text{Conc. NaOH}} & RCH{=}CHCH_3 \quad E \end{cases}$$

A more complete discussion of these two competing mechanisms may be found in Chapter II-2.

# FORMATION OF ORGANOMETALLIC COMPOUNDS

We have previously seen that alkyl halides react with sodium metal to produce hydrocarbon coupling products (Wurtz reaction). An organosodium intermediate has been postulated for this process, and we refer to such molecules as *organometallic* compounds.

$$RBr + Na \longrightarrow [RNa] \xrightarrow{RBr} R{-}R$$

Although organosodium intermediates are not very stable or long lived, many other organometallic compounds are. We will discuss three such classes in this section: (1) organomagnesium halides, or Grignard reagents; (2) organolithium reagents; and (3) lithium dialkyl copper reagents. Each of these is an extremely valuable synthetic tool in the hands of an experienced organic chemist.

## Organomagnesium Halides— Grignard Reagents

In Chapter I-3 the use of Grignard reagents, RMgX, in hydrocarbon synthesis was illustrated. First discovered in the late nineteenth century by the young French graduate student for whom they are named, they have since become one of the most widely used classes of organic reagents. In order for the reagent to form, ethers must be employed as solvents. Normally ethyl ether is utilized for common alkyl halides, while tetrahydrofuran is required for less reactive halides. Apparently ether is necessary in order to stabilize the electron-deficient organometallic species in solution by utilizing the unshared electron pairs of the ether–oxygen to form coordinate covalent linkages.

$$CH_3CH_2CH_2Br + Mg \xrightarrow{Et\ddot{O}Et} \left[ \begin{array}{c} Et \quad Et \\ \ddot{O} \\ CH_3CH_2CH_2-Mg-Br \\ \ddot{O} \\ Et \quad Et \end{array} \right]$$

*n*-Propylmagnesium bromide
stabilized by ethyl ether

$$CH_2=CHBr + Mg \xrightarrow{\phantom{xx}} \left[ \begin{array}{c} CH_2=CH-\overset{..}{\underset{..}{M}}g-Br \end{array} \right]$$

Vinylmagnesium bromide
stabilized by tetrahydrofuran

Grignard reagents, and other organometallic reagents as well, react rapidly and sometimes explosively with water or other compounds containing active hydrogens, forming RH and Mg(OH)X with $H_2O$, for example. In their reactions, the alkyl group may be considered to have carbanion character, and most reactions of organometallic compounds can be rationalized on this basis. This property may be

$$R^{\delta^-}-Mg^{\delta^+}-X^{\delta^-}$$

$$R^{\delta^-} \quad CH_2-CH=CH_2 \xrightarrow{Ether} R-CH_2CH=CH_2 \quad \text{Coupling}$$
$$Mg \quad Cl \quad \text{Reactive} \qquad + MgXCl\downarrow \qquad \text{product}$$
$$X \qquad \text{allylic halide}$$

utilized in reactions with very reactive halides, such as allylic or benzylic species, to produce coupling products. Although this is a useful hydrocarbon synthesis, by far the most valuable reactions of Grignard reagents involve their addition to carbonyl compounds. We will discuss this reaction in great detail in the next chapter where its synthetic usefulness will become apparent.

$$RMgX + \underset{\substack{\\ \text{Carbonyl} \\ \text{group}}}{\overset{\displaystyle O}{\underset{\displaystyle \|}{C}}} \longrightarrow \underset{\substack{\\ \text{Addition} \\ \text{complex}}}{\overset{\displaystyle OMgX}{-\underset{\displaystyle R}{C}-}} \xrightarrow{H_3O^+} \underset{\substack{\\ \text{An alcohol}}}{\overset{\displaystyle OH}{-\underset{\displaystyle R}{C}-}}$$

## Organolithium Reagents

Although organosodium compounds are unstable, even in solution, organolithium reagents may be readily prepared and are relatively stable in the absence of water. They may be prepared by direct lithiation (1° or 2° halides) in hydrocarbon solvents, or by exchange reactions for less reactive halides.

$$RX + 2\,Li \longrightarrow RLi + LiX\downarrow$$

$$CH_3CH_2CH_2CH_2Br + Li \xrightarrow{\text{Hexane}} CH_3CH_2CH_2CH_2Li$$

n-Butyllithium

$$RLi + R'X \rightleftharpoons RX + R'Li$$

Phenyllithium

Alkyllithium reagents are strong bases and powerful nucleophiles. Although they will generally yield the same reaction products as Grignards, such as alcohols by addition to a carbonyl, their ability to abstract protons from a large variety of organic compounds, or to form new reagents, makes them much more versatile. A few examples shown below will illustrate this versatility.

(a)

$$CH_3CH_2CH_2CH_2Li \xrightarrow[\substack{-20° \\ 2.\ H_3O^+}]{1.\ CH_2=CHCHO} CH_3CH_2CH_2CH_2\overset{\overset{\displaystyle OH}{|}}{C}HCH=CH_2$$

Addition to carbonyl

(b)

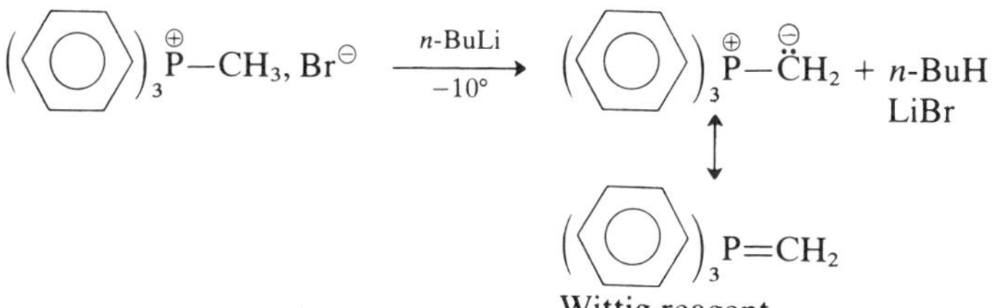

Wittig reagent

(c)

$$CBrCl_3 \xrightarrow{\text{\emph{n}-BuLi}} [Li^{\oplus}\!:\!\overset{\ominus}{C}Cl_3] \xrightarrow{-LiCl} [:CCl_2]$$

Dichlorocarbene

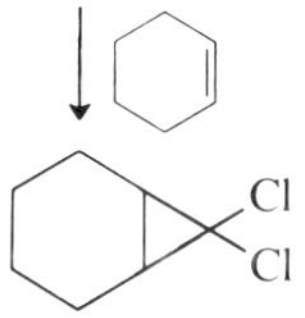

## Lithium Dialkyl Copper Reagents

In recent years a new organometallic reagent, formed from alkyl lithium reagents and cuprous iodide, has essentially replaced many of the more famous hydrocarbon syntheses, such as the Wurtz reaction and Grignard coupling. Lithium dialkyl copper reagents will couple with 1°, 2°, and 3° alkyl halides as well as vinyl or

$$4\,RLi + Cu_2I_2 \longrightarrow 2\,R_2CuLi + 2\,LiI$$

$$CH_3Li + Cu_2I_2 \longrightarrow \underset{\text{Lithium dimethyl copper}}{(CH_3)_2CuLi}$$

aryl halides in good yield. Thus many otherwise difficult hydrocarbon syntheses have been simplified, as some of the following examples illustrate:

(a)

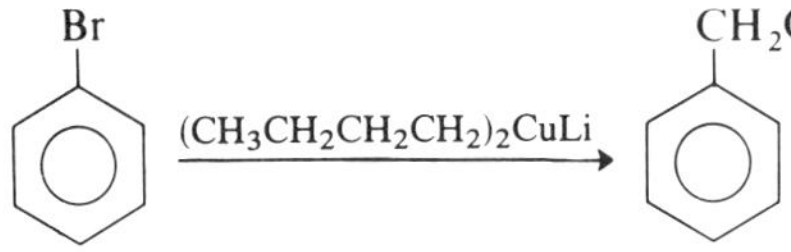


(b)

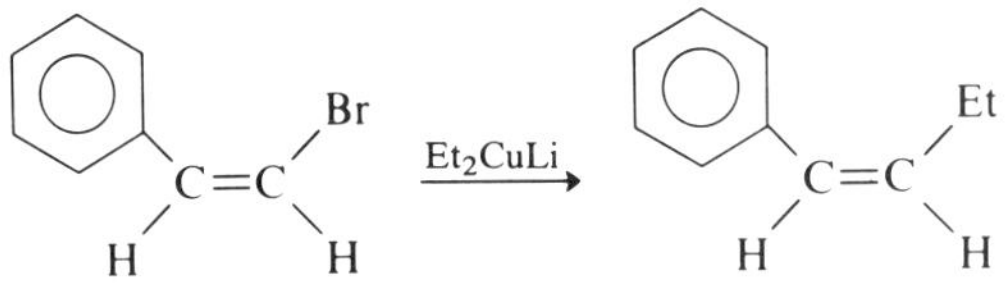


(c)

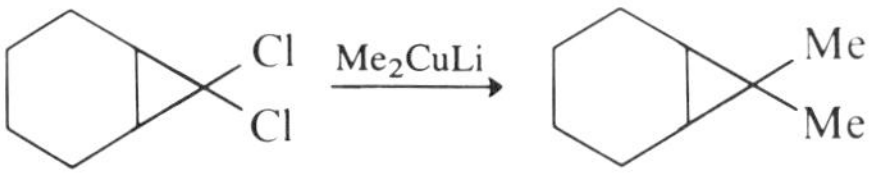


(d)

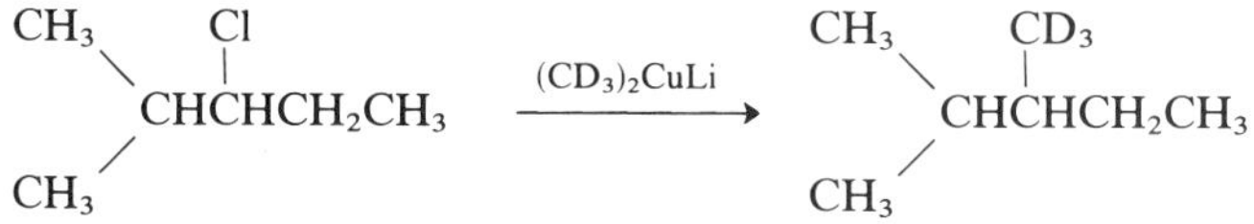


---

**Exercise 7-4:** Predict which of the following compounds might react readily with *n*-BuLi to form new RLi compounds. Consider both the acidity of the proton being abstracted and also the stability of the product.

(a) 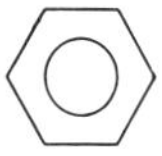

(b) $CH_3CH_3$

(c) 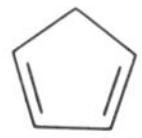

(d) 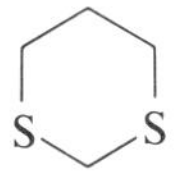


(e) 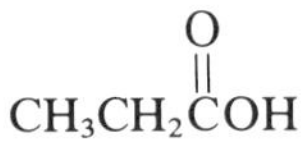


(f) $[(CH_3)_2CH]_2NH$

(g) 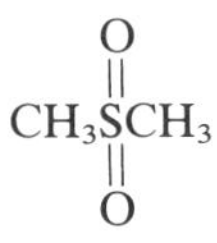


---

Until recent years, very few people realized the extent of our increasing dependence on halo-organic compounds in our everyday life. If asked about their use, most people recall that chloroform ($CHCl_3$) was once used as a general anesthetic, or that carbon tetrachloride ($CCl_4$) was one of the first dry-cleaning solvents. Baseball fans may also know that the "gas" sprayed on minor injuries during a game is really ethylchloride ($CH_3CH_2Cl$), which minimizes pain and swelling by rapidly cooling the injured area through evaporation. Polyvinyl chloride, a vinyl-type plastic, is becoming ubiquitous in our society, replacing metal pipe in plumbing, wood and metal in car dashboards, and finding literally countless other uses. However, there are many other uses and abuses that have only recently been publicized, particularly in relation to our health and our environment. A few of the more important of these will be discussed below.

### Freons

Freon is a general name for the mixed fluorochloromethanes. Their original use was as refrigerating liquids in household refrigerators, replacing the more dangerous liquid ammonia. They are generally prepared by fluorinating carbon tetrachloride. However, in recent years their use has spread alarmingly to include a

$$CCl_4 \xrightarrow[SbF_5]{HF} CCl_3F \longrightarrow CCl_2F_2 \longrightarrow CClF_3 \longrightarrow CF_4$$

$$\text{Freon 11} \quad\quad \text{Freon 12} \quad\quad \text{Freon 13} \quad\quad \text{Freon 14}$$

myriad of spray-can products such as hairspray, deodorant, paint, insecticides, and shave cream, to name but a few. Unfortunately, these propellant gases become part of our atmosphere, and have now been accumulating for several years. We now realize, belatedly, that freons do not degrade rapidly in the environment. The most potentially dangerous consequence of this fact is that freons may react with, and deplete, the ozone ($O_3$) layer in the upper atmosphere. The ozone layer is generally regarded as an essential barrier to ultraviolet radiation from the sun, and its destruction could give rise to severe health problems, such as a significant rise in the occurrence of skin cancers. Several manufacturers have voluntarily abandoned the use of freons in their products, and more recently the FDA has ordered that freon aerosols should be banned for nonessential use in food, cosmetics, and household cleaners.

### Insecticides

Simple organic halides were utilized for many years to repel insects in stored grain or in soil. A few of these simple compounds, sometimes referred to as *first-generation insecticides*, are listed below. However, in general these first-generation insecticides were hit-or-miss (mostly miss) and were ineffectual against such scourges of man as the malaria mosquito, the cotton boll weevil, and the gypsy moth (whose larval form can literally eat every leaf in a forest).

$$\underset{\substack{| \quad | \\ Cl \quad Cl}}{CH_2CH_2} \quad \text{and} \quad \underset{\substack{| \quad | \\ Br \quad Br}}{CH_2CH_2} \quad \text{Used to fumigate grain or soil}$$

1,2-Dichloroethane    1,2-Dibromoethane

Cl—⟨O⟩—Cl     Used to prevent clothes moths and carpet beetles from attacking wool products

para-Dichlorobenzene

$CH_3Br$          Used to fumigate flour mills, grain storage
Methyl bromide     houses, freight cars, and ships

The pressure of war in the South Pacific and in Europe in the 1940s (World War II) led to the development of many *second-generation insecticides*. By far the most important of these, and certainly the most publicized, is DDT. The compound was first utilized by the U.S. Army in Naples, Italy, in 1944 to combat a serious outbreak

Cl—⟨O⟩—C—⟨O⟩—Cl     *D*ichloro*di*phenyl*t*richloroethane,
DDT

of typhus, a disease transmitted by the common body louse. Spraying of the malaria mosquito breeding grounds with DDT almost, but not quite, eradicated the dread disease in a few years. An example of its effectiveness in the United States and in Sri Lanka (Ceylon) will indicate just how effective DDT really was. The World Health Organization (WHO) has estimated that the widespread use of DDT probably saved more than 25,000,000 lives and prevented several hundred million serious cases of illness. (See Table 7-2.)

Table 7-2.    Reported Cases of Malaria in the U.S. and Sri Lanka

| Year | U.S. | Sri Lanka | Comment |
|------|------|-----------|---------|
| 1940 | 78,129 | — | Before DDT use |
| 1946 | — | 2,800,000 | Spraying begun |
| 1955 | 522 | — | DDT in use 8 years |
| 1963 | — | 17 | Spraying stopped |
| 1968 | — | 1,000,000 | No DDT being used |

DDT has proved to be a mixed blessing, however. Many insects, such as the malaria-carrying mosquito, the house fly, and the body louse have developed resistance to the biochemical effects of DDT. Such resistance led to the use of ever-increasing amounts of the insecticide to keep insect populations within tolerable limits. Production of DDT in 1962 reached 167 million pounds/year, and a reasonable estimate of the total amount used in the last 25 years is 3 billion pounds! Unfortunately, DDT persists for an extremely long time once released into the environment. Because of this persistence, DDT has been transported by water systems throughout the world and has even been found in the Antarctic. As DDT progresses up the ecological ladder, it tends to become more concentrated. For example, a typical Coho salmon in Lake Michigan may contain 10–20 ppm DDT, while in the next step up the ladder, a gull may contain up to 3,000 ppm. In 1962, Rachel Carson sounded the alarm in her now famous book *Silent Spring*, and after much public clamor and scientific research, DDT was finally banned from further use.

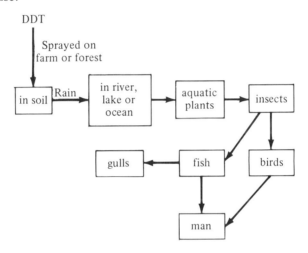

Other organohalogen insecticides took its place. The following brief listing shows a variety of different compounds and structures.

Mirex

Methoxychlor

DDVP
(Shell's No-Pest Strip)

While these new insecticides temporarily solved the insect resistance problem, their effect on the environment is equally bad, and several (Aldrin, Dieldrin) have also been banned from further use.

Chlorine-containing pesticides will be in our environment for a long time. In the United States, it has been estimated that human body fat has already accumulated 5–20 ppm DDT and 1 ppm Dieldrin. They are found in poultry, eggs, cows' milk, and mothers' milk. Yet without their continued use in one form or another, a large percentage of the world's population would probably starve to death. Where to draw the line and where to strike a balance is a problem that future generations must solve.

### Polychlorinatedbiphenyls (PCB's)

Polychlorinatedbiphenyls and polybrominatedbiphenyls (PBB's) have been used as plasticizers and as insulating media in electric transformers for a number of years. Recently it has been recognized that their effects in the environment are similar to those of DDT, and that they follow the same path to man. They have been found in poultry, eggs, milk, and fish. Although their physiological effects have not been well studied, their continued use is being seriously questioned.

Some common polychlorinatedbiphenyls

The main problem with the continued widespread use of the organohalogen compounds in our society is their persistence. Great care must be exercised in the future before any new massive use of this class of chemicals can be safely tolerated.

# REVIEW OF NEW TERMS AND CONCEPTS

Define each term and, if possible, give an example of each:

Organohalogen compound      Nucleophilic displacement of halogen

Dehydrohalogenation      Saytzeff rule

Organometallic compound      Grignard reagent

Organolithium reagent      Lithium dialkyl copper reagent

Grignard coupling reaction      Freon

First-generation insecticide      Second-generation insecticide

PCB

## PROBLEMS: CHAPTER I-7

**1.** Name the following compounds by the IUPAC and common systems were possible:

(a) $CH_2$=$CHCH_2Br$     (b)     (c)

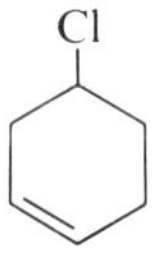

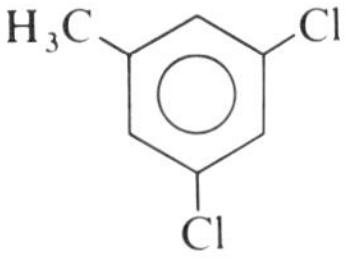

(d) $(CH_3)_2CHCHCH_2CH(CH_3)_2$
                |
                $Cl$

(e) $CH_3CH$=$C-CH_2CH_2-$
               |
               $Br$

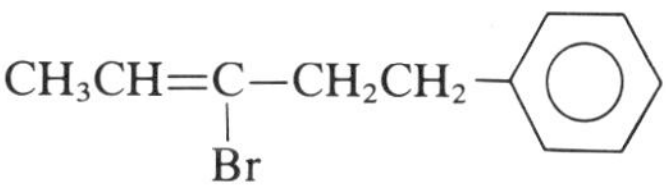

(f) $(CH_3)_2CHCH_2CH_2I$      (g) $ClCH_2CH_2CH_2CH_2CH_2Cl$

**2.** Complete the following reactions showing only organic products. If more than one product is possible, indicate whether they are major or minor.

(a) $(CH_3)_2CHCH_2CH_2OH$ $\xrightarrow{\text{SOCl}_2}$

(b) $CH_3CH$=$CHCH_3$ $\xrightarrow{\text{HBr}}$

(c) $CH_3CH_2CH$=$CH_2$ $\xrightarrow[400°]{\text{Br}_2}$

(d)

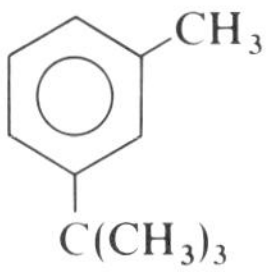

$\xrightarrow[h\nu]{\text{Cl}_2}$

(e) $CH_3CH_2CH_2CH_2Br$ $\xrightarrow{\text{NaOEt}}$

(f) $(CH_3)_2CHMgBr$ $\xrightarrow{CH_2=CHCH_2Cl}$

135

(g) 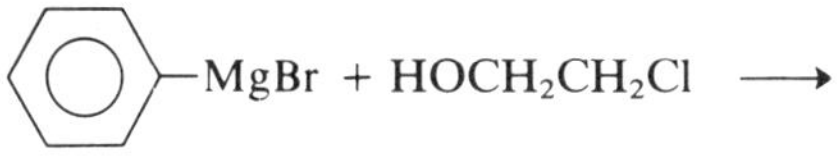   —MgBr + HOCH$_2$CH$_2$Cl  $\longrightarrow$

(h) 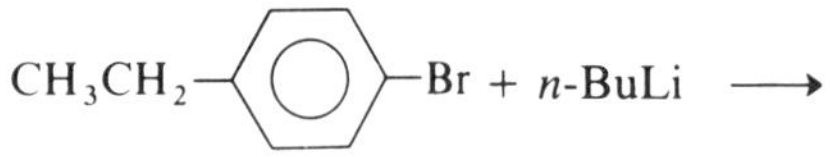  CH$_3$CH$_2$——Br + $n$-BuLi  $\longrightarrow$

(i)

$$\begin{array}{c} Cl \\ | \\ (CH_3)_2CCH_2CH_3 \end{array} \xrightarrow[\Delta]{\text{Alcoholic KOH}}$$

(j)

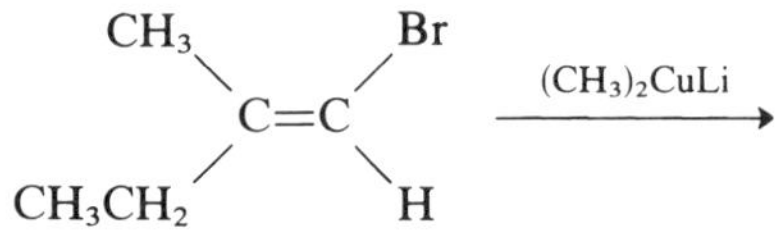

$\xrightarrow{(CH_3)_2CuLi}$

(k) 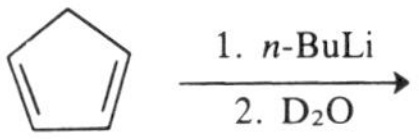   $\xrightarrow[\text{2. D}_2\text{O}]{\text{1. } n\text{-BuLi}}$

**3.** Carry out each of the following conversions utilizing whatever organic or inorganic reagent may be necessary:

(a) 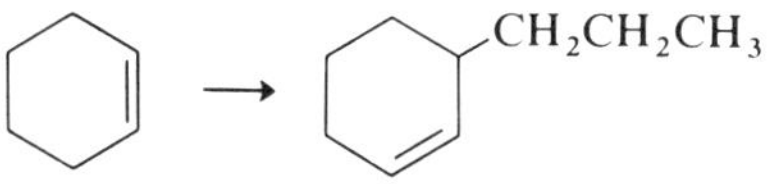   $\rightarrow$  CH$_2$CH$_2$CH$_3$

(b) CH$_3$CH=CHCH$_3$  $\longrightarrow$  CH$_3$C≡CCH$_3$

(c) 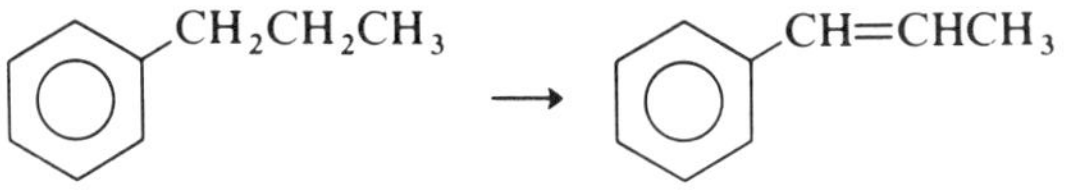  CH$_2$CH$_2$CH$_3$  $\rightarrow$  CH=CHCH$_3$

(d) 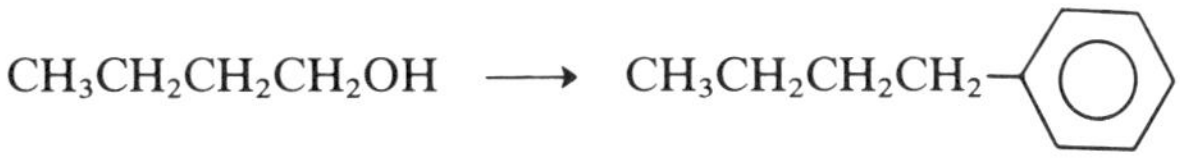 CH$_3$CH$_2$CH$_2$CH$_2$OH  $\longrightarrow$  CH$_3$CH$_2$CH$_2$CH$_2$—

(e)

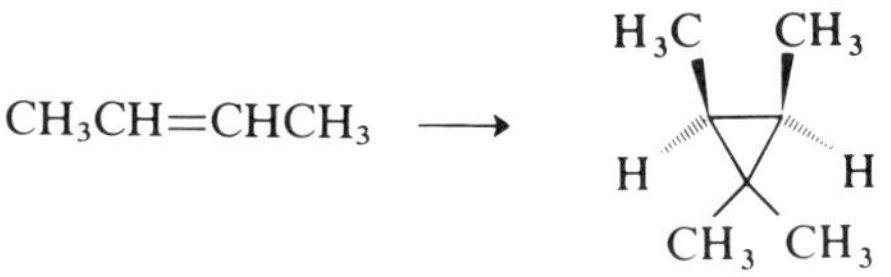

CH$_3$CH=CHCH$_3$  $\longrightarrow$

**4.** Two isomeric unknowns, A and B (C$_6$H$_{11}$Cl), are insoluble in sulfuric acid. Dehydrohalogenation of A gives a single substance C (C$_6$H$_{10}$), while B yields two products of identical formula, D (major) and E (minor). Vigorous oxidation of C yields adipic acid, HOOC(CH$_2$)$_4$COOH. Similarly, upon oxidation D produces

$$\begin{array}{c} CH_3CCH_2CH_2CH_2COOH \\ || \\ O \end{array}$$ and E yields  =O as the only isolable organic product. Suggest structures for A and B and write reactions for all transformations.

**5.** Reaction of CH$_3$CH$_2$CH=CHCH$_2$OH with dilute HCl solution yields two products having identical formulas $C_5H_9Cl$. When these products are separated and hydrolyzed, both yield the same mixture of two alcohols, one of which is identical with the starting material. Suggest reactions which would explain the above observations.

6. Starting with cyclopentene, show how each of the following compounds could be synthesized.

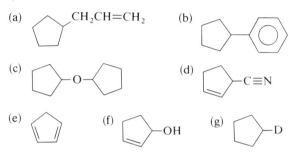

(a)     $CH_2CH=CH_2$     (b)

(c)     (d)     $-C\equiv N$

(e)     (f)     $-OH$     (g)     $-D$

7. The free-radical bromination of cyclohexene is an excellent method of preparing 3-bromocyclohexene. Would the following reaction be an equally effective method of preparing the methyl analog? Explain.

$$\text{(cyclohexene with CH}_3\text{)} \xrightarrow[400°]{Br_2} \text{(cyclohexene with CH}_3\text{ and Br)}$$

8. DDT frequently undergoes slow transformation in the environment to DDE $(C_{14}H_8Cl_4)$. What simple reaction is taking place? Write a structure for DDE.

9. Suggest alternatives for freon aerosol propellants. What advantages or disadvantages would these methods have over freon?

10. Arrange the following compounds in order of increasing reactivity with an aqueous solution of silver nitrate.

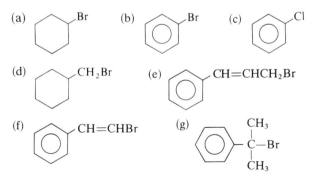

(a)     $Br$     (b)     $Br$     (c)     $Cl$

(d)     $CH_2Br$     (e)     $CH=CHCH_2Br$

(f)     $CH=CHBr$     (g)     $C-Br$ with $CH_3$ groups

# 8

# Alcohols and Ethers

Ethyl alcohol, in the form of wine or other fermented spirits, has been known to man since the earliest recorded times. In fact, it may be that fermentation was the first organic synthesis performed by ancient man, even though he had no knowledge of either the compound itself or of the process involved. Alcohols are compounds containing the *hydroxyl functional group*, —OH, and may be formulated as ROH, where R is any alkyl group. In this chapter we will learn how alcohols may be synthesized and how they react. We will also learn how the solvent properties of alcohol are related to the functional group. Finally, we will learn some important uses of various alcohols in our society.

Ethers are extremely good solvents, although in most people's minds the name "ether" would be associated with hospitals and anesthesia. The ether linkage, R—Ö—R, in which the active hydrogen of an alcohol has been replaced by an alkyl group, is relatively unreactive compared to most other organic functional groups. Together with the coordinating power of the unshared electron pairs on the oxygen atom, this lack of reactivity accounts for the widespread use of ethers as solvents. In this chapter we will explore the preparations of ethers, their use as solvents, and the extent to which their rather limited reactivity yields useful products.

We may consider alcohols and ethers to be derivatives of water in which the active hydrogens have been successively replaced by alkyl groups.

$$HOH \longrightarrow ROH \longrightarrow ROR$$
$$\text{Water} \qquad \text{Alcohol} \qquad \text{Ether}$$

Although this analogy should not be overemphasized, some useful information may be gleaned from the comparison. Water is a highly polar, highly associated liquid. Groups of water molecules are held together by intermolecular hydrogen bonding. The ability of water to solvate both positive and negative charges by association is one of the prime reasons it is a good solvent for ionic or polar compounds.

Water associates by hydrogen-bonding.

Since alcohols also have a polar $\overset{\delta-}{-O}-\overset{\delta+}{H}$ link, one could expect that alcohols also would be associated in the liquid state. Such is the case; however, an alcohol is also an organic compound, and the size and nonpolarity of the alkyl portion affects its ability to associate with itself as well as with other polar or ionic species. Alcohols containing up to three carbons are totally miscible with water, while even $C_4$ and $C_5$ alcohols have a finite solubility (8 g/100 g $H_2O$, and 2 g/100 g $H_2O$, respectively). However, once the organic portion becomes large, solubility in water falls off dramatically and $C_9$, $C_{10}$, and higher alcohols are virtually insoluble.

Alcohols may associate with water or each other by hydrogen-bonding.

Ethers have no active hydrogens, thus they have no tendency for self-association by hydrogen-bonding. However, since they do have unshared electron pairs on the oxygen, they can be hydrogen-bonded to other hydric compounds, such as

139

water or alcohols. Thus ethers are good solvents for polar molecules. Ethers are also good solvents for nonpolar molecules because they themselves are relatively non-polar.

Ethers cannot associate with each other by hydrogen-bonding, but they can associate with other polar compounds containing active hydrogens.

Boiling points of alcohols and ethers demonstrate their relative degrees of association. Alcohols boil considerably higher than alkanes of the same molecular weight, while ethers have boiling points very similar to the corresponding alkanes, as shown in Table 8-1.

Table 8-1.  Comparison of Alcohols, Ethers, and Hydrocarbons of Similar Weight

| Type | Compound | Mol. Wt. | B.p.(°C) |
|---|---|---|---|
| Alkane | $CH_3CH_2CH_2CH_3$ | 58 | −0.5 |
| Ether | $CH_3OCH_2CH_3$ | 60 | 8 |
| Alcohol | $CH_3CH_2CH_2OH$ | 60 | 97 |
| Alkane | $CH_3(CH_2)_3CH_3$ | 72 | 35 |
| Ether | $CH_3CH_2OCH_2CH_3$ | 74 | 36 |
| Alcohol | $CH_3CH_2CH_2CH_2OH$ | 74 | 118 |

## ALCOHOL NOMENCLATURE

In the IUPAC system, alcohols are named as derivatives of the longest carbon chain *containing the OH group*. The chain is then numbered from the end which gives the carbon containing the OH group the lowest possible number. The −*e* ending of the alkane is replaced by the −*ol* and the substituent groups named and numbered in the usual manner. Thus we may also refer to saturated alcohols as *alkanols*. Several examples are shown below:

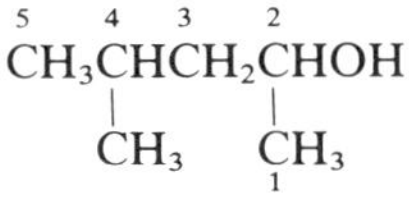

$$\overset{5}{C}H_3\overset{4}{C}H\overset{3}{C}H_2\overset{2}{C}HOH$$

with substituents $CH_3$ (at position 4) and $CH_3$ (at position 1)

Parent hydrocarbon: pentane
Name: 4-Methyl-2-pentanol

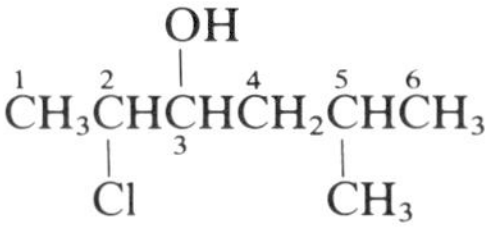

$$\overset{1}{C}H_3\overset{2}{C}H\overset{3}{C}H\overset{4}{C}H_2\overset{5}{C}H\overset{6}{C}H_3$$

with OH (position 2), Cl (position 2... ), $CH_3$

Parent hydrocarbon: hexane
Name: 2-Chloro-5-methyl-3-hexanol

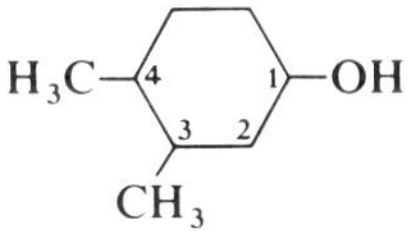

$$H_3C-\!\!\left\langle \overset{4}{\phantom{x}} \quad \overset{1}{\phantom{x}} \right\rangle\!\!-OH$$

with CH₃

Parent hydrocarbon: cyclohexane
Name: 3,4-Dimethylcyclohexanol

*Note*: In the cycloalkanols, it is understood that the C—OH is the number one carbon atom. It is not necessary to include the "1" in the name.

When other functional groups such as the C=C and C≡C also are part of the parent chain, the alcohol naming takes precedence:

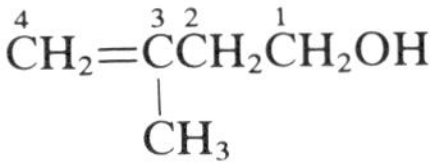

$$\overset{4}{C}H_2=\overset{3}{C}\overset{2}{C}H_2\overset{1}{C}H_2OH$$

with CH₃

3-Methyl-3-buten-1-ol

$$\bigcirc\!\!-\overset{5}{C}\equiv\overset{4}{C}\overset{3}{C}H_2\overset{2}{C}H\overset{1}{C}H_3$$

with OH

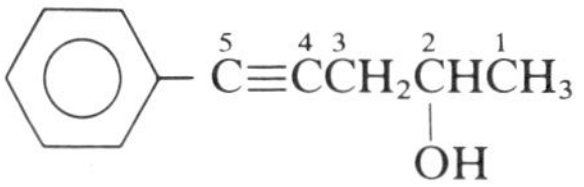

5-Phenyl-4-pentyn-2-ol

Unfortunately, as we discussed in the last chapter, the "common" naming system is still utilized for certain simple organic compounds. Many alcohols are still referred to by this system, which names the alkyl residue followed by the word *alcohol*. Several of the more common examples are listed below with their IUPAC names for comparison.

| Compound | Common name | IUPAC name |
|---|---|---|
| $CH_3OH$ | Methyl alcohol | Methanol |
| $CH_3CH_2OH$ | Ethyl alcohol | Ethanol |
| $CH_3CH_2CH_2OH$ | n-Propyl alcohol | 1-Propanol |
| $CH_3CHCH_3$ with OH | i-Propyl alcohol | 2-Propanol |
| $CH_3CH_2CH_2CH_2OH$ | n-Butyl alcohol | 1-Butanol |
| $(CH_3)_2CHCH_2OH$ | i-Butyl alcohol | 2-Methyl-1-propanol |

| Compound | Common name | IUPAC name |
|---|---|---|
| CH₃CH₂CHCH₃<br>     \|<br>    OH | *s*-Butyl alcohol | 2-Butanol |
| (CH₃)₃COH | *t*-Butyl alcohol | 2-Methyl-2-propanol |
| CH₂=CHCH₂OH | Allyl alcohol | 2-Propen-1-ol |
| 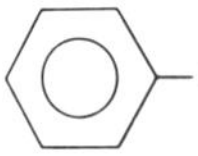 ⬡—CH₂OH | Benzyl alcohol | 1-Phenylmethanol |
| 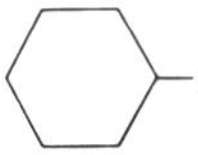 ⬡—OH | Cyclohexyl alcohol | Cyclohexanol |

Molecules containing more than one —OH group are referred to as *diols* (2 OH), *triols* (3 OH), etc., in the IUPAC system. Many of these compounds also bear "trivial" names which are quite unsystematic.

| Compound | IUPAC name | Trivial name |
|---|---|---|
| CH₂CH₂<br>\|  \|<br>OH OH | 1,2-Ethanediol | Ethylene glycol |
| CH₃CHCH₂<br>  \|  \|<br>  HO  OH | 1,2-Propanediol | Propylene glycol |
| CH₂CHCH₂<br> \|  \|   \|<br>HO HO  OH | 1,2,3-Propanetriol | Glycerine (glycerol) |

**Exercise 8-1:** Name the following compounds by the IUPAC system and also by common or trivial names if possible.

(a) 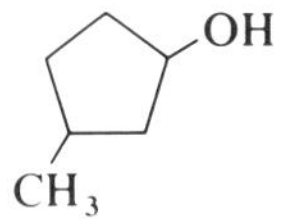  [cyclopentane ring with OH and CH₃ substituents]

(b)  (CH₃)₂C=CHCH₂CH₂OH

(c) 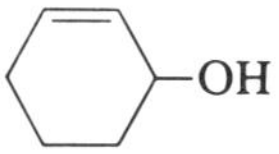  [cyclohexene ring with OH substituent]

(d)  (CH₃)₂CHCH₂CH₂OH

(e) 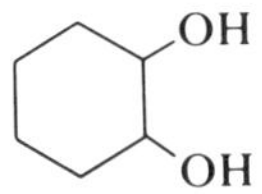  [cyclohexane ring with two OH substituents]

(f)  CH₂=CHC≡CCH₂OH

(g) 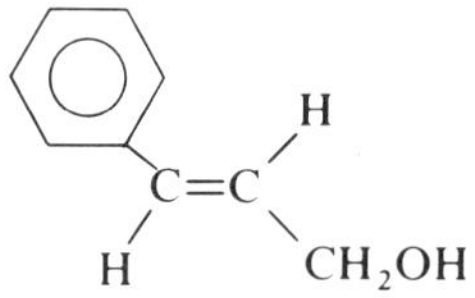  [phenyl ring attached to C=C with H and CH₂OH]

## ETHER NOMENCLATURE

Ethers may be named by either the IUPAC or the common system—both are used with regularity. The alkoxyl group (RO—) is treated as a substituent in the same manner as the halogens in IUPAC naming, the larger of the two groups attached to the oxygen serving as the parent hydrocarbon chain.

$$CH_3CH_2CH_2CH_2\text{—}OCH_3 \qquad \text{IUPAC name: 1-Methoxybutane}$$

Parent hydrocarbon      OCH$_2$CH$_3$   Substituent

$$CH_3CH_2CH_2CHCH_2CH_3 \qquad \text{IUPAC name: 3-Ethoxyhexane}$$

However, simple ethers, and particularly symmetrical (both R groups the same) ethers are more likely to be named by the common system: Both R groups are named followed by the word *ether*. Note that it is not necessary to repeat the R name if the two alkyl groups are identical. Some books, however, insist on this redundancy.

$$CH_3CH_2OCH_3 \qquad \text{Methyl ethyl ether}$$

$$CH_3CH_2OCH_2CH_3 \qquad \text{Ethyl ether (diethyl ether)}$$

Aromatic or mixed aliphatic–aromatic ethers are treated in a similar manner, with the one exception of methoxybenzene, which bears the trivial name *anisole*.

—OCH$_3$    Anisole (methoxybenzene)

—OCH$_2$CH$_3$    Ethoxybenzene

---

**Exercise 8-2:** Name the following ethers by both IUPAC and common names. Which name do you feel would be the most commonly used?

(a) —OCH$_2$CH$_3$

(b) $(CH_3)_3COCH_3$      (c) $CH_3(CH_2)_8OCH_3$

(d) $CH_3O$——Br      (e) $(CH_3)_2CHOCH(CH_3)_2$

---

# CLASSIFICATION OF ALCOHOLS

In many reactions, alcohols differ markedly in their reactivity according to the degree of substitution of the carbon attached to the OH group. It is convenient to thus distinguish between these various types as primary (1°), secondary (2°), and tertiary (3°) alcohols. In this system, methanol is sometimes referred to as a "super-primary" alcohol, since the single carbon is attached only to hydrogen in addition to the OH.

$$
\begin{array}{l}
\text{H} \\
| \\
\text{R}-\text{C}-\text{OH} \\
| \\
\text{H}
\end{array}
$$

1° alcohol: OH attached to primary C
Example: $CH_3CH_2CH_2OH$

$$
\begin{array}{l}
\text{R}-\text{CH}-\text{OH} \\
\quad\quad | \\
\quad\quad \text{R}
\end{array}
$$

2° alcohol: OH attached to secondary C

OH
|
Example: $CH_3CH_2CHCH_3$

$$
\begin{array}{l}
\quad\quad \text{R} \\
\quad\quad | \\
\text{R}-\text{C}-\text{OH} \\
\quad\quad | \\
\quad\quad \text{R}
\end{array}
$$

3° alcohol: OH attached to tertiary C
Example:

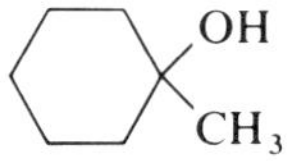

# CLASSIFICATION OF ETHERS

Ethers are generally classified as being either symmetrical or unsymmetrical. Although no chemical differences usually exist between symmetrical and unsymmetrical ethers, the classification is useful in terms of how they may be synthesized.

R—Ö—R   Symmetrical ethers—both R groups identical
Example: $CH_3CH_2CH_2OCH_2CH_2CH_3$

R—Ö—R′   Unsymmetrical ethers—R and R′ groups different
Example: $CH_3CH_2OCH_3$

---

**Exercise 8-3:** Write structures for all possible compounds corresponding to the formula $C_6H_{14}O$. Classify each alcohol as 1°, 2°, or 3° and each ether as symmetrical or unsymmetrical.

---

# INDUSTRIAL PREPARATIONS OF ALCOHOLS

Many alcohols are important industrial chemicals, either as solvents or as chemical intermediates. Some of the more important ones will be discussed.

144

*Methanol*, $CH_3OH$, was once produced commercially by the destructive distillation (heating in the absence of air) of wood, hence its old name, *wood alcohol*. It is produced now by combining carbon monoxide and hydrogen at elevated temperatures and under pressure in the presence of an appropriate catalyst. Much of the

$$CO + 2H_2 \xrightarrow[\text{300°, 300 atm.}]{\text{ZnO}-\text{Cr}_2\text{O}_3} CH_3OH$$

Methanol

methanol produced by this method (about 50%) is oxidized to formaldehyde (see Chapter I-10), a prime ingredient in the preparation of phenolic resins. The balance of the methanol is used as a solvent for shellac and varnish, as a cheap antifreeze, and as a denaturant for ethanol. It has recently been demonstrated that methanol/gasoline and ethanol/gasoline mixtures burn efficiently in car engines. The use of this fuel mixture is under serious consideration by oil companies and the government as a method of stretching our oil reserves.

*Ethanol* may be produced by the acid-catalyzed hydration of ethylene or by *fermentation* of sugars and starches.

$$CH_2{=}CH_2 \xrightarrow{H_3O^+} CH_3CH_2OH \quad \text{Ethanol}$$

$$(C_6H_{10}O_5)_n \xrightarrow[\text{(enzyme)}]{\text{Diastase}} C_{12}H_{22}O_{11}$$
$$\text{Starch} \qquad\qquad\qquad \text{Maltose}$$

$$\downarrow \text{Maltase (enzyme)}$$

$$CO_2{\uparrow} + CH_3CH_2OH \xleftarrow[\text{(enzyme)}]{\text{Zymase}} C_6H_{12}O_6$$
$$\text{Glucose}$$

Grape juice (glucose) $\longrightarrow$ Wine
Potatoes (starch) $\longrightarrow$ Vodka
Barley and hops $\longrightarrow$ Beer

The fermentation process, however, only produces a solution containing approximately 12% ethanol, since concentrations higher than this kill the bacteria or yeasts responsible for the fermentation. In order to make whiskey (Scotch, bourbon, or rye), rum, vodka, gin, or other beverages of high alcoholic content (40–60%), the crude fermentation product must undergo a very careful *distillation* process in order to concentrate the more volatile ethanol. Thus we refer to the manufacturers as "distillers," the plants as "distilleries," and the backwoods home-brew setups as "stills."

Ethanol produced by fermentation is often referred to as *grain alcohol*, and is sold in terms of its "proof," a measure of the actual quantity of ethanol in the solution.

$$100\% \text{ ethanol} = 200 \text{ proof spirit}$$

$$50\% \text{ EtOH, } 50\% \text{ H}_2\text{O} = 100 \text{ proof spirit}$$

Most ethanol used for solvent purposes is 95% ethanol, 5% $H_2O$. Pure ethanol (100%) may be obtained from this by azeotropic distillation with benzene or by treatment with calcium oxide, and is called *absolute ethanol.*

Although we consume a great deal of alcohol produced by fermentation processes (wine, beer, etc.) in the United States, more ethanol is produced by the hydration of ethylene (more than 1 billion lb in 1977 alone) than by all the fermentation processes combined.

*2-Propanol* is produced by the acid-catalyzed hydration of propene. Much of the industrial production is used as a solvent or in the production of acetone, although the man in the street would more likely recognize it in the form of rubbing alcohol for tired, sore muscles.

$$CH_3CH=CH_2 \xrightarrow{H_3O^+} CH_3\overset{\overset{\displaystyle OH}{|}}{C}HCH_3$$

2-Propanol
Isopropyl alcohol

The four butyl alcohols are also produced by hydration of appropriate olefins and are utilized as solvents and plasticizers and to make esters (RCOOButyl). More recently, the Oxo process has been used to make 1-butanol as well as other primary alcohols.

$$CH_3CH=CH_2 + CO + H_2 \xrightarrow[\Delta, P]{Catalyst} CH_3CH_2CH_2CH_2OH$$

$$RCH=CH_2 + CO + H_2 \longrightarrow RCH_2CH_2CH_2OH$$

Oxo process

1,2-Ethanediol (ethylene glycol) is produced industrially from ethylene oxide which in turn is produced from ethylene. An aqueous solution of ethylene glycol is

$$CH_2=CH_2 \xrightarrow[O_2, 300°]{Ag} H_2C\overset{\diagdown\;\;\diagup}{\underset{O}{\quad}}CH_2 \xrightarrow{H_3O^+} \overset{\displaystyle CH_2-CH_2}{\underset{\displaystyle OH\;\;\;OH}{|\quad\;\;|}}$$

Ethylene                Ethylene oxide                Ethylene glycol

sold as antifreeze for car engines (Prestone, Zerex, etc), but over the past 10 years its use in the manufacture of Dacron has increased significantly.

Glycerine, or 1,2,3-propanetriol, is utilized as a moisture-retaining agent in both the tobacco and cosmetic industries. It is both viscous and sweet. It has widespread applications in the plastics industry and is the starting material for the production of nitroglycerine (glyceryltrinitrate) and dynamite. Most fatty acids in both plants and animals are tied up as glycerides, esters formed from one equivalent of glycerine and three equivalents of fatty acids. More will be said about these in Chapter I-12.

$$
\begin{array}{c}
\text{CH}_2-\text{O}-\overset{\displaystyle\text{O}}{\overset{\displaystyle\|}{\text{C}}}\!\!\left(\text{CH}_2\right)_{\!x}\text{CH}_3 \\[4pt]
\text{CH}\ -\text{O}-\overset{\displaystyle\text{O}}{\overset{\displaystyle\|}{\text{C}}}\!\!\left(\text{CH}_2\right)_{\!y}\text{CH}_3 \\[4pt]
\text{CH}_2-\text{O}-\overset{\displaystyle\text{O}}{\overset{\displaystyle\|}{\text{C}}}\!\!\left(\text{CH}_2\right)_{\!z}\text{CH}_3
\end{array}
$$

A triglyceride—an ester composed of one glycerine unit
and three acid units

---

**Exercise 8-4:**  Propylene glycol can be prepared from propylene by the following procedure:

$$
\text{CH}_3\text{CH}=\text{CH}_2 \xrightarrow{\text{HOCl}} \text{CH}_3\overset{\text{OH}}{\underset{\displaystyle|}{\text{CH}}}\text{CH}_2\text{Cl} \xrightarrow[\text{NaOH}]{\text{Dilute}} \text{CH}_3\,\overset{\text{OH}}{\underset{\displaystyle|}{\text{CH}}}\text{CH}_2\text{OH}
$$

Glycerine is also prepared industrially from propylene. Suggest a possible synthesis based on reactions we have already discussed in previous chapters.

---

## LABORATORY PREPARATION OF ALCOHOLS

In previous chapters we have already discussed several methods by which alcohols may be synthesized in the laboratory. In this section we will review these and then discuss some of the more important methods, such as the Grignard synthesis and hydroboration.

### Hydration of Alkenes

Acid-catalyzed addition of water to a carbon–carbon double bond, previously discussed in detail in Chapter I-5, may be used to prepare alcohols in the laboratory as well as in industry. The addition follows Markownikoff's rule, proceeding through the most stable carbonium ion.

$$
\underset{\text{H}}{\overset{\text{R}}{\phantom{.}}}\!\!\!C\!\!=\!\!C\!\!\underset{\text{H}}{\overset{\text{H}}{\phantom{.}}} \;\underset{}{\overset{\text{H}^+}{\rightleftharpoons}}\; [\text{R}\overset{\oplus}{\text{C}}\text{H}-\text{CH}_3] \;\xrightarrow{\text{H}_2\ddot{\text{O}}:}\; [\text{R}-\overset{\overset{\displaystyle\oplus}{\text{OH}_2}}{\underset{\displaystyle|}{\text{C}}}\text{HCH}_3]
$$

$$
\downarrow -\text{H}^{\oplus}
$$

$$
\overset{\text{OH}}{\underset{\displaystyle|}{\text{R}}}\text{CHCH}_3
$$

Examples:

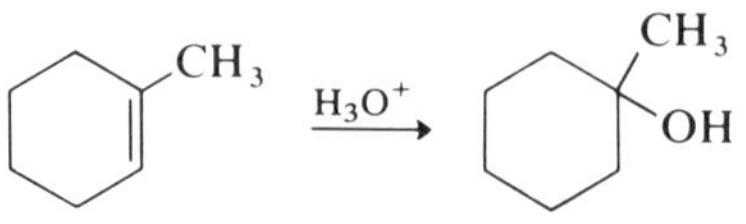


$$(CH_3)_2C{=}CHCH_3 \xrightarrow{H_3O^+} CH_3C(OH)(CH_3)CH_2CH_3$$

### Hydrolysis of Alkyl Halides—Nucleophilic Displacement of Halogen by Hydroxide

We discussed the nucleophilic replacement of halides by various nucleophiles in Chapter I-7. Alcohols may be prepared by treating alkyl halides with cold, dilute sodium hydroxide; the process is referred to as a nucleophilic substitution, or $S_N$ reaction.

$$RCH_2X \xrightarrow[\text{NaOH}]{\text{Cold, dil.}} RCH_2OH \quad 1° \text{ Alcohol}$$

$$RCH(X)R' \xrightarrow{\text{NaOH}} RCH(OH)R' \quad 2° \text{ Alcohol}$$

$$R{-}C(X)(R'')){-}R' \xrightarrow{\text{NaOH}} R{-}C(OH)(R''){-}R' \quad 3° \text{ Alcohol}$$

Examples:

$$CH_3CH(CH_3)CH_2CH_2Br \xrightarrow{\text{NaOH}} CH_3CH(CH_3)CH_2CH_2OH$$

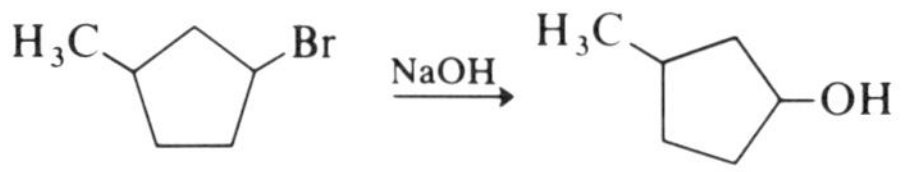


$$CH_3CH_2C(CH_3)_2{-}Cl \xrightarrow{\text{NaOH}} CH_3CH_2C(CH_3)_2{-}OH$$

### Grignard Addition to Carbonyl Compounds

Grignard reagents react with aldehydes and ketones to produce 1°, 2°, and 3° alcohols. This synthetic method is usually the first choice of laboratory chemists because of the ready availability of both organic halides and carbonyl compounds. The aldehyde or ketone is usually added to the Grignard in ether solution, forming an addition complex from which the alcohol is liberated by acid hydrolysis.

$$
RMgX \xrightarrow{R''-\overset{\overset{\displaystyle O}{\|}}{C}-R'} \quad \left[ R-\overset{\overset{\displaystyle OMgX}{|}}{\underset{\underset{\displaystyle R'}{|}}{C}}-R'' \right] \xrightarrow{H_3O^+} \quad R-\overset{\overset{\displaystyle OH}{|}}{\underset{\underset{\displaystyle R''}{|}}{C}}-R'
$$

RMgX
In ether
solution

Addition complex

$+Mg^{+2}, X^-$

R = alkyl, aryl, alkenyl, or alkynyl
R′, R″ = H, alkyl, alkenyl or alkynyl

1°, 2° and 3° alcohols may be prepared by this method, and R, R′ and R″ may be varied almost indefinitely.

$$RMgX + \quad CH_2{=}O \quad \longrightarrow \quad [RCH_2OMgX] \xrightarrow{H_3O^+} RCH_2OH$$

Formaldehyde
1° Alcohol

$$RMgX + R'-\overset{\overset{\displaystyle O}{\|}}{C}-H \quad \longrightarrow \quad \left[ \overset{\overset{\displaystyle OMgX}{|}}{RCHR'} \right] \xrightarrow{H_3O^+} \overset{\overset{\displaystyle OH}{|}}{RCHR'}$$

Any other
aldehyde
2° Alcohol

$$RMgX + R'-\overset{\overset{\displaystyle O}{\|}}{C}-R' \quad \longrightarrow \quad \left[ R-\overset{\overset{\displaystyle OMgX}{|}}{\underset{\underset{\displaystyle R''}{|}}{C}}-R' \right] \xrightarrow{H_3O^+} R-\overset{\overset{\displaystyle OH}{|}}{\underset{\underset{\displaystyle R''}{|}}{C}}-R'$$

Ketone
3° Alcohol

Several examples are outlined below:

$$CH_3CH_2CH_2\overset{\overset{\displaystyle O}{\|}}{C}H \xrightarrow[\text{2. H}_3\text{O}^+]{\text{1. CH}_3\text{CH}_2\text{MgBr}} CH_3CH_2CH_2\overset{\overset{\displaystyle OH}{|}}{C}HCH_2CH_3$$

3-Hexanol

$$\text{(phenyl)}-\overset{\overset{\displaystyle O}{\|}}{C}CH_2CH_3 \xrightarrow[\text{2. H}_3\text{O}^+]{\text{1. CH}_2{=}\text{CHMgBr}} \text{(phenyl)}-\overset{\overset{\displaystyle OH}{|}}{\underset{\underset{\displaystyle CH_2CH_3}{|}}{C}}-CH{=}CH_2$$

3-Phenyl-1-penten-3-ol

$$(CH_3)_2CHCH_2MgCl \xrightarrow[\text{2. H}_3\text{O}^+]{\text{1. CH}_2{=}\text{O}} (CH_3)_2CHCH_2CH_2OH$$

3-Methyl-1-butanol

There are many ways to plan an alcohol synthesis via Grignard addition to a carbonyl. For secondary alcohols, two options are available, but for 3° alcohols,

there are three possible routes. In the following examples these possible combinations are illustrated. *Remember* that the carbon containing the —OH group must
come from the carbonyl compound!

Example 1:
Prepare

$$\underset{\displaystyle CH_3CH_2CH_2CH-\!\!\langle\bigcirc\rangle}{\overset{\displaystyle OH}{|}}$$

Two routes are possible:

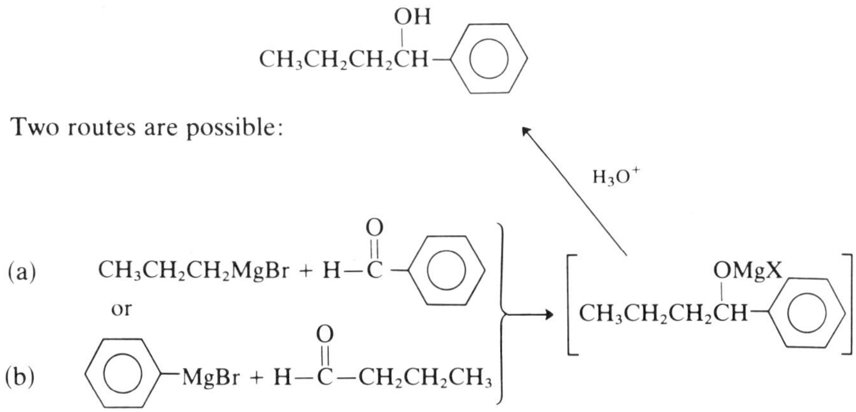

(a)    $CH_3CH_2CH_2MgBr + H-\overset{\overset{\displaystyle O}{\|}}{C}-\!\!\langle\bigcirc\rangle$

     or

(b)    $\langle\bigcirc\rangle\!-MgBr + H-\overset{\overset{\displaystyle O}{\|}}{C}-CH_2CH_2CH_3$

$\xrightarrow{\quad}$ $\left[ \underset{\displaystyle CH_3CH_2CH_2CH-\!\!\langle\bigcirc\rangle}{\overset{\displaystyle OMgX}{|}} \right]$

$\xrightarrow{H_3O^+}$

Example 2:
Prepare

$$\underset{\displaystyle \underset{\displaystyle \bigcirc}{|}}{\underset{\displaystyle CH_3CH_2-\overset{\displaystyle OH}{\underset{\displaystyle |}{C}}-CH=CH_2}{}}$$

Three routes are possible:

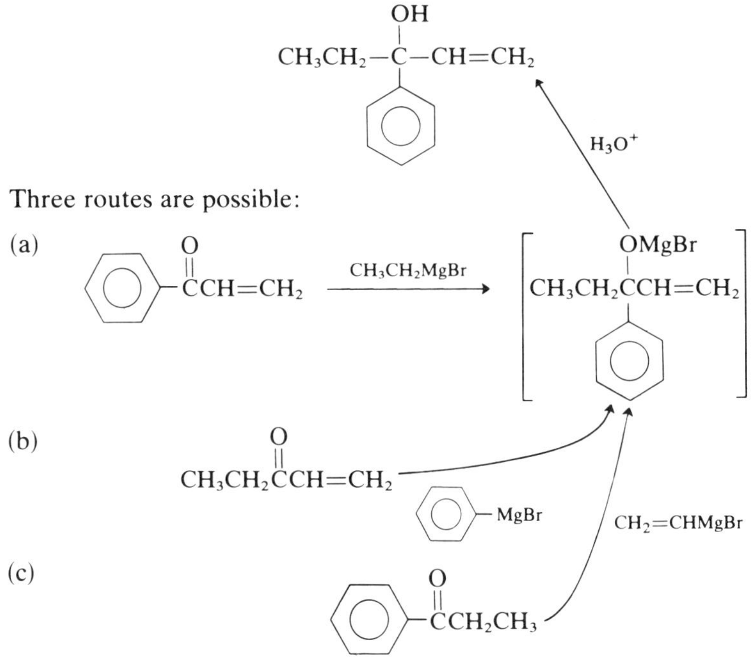

(a)    $\langle\bigcirc\rangle\!-\overset{\overset{\displaystyle O}{\|}}{C}CH=CH_2 \xrightarrow{CH_3CH_2MgBr} \left[ \underset{\displaystyle CH_3CH_2\overset{\displaystyle |}{C}CH=CH_2}{\overset{\displaystyle OMgBr}{}} \right]$

$\xrightarrow{H_3O^+}$

(b)    $CH_3CH_2\overset{\overset{\displaystyle O}{\|}}{C}CH=CH_2$

$\langle\bigcirc\rangle\!-MgBr$

$CH_2=CHMgBr$

(c)    $\langle\bigcirc\rangle\!-\overset{\overset{\displaystyle O}{\|}}{C}CH_2CH_3$

In practice, when more than one route to the derived alcohol is possible, the availability and cost of reagents is usually the determining factor in the synthesis.

---

**Exercise 8-5:** Suggest possible syntheses for the following alcohols. Remember that more than one route may be possible.

(a) $\langle\bigcirc\rangle$—$CH_2CH_2CH_2OH$

(b) 
$$\overset{\quad\quad OH}{CH_2{=}CHCHCH_2CH{=}CH_2}$$

(c) 
$$\overset{\quad\quad OH}{HC{\equiv}CCHCH_2CH_3}$$

(d) 
$$\overset{\quad\quad\quad OH}{(CH_3)_2CHCCH_2CH_3}$$
$$\overset{\quad\quad\quad\quad}{\quad\quad CH_3}$$

---

## Hydroboration of Alkenes

Pure primary alcohols may be prepared by a procedure involving an initial addition of $BH_3$ to a carbon–carbon double bond to yield a trialkylboron derivative, oxidation of the trialkylboron to a trialkylborate with hydrogen peroxide, and

$$RCH{=}CH_2 \xrightarrow[\text{from } B_2H_6]{[BH_3]} [RCH_2CH_2]_3B$$

$$\downarrow \begin{array}{l} H_2O_2 \\ \text{(oxidation)} \end{array}$$

$$\{RCH_2CH_2OH\} + B(OH)_3 \xleftarrow[H_2O]{OH^-} [RCH_2CH_2O]_3B$$

1° Alcohol                    (Hydrolysis)

finally, alkaline hydrolysis. This three-step procedure is referred to as *hydroboration*. The net result of this process is an apparent anti-Markownikoff addition of $H_2O$ to the original double bond. Thus two different alcohols may be obtained from the same olefin.

$$RCH{=}CH_2 \begin{cases} \xrightarrow{H_3O^+} \overset{OH}{RCHCH_3} \\ \xrightarrow[\text{(3 steps)}]{\text{Hydroboration}} RCH_2CH_2OH \end{cases}$$

Example:

$$(CH_3)_2CHCH_2CH_2CH{=}CH_2 \xrightarrow{\text{[BH}_3\text{]}} [(CH_3)_2CH(CH_2)_4]_3B$$

$$\downarrow \begin{array}{l} 1.\ H_2O_2 \\ 2.\ OH^-/H_2O \end{array}$$

$$(CH_3)_2CH(CH_2)_4OH$$

---

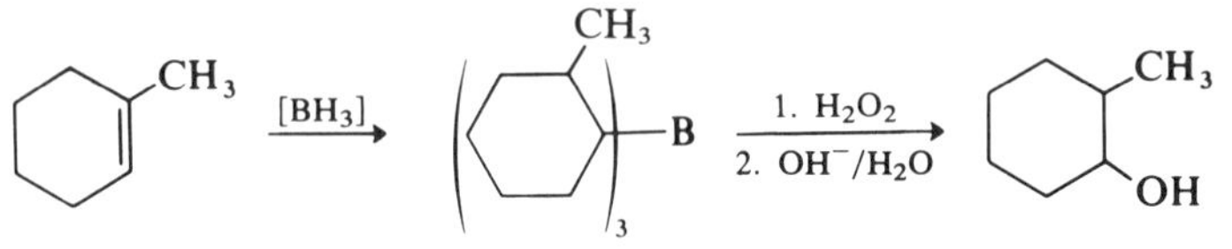

**Exercise 8-6:**  In the preceding two examples, what products would be produced from acid-catalyzed addition of water to the two olefins? Are they different from the hydroboration products?

---

## PREPARATION OF ETHERS

Both symmetrical (ROR) and unsymmetrical (ROR′) ethers are prepared from alcohols. Acid-catalyzed dehydration of alcohols in sulfuric acid at elevated temperatures (150–180°) produces the former. In general practice, the alcohol is added to sulfuric acid at room temperature and the mixture heated until the volatile product distills from the reaction mixture.

$$2\ RCH_2OH \xrightarrow[150-180°]{H_2SO_4} RCH_2OCH_2R$$

$$CH_3CH_2CH_2OH \xrightarrow[150-180°]{H_2SO_4} CH_3CH_2CH_2OCH_2CH_2CH_3$$
$$\text{Symmetrical ethers} \qquad\qquad\qquad \textit{n-}\text{Propyl ether}$$

Unsymmetrical ethers are prepared by a reaction sequence known as the *Williamson synthesis*. Alcohols react with sodium metal to produce sodium alkoxide. Alkoxides are strong bases and good nucleophiles. Under mild conditions,

$$ROH \xrightarrow{\text{Na}} RO^{\ominus}Na^{\oplus} + \tfrac{1}{2}\,H_2$$
$$\text{Sodium}$$
$$\text{alkoxide}$$

then, alkoxides will displace halide ions from alkyl halides in typical $S_N$ fashion to yield ethers. Both symmetrical and unsymmetrical ethers may be produced by this method.

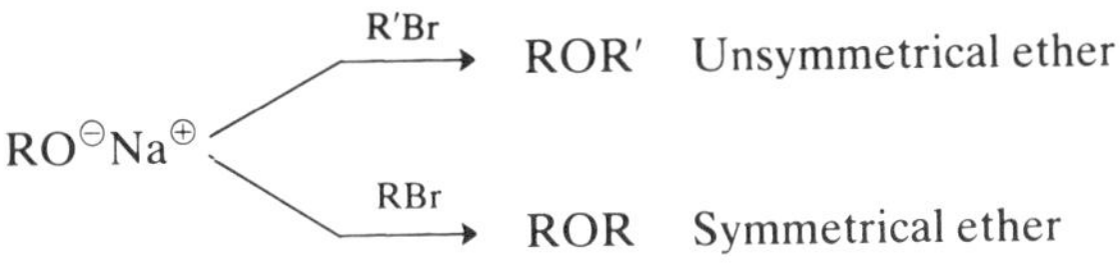

$$RO^{\ominus}Na^{\oplus} \begin{cases} \xrightarrow{\text{R'Br}} & ROR' \quad \text{Unsymmetrical ether} \\ \xrightarrow{\text{RBr}} & ROR \quad \text{Symmetrical ether} \end{cases}$$

Williamson Synthesis

---

**Exercise 8-7:** Suggest methods for preparing the following ethers via Williamson syntheses:

(a) $CH_3CH_2OCH_2CH_3$

(b) $CH_3CH_2OCH(CH_3)_2$

(c) $CH_3CH_2CH_2CH_2OCH_2CH_2CH_3$

---

Aromatic ethers are usually prepared from sodium phenoxide and the corresponding alkyl halide.

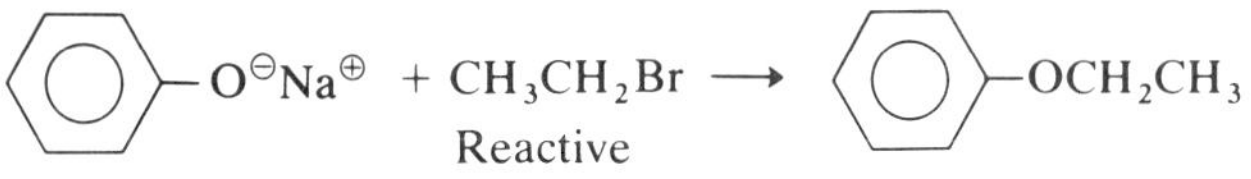

$$\text{Ph-}O^{\ominus}Na^{\oplus} + CH_3CH_2Br \longrightarrow \text{Ph-}OCH_2CH_3$$
Reactive

$$CH_3CH_2O^{\ominus}Na^{\oplus} + \text{Ph-}Br \longrightarrow \text{N.R.}$$
Unreactive

However, the reaction of sodium ethoxide and bromobenzene does not yield the desired product because of the lack of reactivity of aryl halides toward nucleophilic displacement.

## CYCLIC ETHERS

Cyclic ethers are related to acyclic ethers in the same manner as cycloalkanes are related to the straight-chain alkanes. Their reactions and physical properties are similar but their preparations involve special techniques. By far the most important cyclic ether is ethylene oxide, formed from ethylene by vapor-phase oxidation.

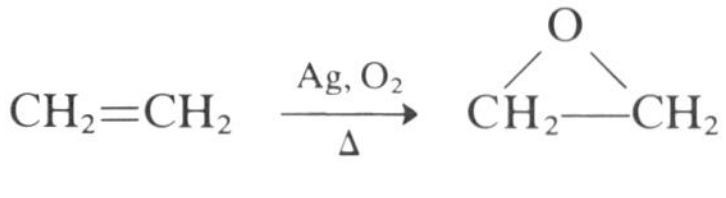

$$CH_2{=}CH_2 \xrightarrow[\Delta]{Ag, O_2} \underset{\overset{\diagup\ \diagdown}{CH_2{-}CH_2}}{O} \text{ Ethylene oxide}$$
$$\text{(oxirane)}$$

The 3-membered oxirane ring is most commonly referred to as an *epoxide*. Epoxides may be prepared from any olefin by oxidation with a peracid, $RCO_3H$, such as performic ($HCO_3H$) or peracetic ($CH_3CO_3H$) acid.

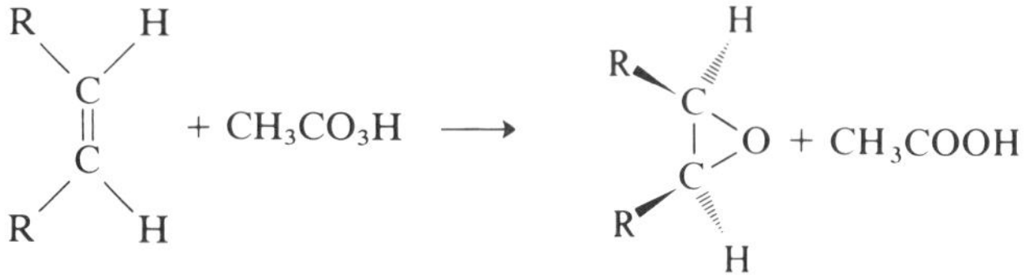

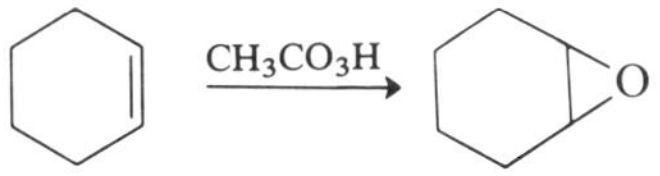

Other cyclic ethers may be formed by dehydration of appropriate diols:

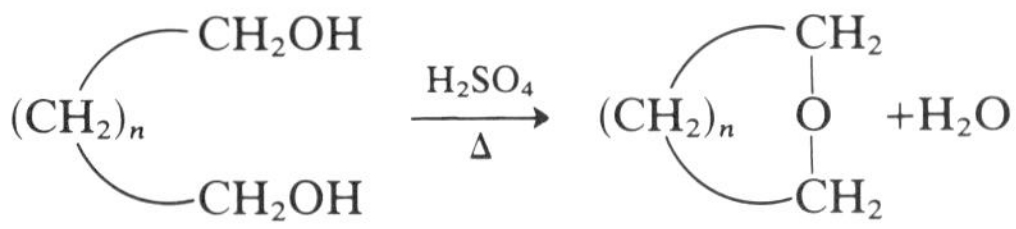

## REACTIONS OF ALCOHOLS AND ETHERS

The reactions of alcohols can be broken down into two main classes: (a) reactions in which the O—H bond is broken, and (b) reactions in which the C—O bond breaks.

We have already seen an example of the first class in the Williamson ether synthesis:

$$RO{-}H \xrightarrow[\text{Li, Na, K, etc.}]{\text{Active metal}} RO^{\ominus}M^{\oplus} + \tfrac{1}{2}H_2$$

### Reactions Involving RO⊢H Scission

In this reaction the O—H bond behaves much like the OH bond in water. Alkoxides are strong bases, as is the hydroxide ion, and are very useful in organic reactions as basic catalysts, usually in alcohol solution.

Alcohols also form esters with many organic acids.

$$RO{-}H + HO{-}\overset{\overset{\textstyle O}{\|}}{\underset{\underset{\textstyle O}{}}{S}}OH \quad\xrightarrow{-H_2O}\quad R{-}O{-}\overset{\overset{\textstyle O}{\|}}{\underset{\underset{\textstyle O}{}}{S}}{-}OH \quad\xrightarrow[-H_2O]{+ROH}\quad R{-}O{-}\overset{\overset{\textstyle O}{\|}}{\underset{\underset{\textstyle O}{}}{S}}{-}O{-}R$$

Sulfuric       Alkyl hydrogen       Dialkyl<br>
acid       sulfate       sulfate

$$RO{-}H + H{-}O{-}\overset{\overset{\textstyle O}{\|}}{\underset{\underset{\textstyle OH}{|}}{P}}{-}OH \quad\xrightarrow{-H_2O}\quad RO{-}\overset{\overset{\textstyle O}{\|}}{\underset{\underset{\textstyle OH}{|}}{P}}{-}OH \quad\xrightarrow[-H_2O]{+ROH}\quad RO{-}\overset{\overset{\textstyle O}{\|}}{\underset{\underset{\textstyle OH}{|}}{P}}{-}OR$$

Phosphoric       Alkyl dihydrogen       Dialkyl hydrogen<br>
acid       phosphate       phosphate

$$RO{-}H + HO{-}NO_2 \quad\xrightarrow{-H_2O}\quad RONO_2$$

$$RO{-}\overset{\overset{\textstyle O}{\|}}{\underset{\underset{\textstyle OR}{|}}{P}}{-}OR \quad\overset{+ROH}{\underset{-H_2O}{\nwarrow}}$$

Alkyl       Trialkyl<br>
nitrate       phosphate

Many of those esters of inorganic acids play important roles in both organic synthesis and biochemistry. Dialkyl hydrogen phosphates are present in nucleic acids, while diphosphate and triphosphate esters are biochemically formed in living cells by enzymatic catalysis and play major roles in energy storage and release at the

$$RO\overset{\overset{\textstyle O}{\|}}{\underset{\underset{\textstyle OH}{|}}{P}}{-}O{-}\overset{\overset{\textstyle O}{\|}}{\underset{\underset{\textstyle OH}{|}}{P}}{-}OH \qquad\qquad RO\overset{\overset{\textstyle O}{\|}}{\underset{\underset{\textstyle OH}{|}}{P}}{-}O{-}\overset{\overset{\textstyle O}{\|}}{\underset{\underset{\textstyle OH}{|}}{P}}{-}O{-}\overset{\overset{\textstyle O}{\|}}{\underset{\underset{\textstyle OH}{|}}{P}}{-}OH$$

Alkyl diphosphate       Alkyl triphosphate

cellular level in such molecules as adenosine triphosphate (ATP).

Adenosine triphosphate (ATP)

Alcohols also form esters with organic acids. These esters and their formation will be discussed in greater detail in Chapter I-12. It is interesting to note that organic esters also play important biochemical roles as flavor and odor enhancers.

$$ROH + R'\overset{\overset{\textstyle O}{\|}}{C}OH \overset{H^+}{\rightleftharpoons} R-O-\overset{\overset{\textstyle O}{\|}}{C}-R' + H_2O$$

Alcohol  Carboxylic                    Organic
         acid                          ester

---

**Exercise 8-8:** Write formulas for the esters that are formed in excess inorganic acid for the following alcohols:

(a) $CH_2OH$
    |
    $CH-OH$  $\xrightarrow{HONO_2}$  Nitroglycerine, a powerful explosive, muscle relaxant, and blood-vessel dilator used in the treatment of angina pectoris
    |
    $CH_2OH$

(b) $CH_3OH$  $\xrightarrow{H_2SO_4}$  A reagent used to "methylate" other active OH groups, particularly phenols
    Methanol

(c) $C(CH_2OH)_4$  $\xrightarrow{HONO_2}$  Pentaerythritol tetranitrate (PETN), a powerful explosive
    Pentaerythritol

---

## Reactions Involving R $\dashv$ OH Scission

We have already discussed the more important reactions falling into this category in previous chapters. Alkyl halides are usually prepared from alcohols in the laboratory and may be briefly summarized below:

1.  $ROH + HX \rightleftharpoons RX + H_2O$
    $X = Cl, Br, I$

2.  $ROH + PX_3 \longrightarrow RX + P(OH)_3$
    $X = Cl, Br$

3.  $ROH + SOCl_2 \longrightarrow RCl + SO_2\uparrow + HCl\uparrow$

Olefins may also be obtained from alcohols by acid-catalyzed dehydration (Chapter I-5), either in solution $(H_2SO_4/H_2O)$ or catalytically in the vapor phase $(Al_2O_3/300-400°)$.

4.

5.

$$CH_3CH_2\underset{\underset{CH_3}{|}}{\overset{\overset{CH_3}{|}}{C}}-OH \xrightarrow[100°]{H_2SO_4} CH_3CH=C\underset{CH_3}{\overset{CH_3}{<}} + CH_3CH_2C\underset{CH_3}{\overset{CH_2}{<}}$$

Major product          Minor product

You will recall that in all acid-catalyzed dehydrations carbonium ions are formed as intermediates and that the more stable olefins are preferentially formed in the reaction (Saytzeff elimination rule).

## Oxidation of Alcohols

Aldehydes (RCHO) and ketones (RCOR) may be obtained from alcohols by chemical oxidation or catalytic dehydrogenation, or enzymatically in both plants and animals.

$$RCH_2OH \underset{\substack{Strong\,[O]}}{\overset{\substack{Mild\,oxidation \\ [O]}}{<}} \quad \overset{\overset{O}{\parallel}}{R-C-H} \text{ An aldehyde} \downarrow[O]$$

$$\overset{\overset{O}{\parallel}}{R-C-OH} \text{ A carboxylic acid}$$

$$\underset{RCHR}{\overset{OH}{|}} \xrightarrow{[O]} \overset{\overset{O}{\parallel}}{R-C-R'} \text{ A ketone}$$

Mild oxidizing agents: $CrO_3$/pyridine; $K_2Cr_2O_7$/DMSO
Strong oxidizing agents: alkaline $KMnO_4$; aqueous $K_2Cr_2O_7$

$$RCH_2OH \xrightarrow[250°]{Cu} \overset{\overset{O}{\parallel}}{R-C-H} + H_2$$

$$\underset{RCHR'}{\overset{OH}{|}} \xrightarrow[250°]{Cu} \overset{\overset{O}{\parallel}}{R-C-R'} + H_2$$

Catalytic dehydrogenation (1° + 2° alcohols)

$$CH_3-\underset{\underset{CH_3}{|}}{\overset{\overset{CH_3}{|}}{C}}-OH \underset{\underset{250°}{Cu}}{\overset{\overset{Chemical}{[O]}}{\rightleftharpoons}} \text{No reaction under normal conditions}$$

3° Alcohol

Oxidation or dehydrogenation of 1° alcohols yields aldehydes. However, aldehydes are more easily oxidized than alcohols, yielding carboxylic acids, and this complication severely limits the number of reagents that can be utilized in this reaction. In recent years it has been recognized that the oxidation may be stopped at the aldehyde stage if the exact stoichiometric quantity of oxidizing agent is used (no excess!) in an aprotic solvent such as dimethylsulfoxide (DMSO—$CH_3SOCH_3$) or pyridine ($C_5H_5N$). Ketones, on the other hand, cannot easily be oxidized further, and much stronger oxidizing media may be used for the oxidation of 2° alcohols. Tertiary alcohols resist oxidation or dehydrogenation and are sometimes used as solvents for oxidation reactions.

Examples:

3-Methyl-2-buten-1-ol          3-Methyl-2-butenal

4-Methylcyclo-          4-Methylcyclo-
hexanol          hexanone

2,3-Dimethyl-1-butanol          2,3-Dimethylbutanal

We will discuss the topic of oxidation again in the next chapter in more detail, including a review of electron transfer and the balancing of equations.

# REACTIONS OF ETHERS

Ethers are inert to most laboratory reagents. However, many organic compounds are very soluble in aliphatic ethers, while inorganic compounds are not. For these reasons, ethers, and particularly ethyl ether, are widely used as sovents in organic reactions (Grignard) and to extract soluble organic compounds from inorganic or insoluble material. In fact, the only chemical reaction which distinguishes ethers from hydrocarbons is their solubility in strong acids. If concentrated mineral acids

$$R\ddot{O}R \underset{}{\overset{H^+}{\rightleftharpoons}} R\overset{H}{\ddot{O}}{}^{\oplus}R \quad \text{An oxonium ion}$$

are used, this property may be used to hydrolyze the ether completely. Thus ethers *cannot* be used as solvents in strongly acidic media.

$$R\ddot{O}R \overset{HI}{\rightleftharpoons} R\overset{H}{\ddot{O}}R^{\oplus}, I^{\ominus} \longrightarrow RI + ROH$$

One particular ether does display inordinate reactivity—ethylene oxide. Ring opening of an epoxide may be either acid or base-catalyzed, thus ethylene oxide will react with either electrophiles or nucleophiles to yield difunctional intermediates, many of which have important uses.

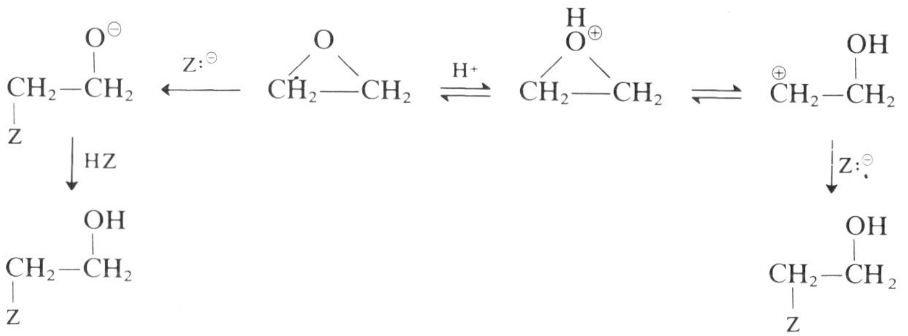

Examples:

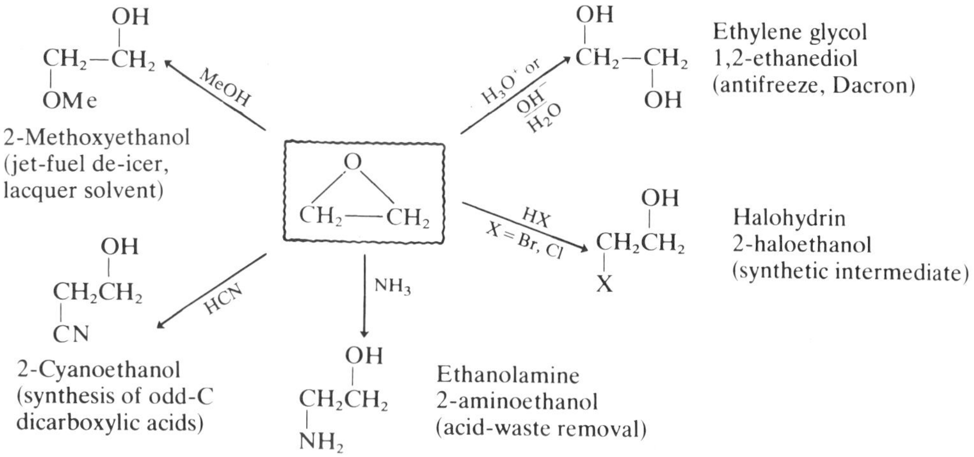
MeOH
2-Methoxyethanol
(jet-fuel de-icer,
lacquer solvent)
$H_3O^+$ or $OH^-$/$H_2O$
Ethylene glycol
1,2-ethanediol
(antifreeze, Dacron)
HX
X = Br, Cl
Halohydrin
2-haloethanol
(synthetic intermediate)
HCN
2-Cyanoethanol
(synthesis of odd-C
dicarboxylic acids)
$NH_3$
Ethanolamine
2-aminoethanol
(acid-waste removal)

Ethylene oxide also reacts with Grignard reagents to produce 1° alcohols.

$$\text{R---MgX} + \underset{\text{O}}{\overset{/\ \backslash}{CH_2CH_2}} \longrightarrow [RCH_2CH_2OMgX] \overset{H_3O^+}{\longrightarrow} RCH_2CH_2OH$$

---

**Exercise 8-9:** Epichlorohydrin is an epoxide widely used in the manufacture of epoxide resins. Outline a synthesis of this material from propene.

$$CH_3CH=CH_2 \longrightarrow \ \longrightarrow \ \longrightarrow \underset{\text{O}}{\overset{}{CH_2---CHCH_2Cl}}$$

Epichlorohydrin

---

## ETHER PEROXIDES

Aliphatic ethers have a tendency to form peroxides when left in contact with air for extended periods of time. Most organic peroxides are thermally unstable and explode when heated. Since ethers are widely used as solvents, care must be taken to ensure that no peroxides exist prior to removal of the solvent by distillation. The

$$CH_3CH_2OCH_2CH_3 \xrightarrow[\text{Slow}]{O_2} CH_3CH_2OCHCH_3 \quad \text{Ethyl ether}$$
$$\uparrow \quad \uparrow \qquad\qquad\qquad \underset{H}{\overset{O}{|}}\underset{}{\overset{|}{O}} \qquad \text{peroxide}$$

$$(CH_3)_2CHOCH(CH_3)_2 \xrightarrow[\text{Fast}]{O_2} (CH_3)_2\,C{-}OCH(CH_3)_2 \quad \textit{i}\text{-Propyl ether}$$
$$\uparrow \quad \uparrow \qquad\qquad\qquad\qquad \underset{H}{\overset{O}{|}}\underset{}{\overset{|}{O}} \qquad \text{peroxide}$$

$\alpha$-C—H bonds are
susceptible to
oxidation

Explosive when
heated

presence of ether peroxides may be detected by shaking a sample of the suspect solvent with ferrous sulfate and testing the solution with thiocyanate ion. A deep red color indicates that peroxides are indeed present. Peroxides may be destroyed by passing the ether through a column of activated alumina.

$$Fe^{+2} \xrightarrow[\text{peroxides}]{\text{Ether}} Fe^{+3} \xrightarrow{SCN^{\ominus}} \text{Red color}$$

# ALCOHOLS AND ETHERS IN NATURE

A large variety of both simple and complicated alcohols occur in nature. All straight-chain alcohols up to $C_{10}$ have been isolated from plants, although they usually occur as esters rather than free alcohols. Alcohols and esters are primarily responsible for both the flavor and fragrance of fruit. For example, at least seven common alcohols are found to be flavor constituents in strawberries:

$$CH_3CH_2OH, \quad CH_3CH_2CH_2OH, \quad CH_3(CH_2)_4OH, \quad (CH_3)_2CHCH_2CH_2OH,$$

$$CH_3CH(OH)CH_2CH_2CH_3, \quad (CH_3)_2CHCH_2OH, \quad \text{and} \quad CH_3(CH_2)_5OH,$$

as well as several esters of alcohols.

As an example of an extremely important class of natural products containing an —OH functionality, consider the steroid family. This family is widely distributed in both plants and animals as free sterols and their esters. Although their function in plants is obscure, they are profoundly important to most animal life as hormones, coenzymes, and Vitamin D. A few of the more important sterols are listed below:

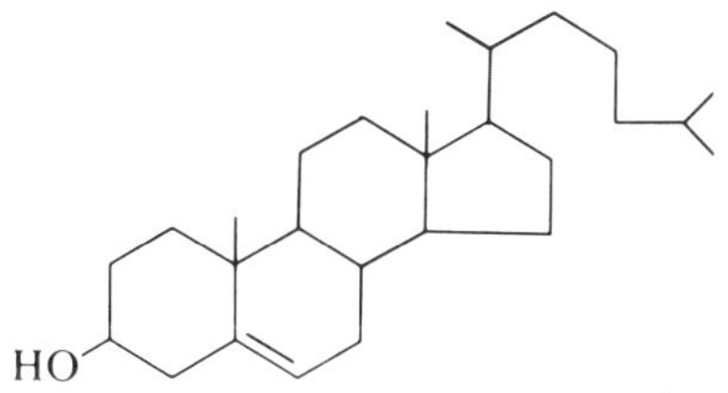

Cholesterol—most common animal steroid—end product of steroid synthesis in humans and normally concentrated in brain. Excess dietary intake has been linked with arteriosclerosis and formation of gallstones.

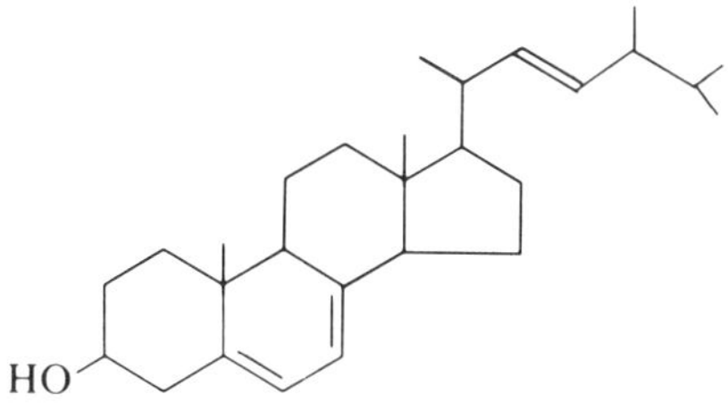

Ergosterol—precursor of Vitamin D

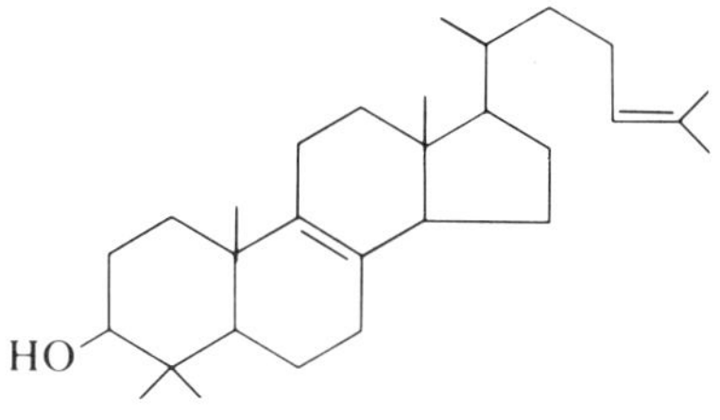

Lanosterol—wool fat, yeast, plants—intermediate in steroid biosynthesis leading to other steroids

Ether functional groups also occur frequently in nature, although not as the main functional group. Methoxy groups dominate the naturally occurring ethers, particularly on aromatic rings, as in mescaline, reserpine, morphine, and quinine.

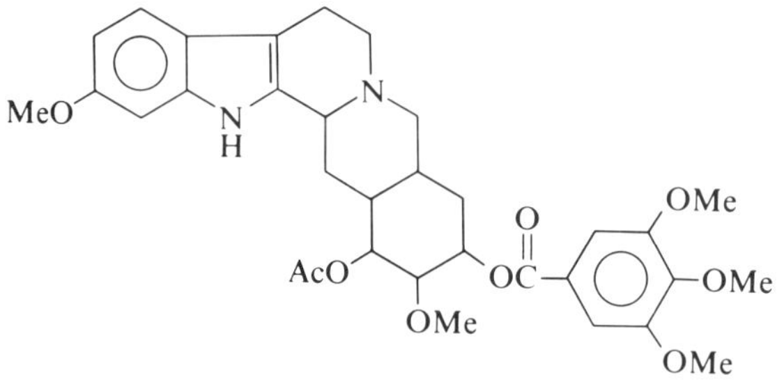

Reserpine—tranquilizer

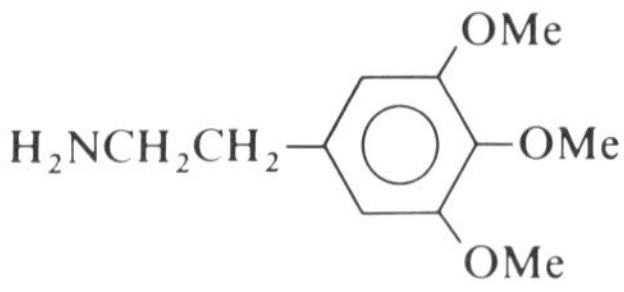

Mescaline—hallucinogen

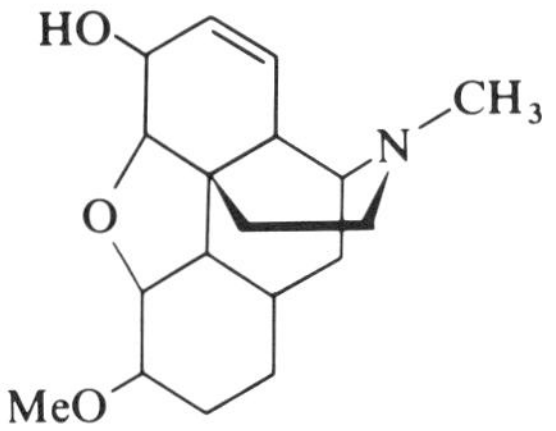

Morphine—narcotic
pain-killer

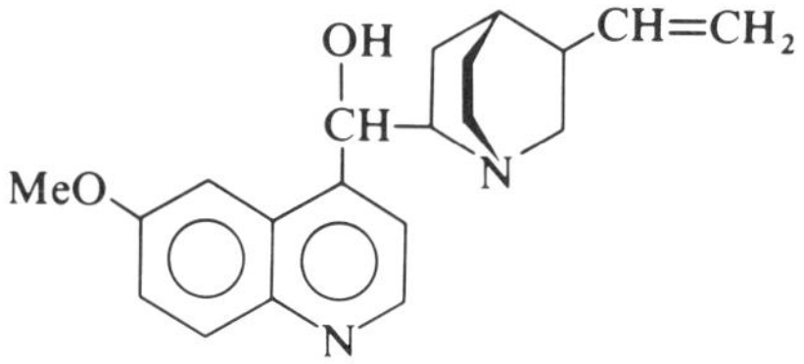

Quinine—cardiac depressant

## REVIEW OF NEW TERMS AND CONCEPTS

Define each term and, if possible, give an example of each:

| | |
|---|---|
| 1° alcohol | 2° alcohol |
| 3° alcohol | Symmetrical ether |
| Unsymmetrical ether | Fermentation |
| Alcohol "proof" | Absolute alcohol |
| Oxo process | Triglyceride |
| Hydroboration | Williamson synthesis |
| Epoxide | Inorganic acid ester |
| Organic acid ester | Sterol |

## PROBLEMS: CHAPTER I-8

1. Name the following compounds by the IUPAC system, and if possible, by common or trivial names. Classify each alcohol as 1°, 2° or 3°.

(a)    OH

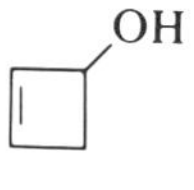

(b)    OCH(CH$_3$)$_2$

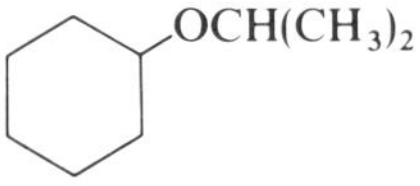

(c) 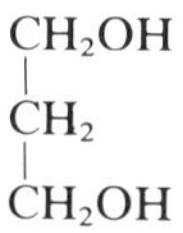 $CH_2OH$
    $CH_2$
    $CH_2OH$

(d) 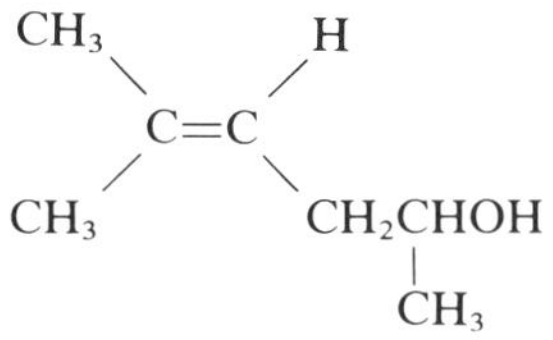 
$$CH_3 \quad\quad\quad H$$
$$C=C$$
$$CH_3 \quad\quad\quad CH_2CHOH$$
$$CH_3$$

(e) 

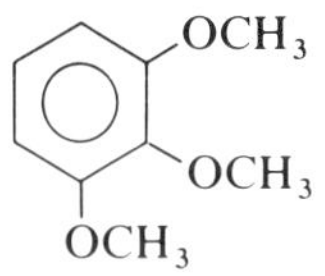

(f) $CH_3CH_2OCH_2CH_2OH$

(g) $(CH_3)_2CHCH_2CH_2OH$

(h) 

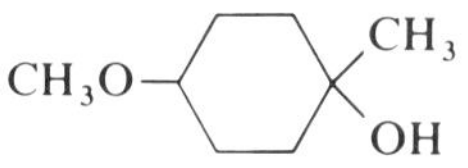

(i) $CH_2=CHCH_2OH$

2. Complete each of the following reactions. Show only organic products.

(a) $CH_3CH_2CH_2CH=CH_2 \xrightarrow[\Delta]{H_3O^+}$

(b) $CH_3CH_2CH_2CH=CH_2 \xrightarrow[\substack{2.\ H_2O_2 \\ 3.\ OH^-}]{1.\ B_2H_6}$

(c) 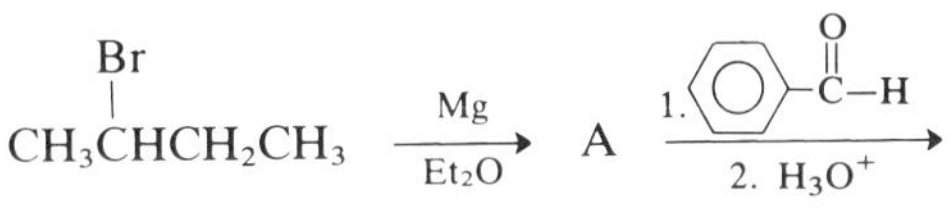 
$$\begin{array}{c} Br \\ | \\ CH_3CHCH_2CH_3 \end{array} \xrightarrow[Et_2O]{Mg} A \xrightarrow[2.\ H_3O^+]{1.}$$

(d) $(CH_3)_2CHCH_2OH \xrightarrow{PBr_3}$

(e) $CH_3CH_2CH_2CH_2CH_2Br \xrightarrow{Na\overset{+}{O}CH(CH_3)_2 \overset{-}{}}$

(f) $(CH_3)_2CHCH_2OCH_2CH_3 \xrightarrow{Excess\ HI}$

(g) 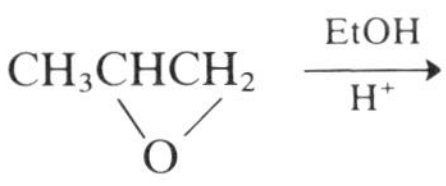 
$$CH_3CHCH_2 \xrightarrow[H^+]{EtOH}$$
$$\diagdown O \diagup$$

(h) 

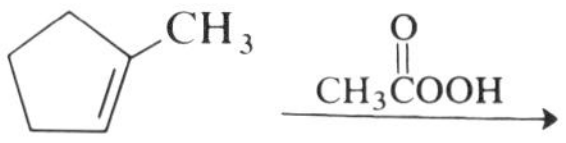

(i) 

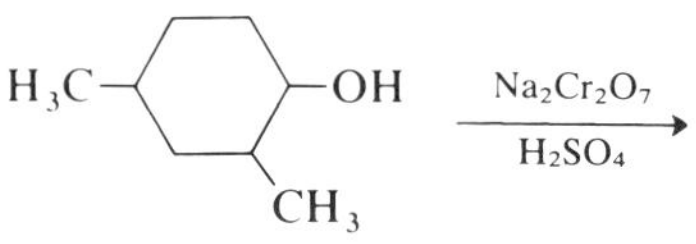

(j) $CH_3CH_2CH=CHCH_2OH \xrightarrow[Pyridine]{CrO_3}$

(k)

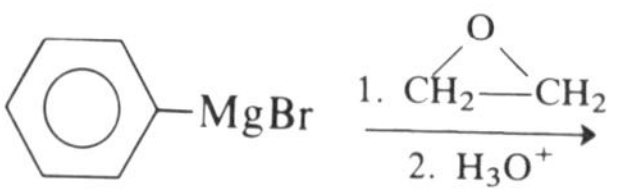

(l)  $CH_3CH_2CH_2CH_2Br$ $\xrightarrow[25°]{Dil.\ NaOH}$

(m) $CH_3CH_2CHCH_3$ $\xrightarrow[100°]{Conc.\ KOH}$
      |
      $Br$

(n)  $CH_2{=}CHCH_2CH_2OH$ $\xrightarrow[250°]{Cu}$

**3.** The following chemicals were mislabeled by a careless laboratory assistant. What simple chemical tests could you perform which would allow you to place the correct labels on each bottle?

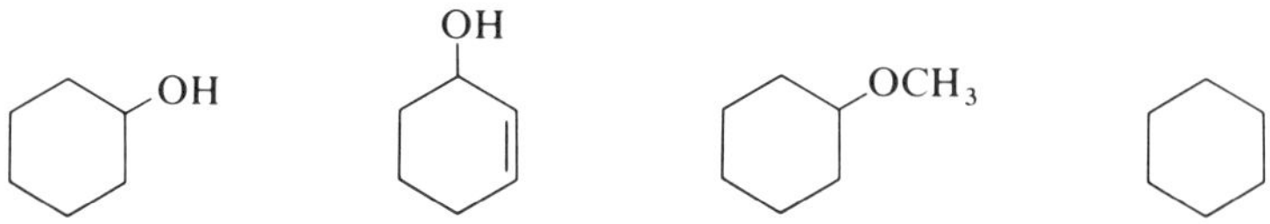

**4.** Carry out each conversion as indicated. You may use any organic or inorganic reagents. Each conversion will require more than one step.

(a)         $OH$                        $OH$
            |                           |
    $CH_3CH_2CHCH_3$ $\longrightarrow$ $CH_3CH_2CCH_3$
                                        |
                                        $CH_2CH_3$

(b) $CH_3CH_2CH{=}CH_2$ $\longrightarrow$ $CH_3CH_2CH_2CH_2OCH_2CH_3$

(c)

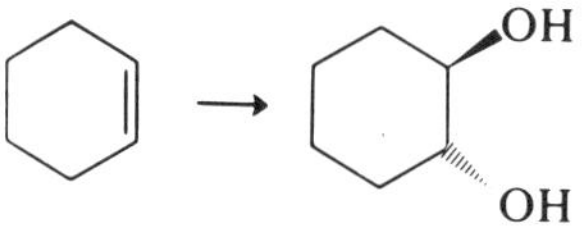

(d)

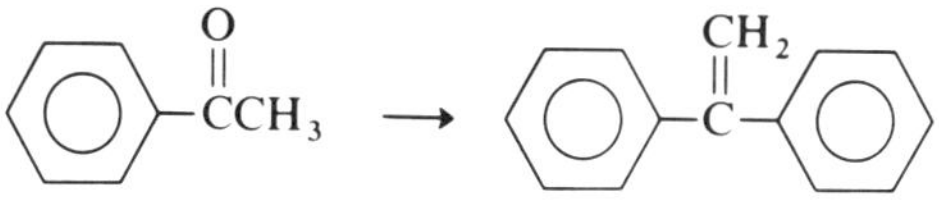

**5.** The Lucas test is often used to distinguish between 1°, 2°, and 3° alcohols. The test depends on the differing rates for the conversion of the alcohol to a chloride using conc. $HCl/ZnCl_2$ (Lucas reagent).

$$ROH \xrightarrow[ZnCl_2]{HCl} RCl$$

A positive test is the appearance of the insoluble alkyl halide which turns the clear solution cloudy.

Allyl, benzyl, 3° $\longrightarrow$ 1–5 minutes

2° $\longrightarrow$ 10–15 minutes

1° $\longrightarrow$ 1 hour

Arrange the following alcohols in order of increasing reactivity toward Lucas reagent.

(a) CH₃CHCH₂CH₃
         |
         OH

(b) (CH₃)₂CHCH₂CH₂OH

(c) CH₃CH=CHCH₂OH

(d) (CH₃)₃CCH₂OH

**6.** Compound A ($C_6H_{14}O$) is soluble in sulfuric acid and gives off hydrogen when treated with sodium metal. When the sulfuric acid solution is heated, a new compound, B ($C_6H_{12}$), is obtained. B decolorizes a solution of bromine in carbon tetrachloride. Ozonolysis of compound B produces a single substance, C ($C_3H_6O$). Treatment of C with isopropyl magnesium bromide followed by hydrolysis yields the original compound, A. Assign structures to A, B, and C.

**7.** Symmetrical ethers may be prepared by acid-catalyzed dehydration of an appropriate alcohol. However, this is a poor method for producing an unsymmetrical ether such as 1-methoxybutane. Explain.

$$CH_3OH + HOCH_2CH_2CH_2CH_3 \xrightarrow{H_2SO_4} CH_3OCH_2CH_2CH_2CH_3$$

**8.** Base-catalyzed ring-opening of cyclopentane epoxide yields only *trans*-1,2-pentane diol. Explain why no *cis* isomer is obtained.

**9.** A graduate student attempted the following reaction as part of his thesis research. It did not work and he recovered only starting material. Why did his reaction fail? Is there anything he could do to make it work?

$$\underset{\substack{|\\OH\\ \text{1 Mole}}}{CH_3CHCH_2CH_2}\overset{\overset{O}{\|}}{C}CH_3 + \underset{\text{1 Mole}}{CH_3MgBr} \;\not\longrightarrow\; \underset{\substack{|\\OH\\}}{CH_3}\underset{}{CHCH_2CH_2}\underset{\substack{|\\CH_3}}{\overset{\overset{OH}{|}}{C}}CH_3.$$

# 9

# Phenols

There is an old saying that maintains that if something looks like a duck, quacks like a duck, and walks like a duck, then it is reasonable to assume that it is indeed a duck. Phenols *look* like alcohols when you write their formula (Ar—OH), but they don't *act* like alcohols in the lab, which is why we are discussing them in a separate chapter. We will learn why it is not correct to think of phenols as "aromatic alcohols," and why they are important in their own right. Phenols are much more acidic than alcohols, and the reason behind this acidity will also tell us something about the remarkable reactivity of the aromatic ring toward electrophilic substitution. Finally, we will explore the extent to which phenolic products affect our everyday lives.

## THE STRUCTURE AND ACIDITY OF PHENOL

When we compare the autopyrolysis constants for water and typical aliphatic alcohols, it is easy to comprehend their similarity. Alcohols are generally less acidic

$$H_2O + H_2O = H_3O^+ + OH^- \qquad K = 10^{-14}$$

$$ROH + ROH = R\overset{+}{O}H_2 + RO^- \qquad K = 10^{-15} \text{ to } 10^{-20}$$

than water, and one can rationalize the small difference on the basis of a combination of polarization and solvation effects in solution. However, phenol itself is more than $10^6$ times more acidic than the corresponding saturated alcohol, cyclohexanol!

$$C_6H_5OH \rightleftharpoons C_6H_5O^{\ominus} + H^{\oplus} \quad K_a \sim 10^{-10}$$

$$C_6H_{11}OH \rightleftharpoons C_6H_{11}O^{\ominus} + H^{\oplus} \quad K_a \sim 10^{-16}$$

$$\frac{K_a \text{ (phenol)}}{K_a \text{ (cyclohexanol)}} \approx 10^6$$

The explanation for the large acidity difference lies in considering localization versus delocalization of negative charge in the two anions. In our previous discussions of resonance stabilization, we learned that delocalization of charge equates with added stability—the greater the delocalization, the greater the stability.

$$R\ddot{O}H \rightleftharpoons R\ddot{O}:^{\ominus} + H^{\oplus} \quad \text{Charge localized on O}$$

Charge delocalized over O and aromatic ring

The phenoxide ion, by virtue of the delocalization of the negative charge over the aromatic ring, is more stable than the cyclohexanol anion.

Phenols are strongly hydrogen-bonded and have high boiling points. In general, their solubility in water is not great, although the anions are completely soluble.

## SUBSTITUENT EFFECTS ON PHENOL ACIDITY

If the acidity of phenols is really derived from electron delocalization in the anion, then one should expect that substituent groups which either add or subtract electron density will affect this acidity. Let us examine the effects of a strong electron-withdrawing group ($-NO_2$) on phenol acidity, reflected in the $pK_a$ of the p-substituted compound compared with the parent, phenol.

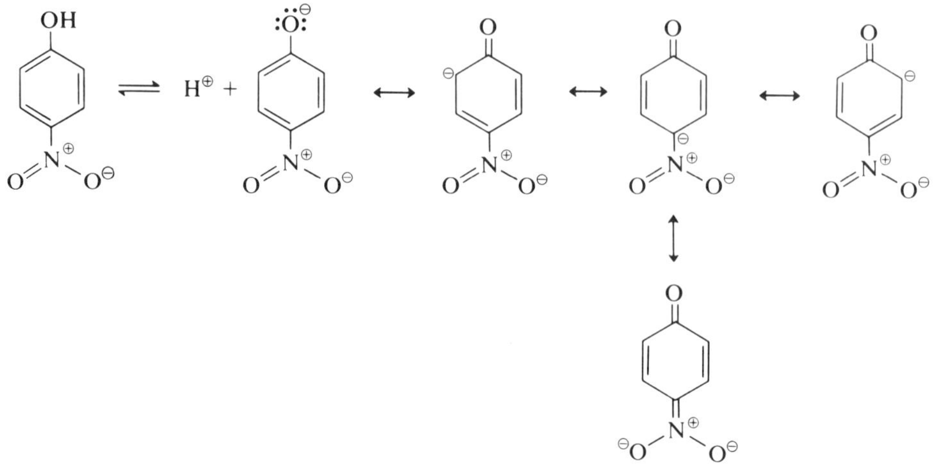

Extra resonance
form due to
$NO_2$ delocalization

The negative charge in the *p*-nitrophenoxide anion *is* delocalized to a greater extent than is the charge on the parent phenoxide, which *is* reflected in the $pK_a$'s. Other groups similarly affect the acidity of phenols.

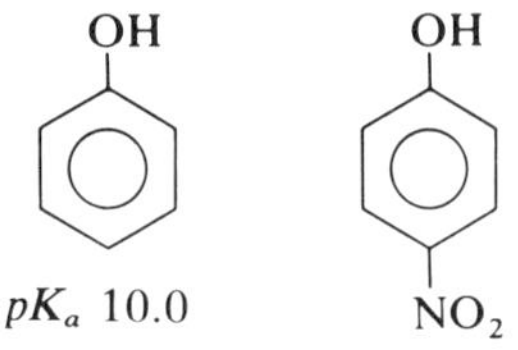

$pK_a$ 10.0        $pK_a$ 7.2     Stronger acid

---

**Exercise 9-1:**  Predict the acidities of the following phenols relative to the parent (greater or less):

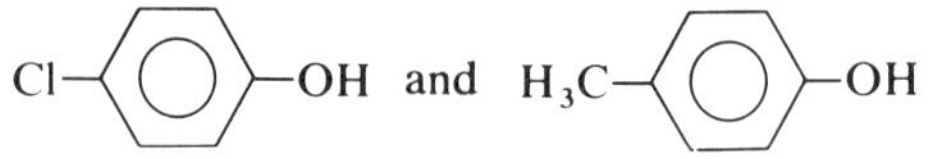

---

## PHENOL NOMENCLATURE

Phenols are most often named as derivatives of the parent compound, phenol. The —OH is positioned at C-1, and the substituent groups given the lowest possible numbering. The use of *o*, *m*, and *p* designations to locate substituents on the ring is also still prevalent in naming simple phenols.

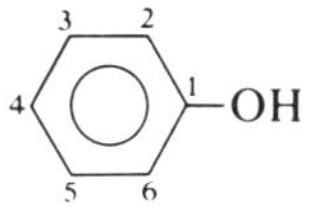

Phenol

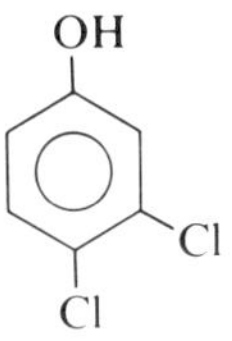

3,4-Dichlorophenol

4-Nitrophenol
(*p*-nitrophenol)

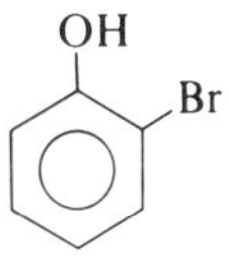

2-Bromophenol
(*o*-bromophenol)

Many phenolic compounds have common names which are still widely used in the chemical literature.

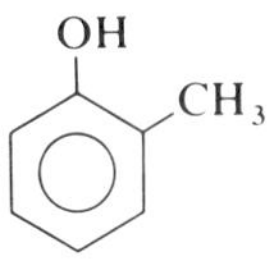

*o*-Cresol

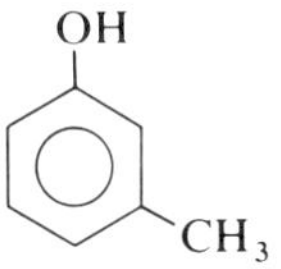

*m*-Cresol

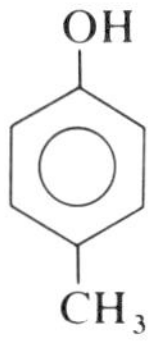

*p*-Cresol

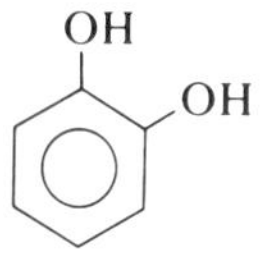

Catechol

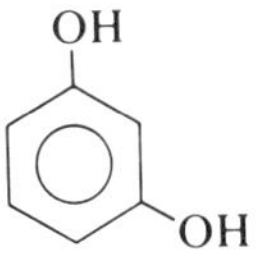

Resorcinol

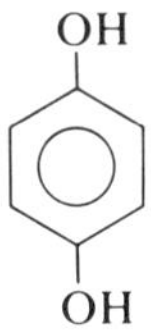

Hydroquinone

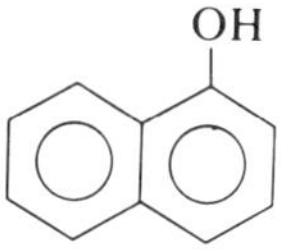

*α*-Naphthol
(1-naphthol)

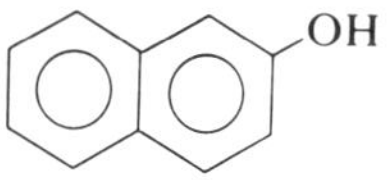

*β*-Naphthol
(2-naphthol)

**Exercise 9-2:**   Name the following compounds by the IUPAC system and by common names where possible:

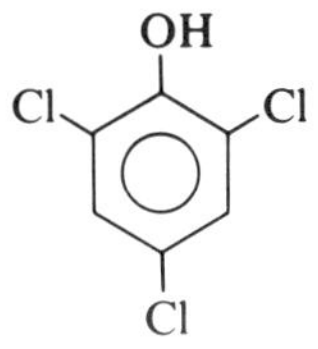

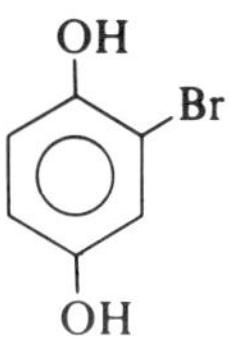

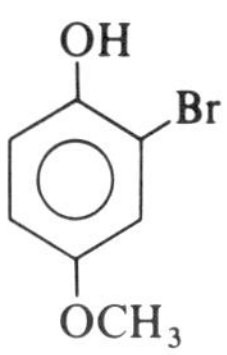

## INDUSTRIAL PREPARATION OF PHENOL

Two processes currently dominate the industrial production of phenol. The *Dow process* converts chlorobenzene into sodium phenoxide by high pressure heating with sodium hydroxide. The *cumene hydroperoxide process* involves air-oxidation of *i*-propylbenzene followed by acid-catalyzed rearrangement of the intermediary cumene hydroperoxide. The latter process produces more than 70% of the phenol made in the United States.

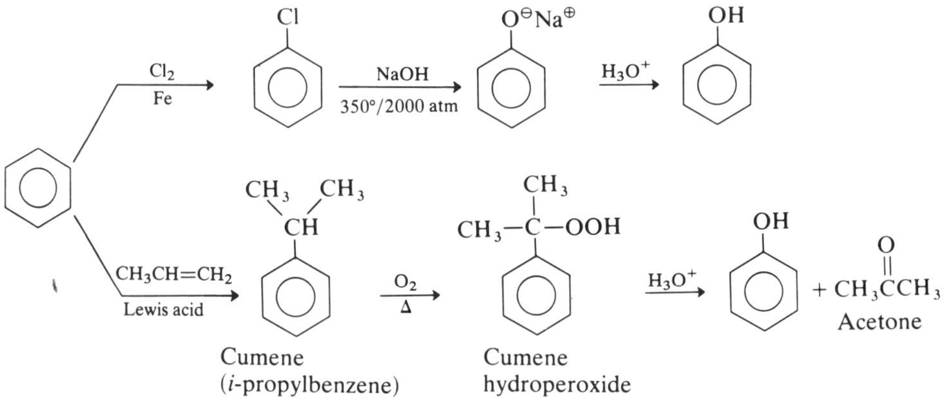

## LABORATORY PREPARATION OF PHENOLS

Phenols are normally prepared in the laboratory by fusing sodium benzene-sulfonates with sodium hydroxide, or by warming benzenediazonium halides in water.

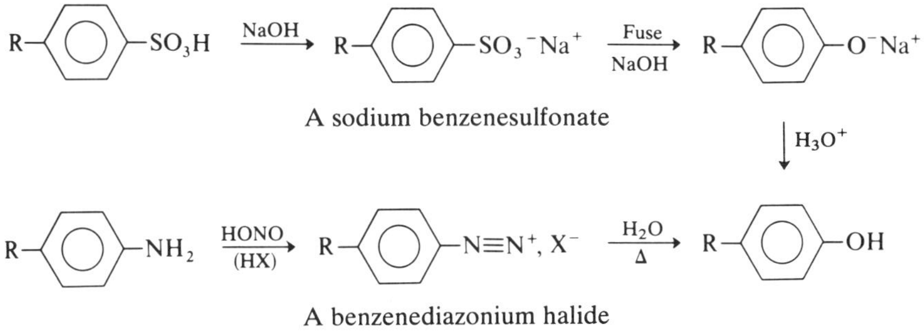

A sodium benzenesulfonate

A benzenediazonium halide

The fusion method is quite harsh, and cannot be utilized if the R group contains any functional group that is unstable to either strong base or heat. For this reason, the diazonium salt route is preferable when sensitive groups are attached to the ring.

170

Example: Prepare

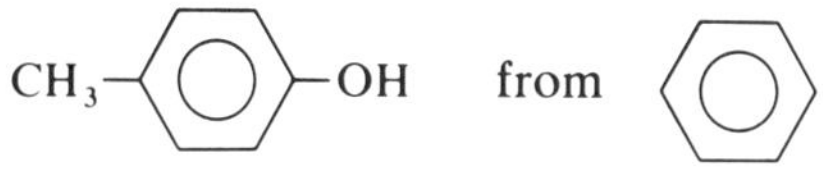

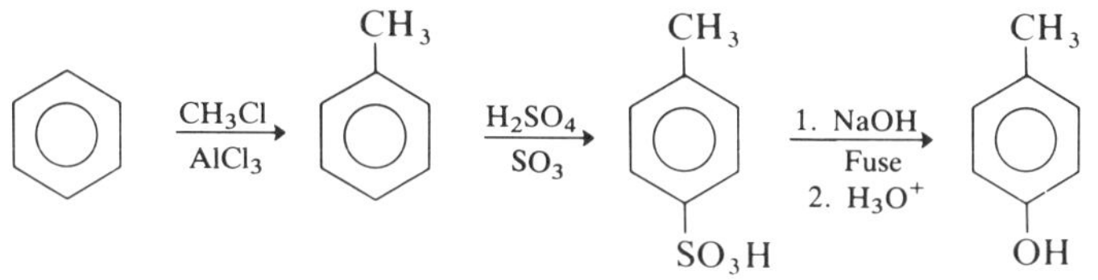

$CH_3$—⟨benzene ring⟩—OH   from   ⟨benzene ring⟩

⟨benzene⟩ $\xrightarrow[\text{AlCl}_3]{\text{CH}_3\text{Cl}}$ ⟨toluene, $CH_3$⟩ $\xrightarrow[\text{SO}_3]{\text{H}_2\text{SO}_4}$ ⟨$CH_3$ ... $SO_3H$⟩ $\xrightarrow[\substack{\text{Fuse} \\ 2.\ \text{H}_3\text{O}^+}]{1.\ \text{NaOH}}$ ⟨$CH_3$ ... OH⟩

Example: Prepare

$HOOC$—⟨benzene ring⟩—OH   from   $CH_3$—⟨benzene ring⟩

⟨$CH_3$⟩ $\xrightarrow[\text{H}_2\text{SO}_4]{\text{HONO}_2}$ ⟨$CH_3$ ... $NO_2$⟩ $\xrightarrow[\Delta]{\text{KMnO}_4}$ ⟨$COOH$ ... $NO_2$⟩ $\xrightarrow[\text{HCl}]{\text{Sn}}$ ⟨$COOH$ ... $NH_2$⟩

$\swarrow$ 1. HONO
2. H$_2$O, Δ

⟨$COOH$ ... OH⟩

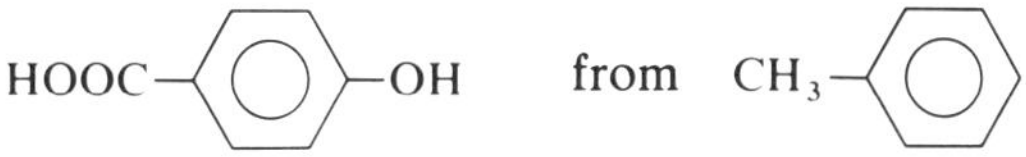

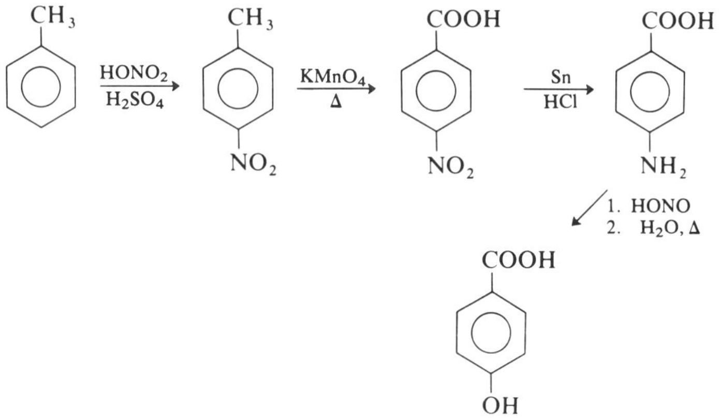

# REACTIONS OF PHENOLS

You will recall that alcohols participate in a number of reactions involving the breaking of the O—H bond. The phenolic O—H bond is considerably more labile, as discussed previously in this chapter under the heading of acidity. Sodium phenoxide, for example, is easily formed by dissolving phenol in dilute hydroxide solution, while sodium alkoxides must be formed from the alkali metal. Many of the reactions of

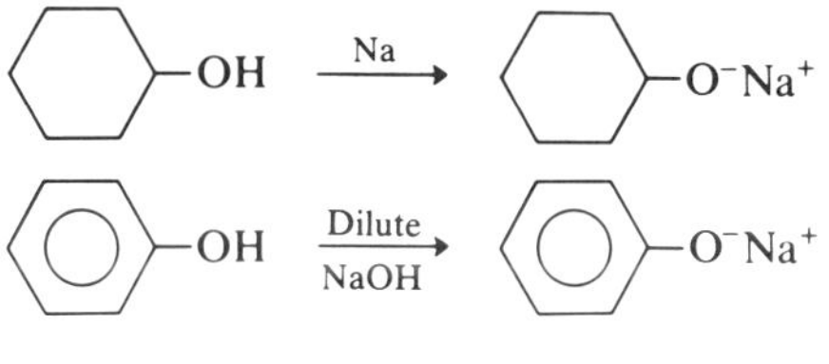

⟨cyclohexane⟩—OH $\xrightarrow{\text{Na}}$ ⟨cyclohexane⟩—O⁻Na⁺

⟨benzene⟩—OH $\xrightarrow[\text{NaOH}]{\text{Dilute}}$ ⟨benzene⟩—O⁻Na⁺

phenoxide ion are similar to the reactions of alkoxides, such as ether and ester formation. However, many reactions of alcohols involve the breaking of the C—OH

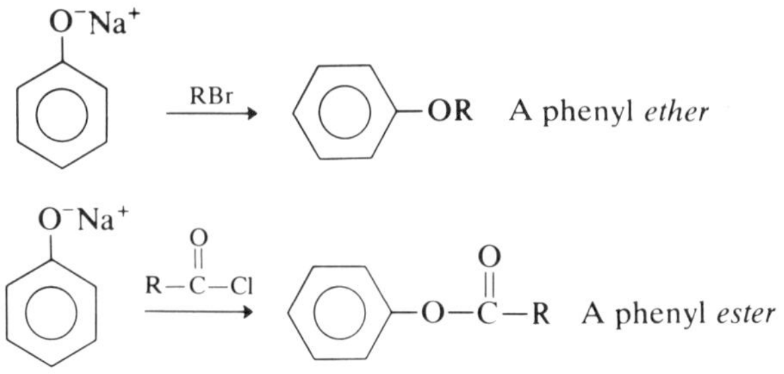

Examples:

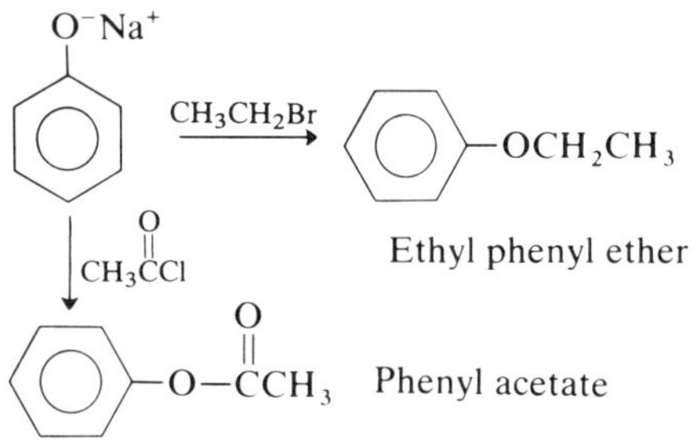

Ethyl phenyl ether

Phenyl acetate

bond, for which there are *no* counterparts in phenol chemistry. For example, aliphatic alcohols react with HBr to produce alkyl bromides, but HBr does not react with phenol.

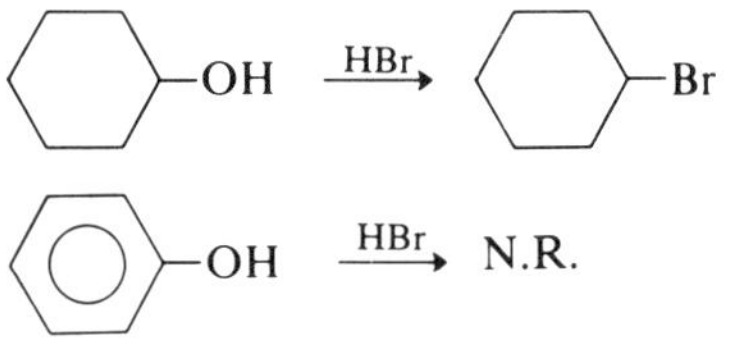

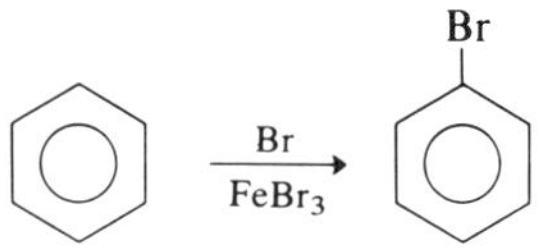

Most of the interesting chemistry of phenols involves ring substitution. The OH group is highly activating toward electrophilic attack, and the incoming electrophile is directed to the ortho- and para- positions. The ring is activated to the extent that monosubstitution is sometimes difficult to achieve. For example, bromination of phenol in aqueous solution yields 2,4,6-tribromophenol under conditions which are much milder than the monobromination of benzene. Similarly, nitration of phenol may be carried out in dilute nitric acid, while nitration of benzene requires a mixture of concentrated nitric acid and sulfuric acid. From these observations and also from our previous discussions of phenol acidity, one may conclude

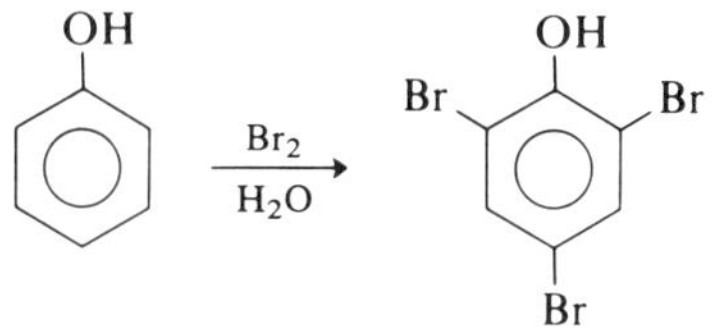

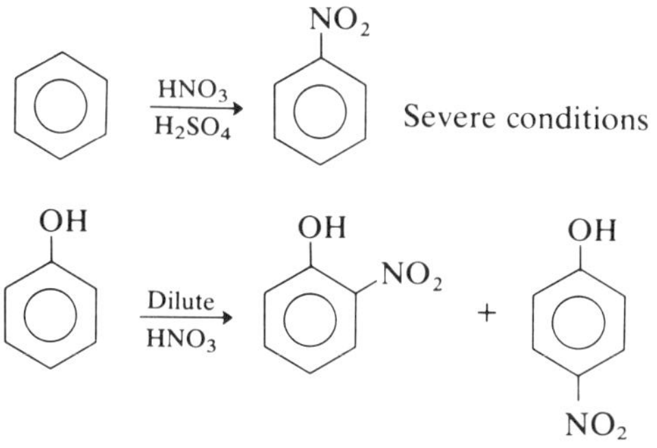

that the $\pi$ electron cloud of the phenol ring is electron-rich and hence much more susceptible to electrophilic substitution.

---

**Exercise 9-3:**   Suggest a reasonable preparation of 4-*tert*-butylphenol from benzene.

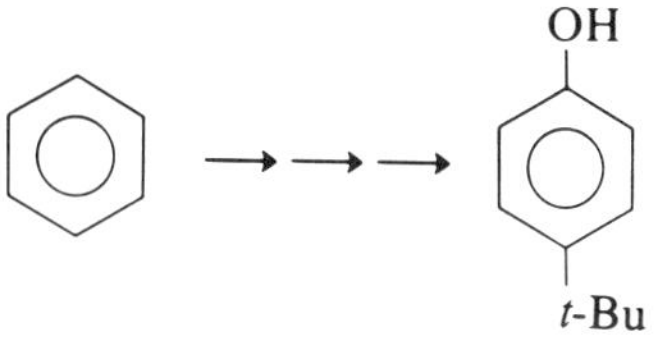

---

## PRACTICAL APPLICATIONS OF PHENOL DERIVATIVES

### Phenols as Germicides

Joseph Lister was one of the first doctors to recognize the germ-killing properties of phenols. In 1867 he used an aqueous solution of phenol to sterilize his instruments and hands prior to surgery and was thus able to prevent postoperative infection. Since then, many phenolic derivatives have been discovered which have germicidal properties, and these have been sold commercially as disinfectants, mouthwashes, cough lozenges, wood preservatives, and deodorants. Several of these useful compounds are shown below:

OH

HO    CH$_2$(CH$_2$)$_4$CH$_3$

*n*-Hexyl resorcinol

Antiseptic and germicide—used in Sucrets and various mouth washes

OH

Cl      Cl

Cl      Cl

Cl

Pentachlorophenol

Active against both bacteria and fungi—used mainly as a wood preservative

OH

*o*-Phenyphenol

Disinfectant and germicide—present in Lysol

CH$_3$

OH

Thymol

Antiseptic flavoring agent found in the herb thyme—used in many mouthwashes because of its pleasant flavor

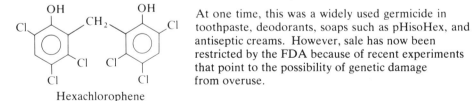

OH          CH$_2$         OH

Cl                            Cl

Cl          Cl

Cl                    Cl

Hexachlorophene

At one time, this was a widely used germicide in toothpaste, deodorants, soaps such as pHisoHex, and antiseptic creams. However, sale has now been restricted by the FDA because of recent experiments that point to the possibility of genetic damage from overuse.

The ability of a phenol derivative to kill bacteria is measured by its effectiveness compared with that of the parent phenol. The lowest concentration of the derivative which can kill a test sample of bacteria, divided by the concentration of phenol having equivalent killing power, is referred to as the *phenol coefficient* (PC). *n*-Hexylresorcinol has a PC of 50, while that of hexachlorophene is 125. Thus hexachlorophene solutions which are 125 times more dilute than given phenol solutions will have equivalent germicidal properties.

## Phenolic Resins

Phenol reacts with formaldehyde ($CH_2O$) in either acidic or basic solutions to produce Bakelite, one of the first commercially successful synthetic polymers. The reaction occurs stepwise, with the formaldehyde reacting with the ortho and para positions of the ring.

Intermediates in polymer formation

Two of the intermediary hydroxymethylated phenols may also react together by eliminating water and forming a methylene bridge between two aromatic rings. In this manner, a huge three-dimensional polymeric chain of phenol rings linked by

methylene bridges results. Many early molded products such as telephone casings and light switches were made from Bakelite, and similar resins are utilized today as protective coatings and in plywood manufacture.

Bakelite polymer

# REVIEW OF NEW CONCEPTS AND TERMS

Define and explain each term and, if possible, give a specific example of each:

Acid properties of phenol

Cumene hydroperoxide process

Bakelite

Dow process

Phenol coefficient

## PROBLEMS: CHAPTER I-9

**1.** Name the following compounds by the IUPAC system and by common names where possible:

(a)

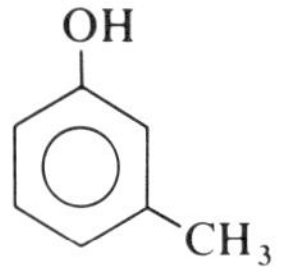

(b)

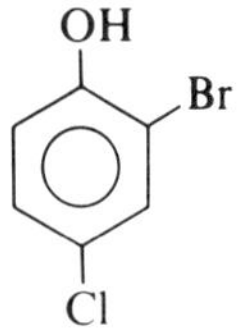

(c)

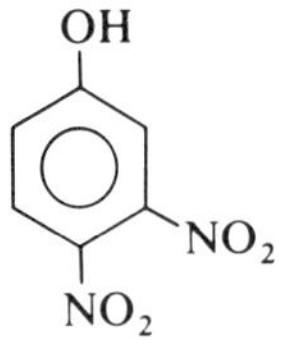

(d)

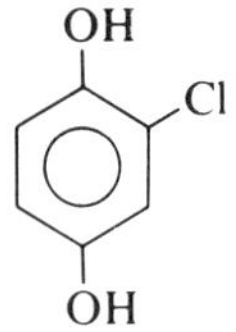

(e)

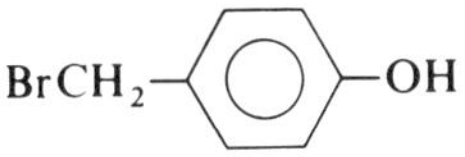

$BrCH_2$—⟨O⟩—OH

**2.** Complete the following reactions, showing only organic products:

(a)

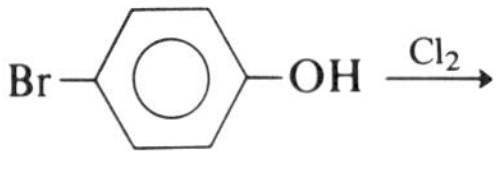

$Br$—⟨O⟩—OH $\xrightarrow{Cl_2}$

(b)

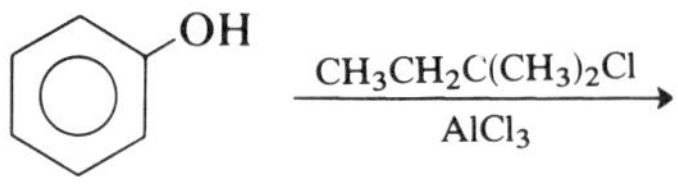

⟨O⟩—OH $\xrightarrow[AlCl_3]{CH_3CH_2C(CH_3)_2Cl}$

(c)

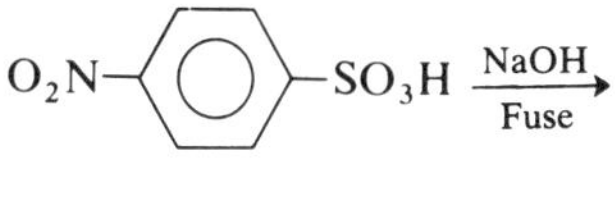

$O_2N$—⟨O⟩—$SO_3H$ $\xrightarrow[Fuse]{NaOH}$

(d)

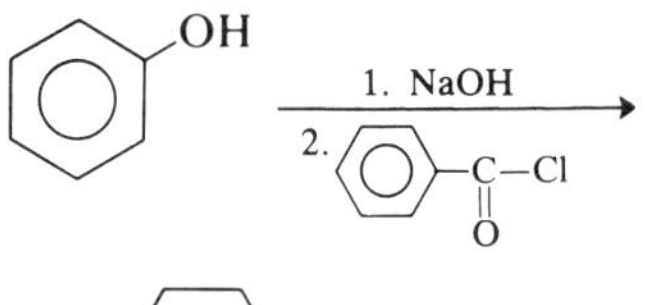

⟨O⟩—OH $\xrightarrow[2.\ \text{⟨O⟩—C—Cl}]{1.\ NaOH}$

(e)

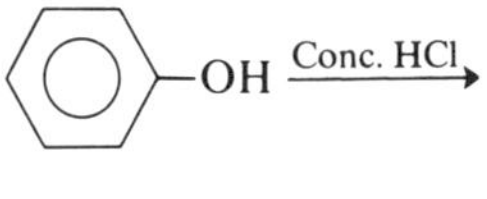

⟨O⟩—OH $\xrightarrow{Conc.\ HCl}$

(f)

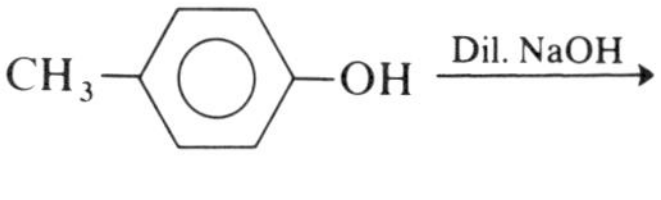

$CH_3$—⟨O⟩—OH $\xrightarrow{Dil.\ NaOH}$

(g)

⟨O⟩ $\xrightarrow[2.\ CH_2CHCH_2Cl]{1.\ NaOH}$

176

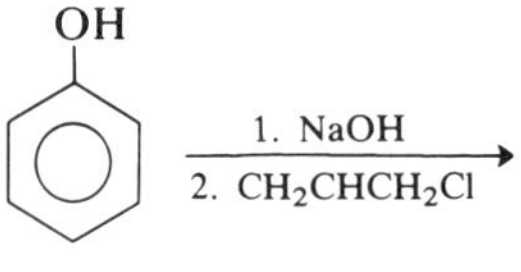

**3.** Carry out the following conversions from phenol and any other inorganic or organic reagents:

(a)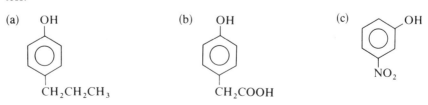

(b)

(c)

**4.** Utilizing simple chemical tests, how would you distinguish between the following compounds?

(a)   OH            OH          (b)   OH            OH

      and                            and

(c)   OH

      and   $CH_3CH_2CH_2OH$

**5.** In this chapter we have learned that phenol does not undergo any of the reactions involving C—OH scission common to alcohols. Why? Consider as an example, the lack of reaction with concentrated HBr:

$\longrightarrow$—OH $\xrightarrow{\text{HBr}}$ N.R.;   $\longrightarrow$—OH $\xrightarrow{\text{HBr}}$ $\longrightarrow$—Br

**6.** Outline reasonable syntheses of the following compounds from either benzene or toluene and any inorganic reagents or organic alcohols containing four carbons or less.

(a)   OH                    (b)   OH                    (c)        OH

      $CH_2CH_2CH_3$              $CH_2COOH$                    $NO_2$

**7.** Could a phenolic resin similar to Bakelite be prepared from the following phenol derivatives? If your answer is *yes*, would there be any important structural differences? If your answer is *no*, explain why no polymer would be formed.

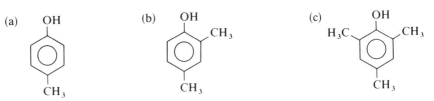

# 10

# Carbonyl Compounds:
## Aldehydes and Ketones

The chemistry of aldehydes $(R-\overset{\overset{O}{\|}}{C}-H)$ and ketones $(R-\overset{\overset{O}{\|}}{C}-R)$ is determined by the nature of the carbonyl group $(-\overset{\overset{O}{\|}}{C}-)$. This simple functional group is responsible for more interesting chemistry than is any other functionality in organic chemistry, and is probably the favorite of synthetic organic chemists. Carbonyl compounds are extremely valuable synthetic intermediates and are found as odor and flavor agents in nature. In this chapter we will learn how carbonyl compounds may be prepared and identified, and how the reactions of aldehydes and ketones may be understood in terms of the structure of the carbonyl group. Finally, we will examine the importance of volatile carbonyl compounds in everyday life.

## ALDEHYDES AND KETONES

The major difference between an aldehyde and a ketone is the nature of the groups attached to the carbonyl functional group. Simply, an aldehyde must have at least *one* hydrogen attached to the carbonyl. Ketones, on the other hand, have aryl or alkyl groups attached to the carbonyl group. Examples of both are shown below, as well as in Fig. 10-1.

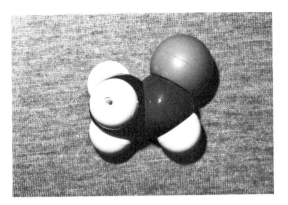 

$$\underset{\substack{\| \\ CH_3C-H}}{O}$$  $$\underset{\substack{\| \\ CH_3CCH_3}}{O}$$

**Figure 10-1.** Models of acetaldehyde, a typical aldehyde, and acetone, a typical ketone.

$$\underset{\substack{\| \\ / C \backslash}}{:O:}$$  Carbonyl group

$$\underset{\substack{\| \\ H-C-H}}{O}$$
$$\underset{\substack{\| \\ R-C-H}}{O}$$ }  Aldehydes—have at least one H attached to carbonyl group

$$\underset{\substack{\| \\ R-C-R}}{O}$$
$$\underset{\substack{\| \\ R-C-Ar}}{O}$$
$$\underset{\substack{\| \\ Ar-C-Ar}}{O}$$ }  Ketones—have alkyl or aryl groups but *no* hydrogens attached to the carbonyl group

**Exercise 10-1:** Identify each of the following as either an aldehyde or a ketone:

(a)  $$\underset{\substack{\| \\ CH_3CCH_2CH_3}}{O}$$

(b)  $$H_3C-\langle\bigcirc\rangle-\underset{\substack{\| \\ CH}}{O}$$

(c)  $$\underset{\substack{\| \\ CH_2=CHCCH_3}}{O}$$

# THE NATURE OF THE CARBONYL GROUP

The carbonyl group structure may be compared with that of the carbon–carbon double bond, except that an oxygen atom replaces one of the carbons.

Since oxygen is more electronegative than carbon, the carbonyl group is polarized toward oxygen.

$$\overset{\delta^+}{C}=\overset{\delta^-}{O}$$

Thus the carbonyl group may be considered to be a polar functional group, but not as associated as alcohols of comparable size as evidenced by boiling points or solubility in polar solvents. However, the inherent polarity of the carbonyl group has a pronounced effect on the chemistry of aldehydes and ketones, as we shall see later in this chapter.

| Compound | B.p. (°C) | Solubility in $H_2O$ (g/100 m$\ell$) |
|---|---|---|
| (benzene ring) | 80 | 0.08 |
| (benzaldehyde: ring—C(=O)—H) | 178 | 0.3 |
| (benzyl alcohol: ring—$CH_2OH$) | 205 | 4. |

# NOMENCLATURE OF ALDEHYDES AND KETONES

## Aldehydes

The IUPAC names for aldehydes are derived from the parent hydrocarbon by identifying the longest chain containing the aldehyde group ($-\overset{\|}{\underset{O}{C}}-H$) and replacing the *-ane* ending with *-al*. The aldehyde carbon is designated C-1 and substituents numbered accordingly.

Example:

$$\underset{5}{CH_3}\underset{4}{CH}\underset{3}{CH_2}\underset{2}{CH_2}\underset{1}{C}-H$$

with $CH_3$ on C-4 and O double-bonded to C-1

Parent hydrocarbon chain: pentane
IUPAC name: 4-methylpentanal

Example:

$$\underset{4}{CH_3}\underset{3}{CH}=\underset{2}{CH}\underset{1}{C}-H$$

with O double-bonded to C-1

Parent hydrocarbon chain: butene
IUPAC name: 2-butenal

Aromatic aldehydes are named as derivatives of the parent benzaldehyde, with the aldehyde ring position being designated as the 1-position.

CHO (on benzene ring) — Benzaldehyde

CHO (on benzene ring with positions 1, 2, 3, 4 and Br at position 4) — 4-Bromobenzaldehyde or *p*-Bromobenzaldehyde (common)

Aldehydes may also be named by a common system related to the common names of carboxylic acids of similar carbon content and chain structure (see Chapter I-12). The IUPAC and common names for simple aldehydes are shown in Table 10-1.

## Ketones

In the IUPAC system, ketones are named by first identifying the longest chain containing the carbonyl group and replacing the *-ane* with *-one*. However, since the carbonyl group is not at the chain terminus in ketones, we must number the chain so as to give the carbonyl group the lowest possible number.

Example:

$$\underset{5}{CH_3}\underset{4}{CH_2}\underset{3}{CH_2}\underset{2}{C}\underset{1}{CH_3}$$

with O double-bonded to C-2

Parent chain: pentane
IUPAC name: 2-pentanone

Example:

$$\underset{1}{CH_2}=\underset{2}{CH}\underset{3}{CH_2}\underset{5}{CH_2}\underset{6}{CH_3}$$

with O double-bonded to C-3

1-Hexen-3-one

Cyclic ketones are named in a similar fashion, with the carbonyl group occupying the 1-position in the ring. As in the case of aldehydes, it is not necessary to include the number 1 in the final name.

Table 10-1.   Names of Simple Aldehydes

| Structure | IUPAC name | Common name |
|---|---|---|
| $H-\overset{\overset{\displaystyle O}{\|\|}}{C}-H$ | Methanal | Formaldehyde |
| $CH_3\overset{\overset{\displaystyle O}{\|\|}}{C}-H$ | Ethanal | Acetaldehyde |
| $CH_3CH_2\overset{\overset{\displaystyle O}{\|\|}}{C}-H$ | Propanal | Propionaldehyde |
| $CH_3CH_2CH_2\overset{\overset{\displaystyle O}{\|\|}}{C}-H$ | Butanal | Butyraldehyde |
| $CH_3(CH_2)_3\overset{\overset{\displaystyle O}{\|\|}}{C}-H$ | Pentanal | Valeraldehyde |
| $CH_2{=}CH\overset{\overset{\displaystyle O}{\|\|}}{C}-H$ | Propenal | Acrolein |
| $CH_3CH{=}CH\overset{\overset{\displaystyle O}{\|\|}}{C}-H$ | 2-Butenal | Crotonaldehyde |
| $CH_3\underset{\underset{\displaystyle CH_3}{\|}}{CH}-\overset{\overset{\displaystyle O}{\|\|}}{C}-H$ | 2-Methylpropanal | Isobutyraldehyde |

Example:

Parent chain: cyclohexane
IUPAC name: 3-methylcyclohexanone

Although common names for ketones do not occur with the same degree of frequency as for aldehydes, they do exist. A few of the more common examples are shown in Table 10-2.

Table 10-2.   Names of Simple Ketones

| Structure | IUPAC name | Common name |
|---|---|---|
| $CH_3\overset{\overset{\displaystyle O}{\|\|}}{C}CH_3$ | 2-Propanone | Acetone |
| $CH_3\overset{\overset{\displaystyle O}{\|\|}}{C}CH_2CH_3$ | 2-Butanone | Methyl ethyl ketone |
| $CH_2{=}CH\overset{\overset{\displaystyle O}{\|\|}}{C}CH_3$ | 3-Buten-2-one | Methyl vinyl ketone |

183

**Exercise 10-2:**   Name the following compounds by the IUPAC system:

(a)

$$(CH_3)_2CHCH_2CH_2\overset{\overset{\displaystyle O}{\|}}{C}H$$

(b)

$$(CH_3)_2CH\overset{\overset{\displaystyle O}{\|}}{C}CH=CH_2$$

(c)

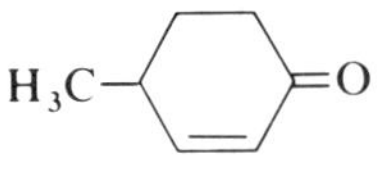

$$H_3C-\langle\ \rangle=O$$

(d)

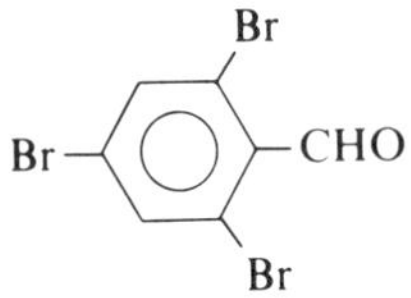

Br—(benzene ring with Br at top, Br at bottom)—CHO

(e)   $(CH_3)_3CCHO$

## PREPARATION OF ALDEHYDES AND KETONES

Both aldehydes and ketones are normally prepared from alcohols by either chemical oxidation or catalytic dehydrogenation.  Aldehydes are produced from 1° alcohols, while 2° alcohols yield ketones.

$$RCH_2OH \xrightarrow[\text{or}\,[-H_2]]{[O]} R-\overset{\overset{\displaystyle O}{\|}}{C}-H \quad \text{Aldehyde}$$

$$\overset{\overset{\displaystyle OH}{|}}{R}CHR \xrightarrow[\text{or}\,[-H_2]]{[O]} R-\overset{\overset{\displaystyle O}{\|}}{C}-R \quad \text{Ketone}$$

### Chemical Oxidation

Strong oxidizing agents such as $KMnO_4$ in basic solution or $K_2Cr_2O_7$ in acidic solution may be used to oxidize 2° alcohols to ketones.  However, these methods are less successful in the preparation of aldehydes because of their susceptibility to further oxidation to carboxylic acids.  Aldehydes may be prepared in good yield by

$$CH_3CH_2\overset{\overset{\displaystyle OH}{|}}{C}HCH_2CH_3 \xrightarrow[\text{or}\,K_2Cr_2O_7/H^+]{KMnO_4/OH-} CH_3CH_2\overset{\overset{\displaystyle O}{\|}}{C}CH_2CH_3$$

Oxidation stops at ketone—
good yields obtained

$$CH_3CH_2CH_2CH_2OH \xrightarrow[\text{or}\,K_2Cr_2O_7/H^+]{KMnO_4/OH-} \left[CH_3CH_2CH_2\overset{\overset{\displaystyle O}{\|}}{C}-H\right]$$

Intermediary aldehyde—
oxidation does not stop,
poor yields obtained

Further oxidation

$$CH_3CH_2CH_2COOH \quad \text{Carboxylic acid}$$

using any one of a number of mild oxidizing agents developed recently, such as $CrO_3$ in pyridine or dichromate in DMSO (dimethylsulfoxide).

$$CH_3CH_2CH_2CH_2OH \xrightarrow[\text{or Na}_2\text{Cr}_2\text{O}_7/\text{DMSO}]{CrO_3/C_5H_5N} CH_3CH_2CH_2\overset{\overset{\displaystyle O}{\|}}{C}-H$$

No further
oxidation

## Catalytic Dehydrogenation

Dehydrogenation of 1° or 2° alcohols may be accomplished by passing the alcohol vapor over a copper catalyst at 250–300°.

$$\overset{\overset{\displaystyle OH}{|}}{RCHR'} \xrightarrow[\Delta]{Cu} R-\overset{\overset{\displaystyle O}{\|}}{C}-R'$$

$$RCH_2OH \xrightarrow[\Delta]{Cu} R-\overset{\overset{\displaystyle O}{\|}}{C}-H$$

Example:

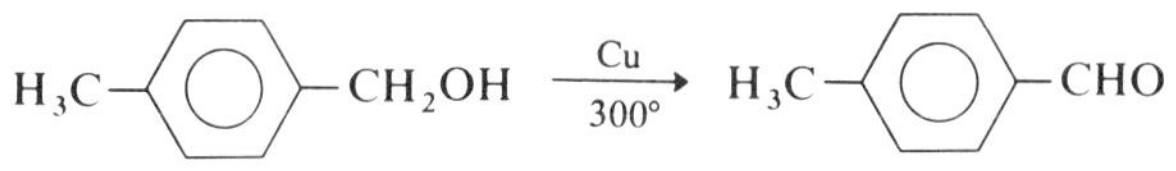

$$H_3C-\!\!\!\left\langle\bigcirc\right\rangle\!\!\!-CH_2OH \xrightarrow[300°]{Cu} H_3C-\!\!\!\left\langle\bigcirc\right\rangle\!\!\!-CHO$$

## Hydrolysis of Dihalides

Dihalides, wherein both halogens are attached to the same carbon, may be hydrolyzed under mild conditions in basic solution to yield the corresponding carbonyl compound. This method is useful for preparing aldehydes and ketones from either alkynes or aromatic hydrocarbons, as shown below:

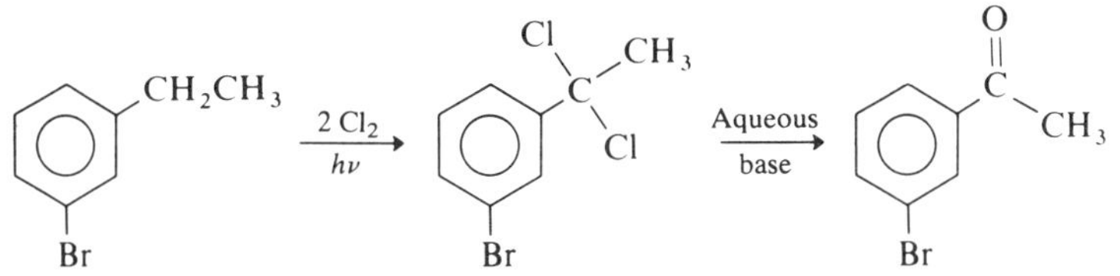

$$CH_3CH_2CH_2C\equiv CH \xrightarrow{2\,HCl} CH_3CH_2CH_2\overset{\overset{\displaystyle Cl}{|}}{\underset{\underset{\displaystyle Cl}{|}}{C}}CH_3 \xrightarrow[\text{base}]{\text{Aqueous}} CH_3CH_2CH_2\overset{\overset{\displaystyle O}{\|}}{C}CH_3$$

## Addition of H₂O to Alkynes

In Chapter I-5, we discussed the addition of water to a triple bond. The intermediary enol that is formed after the addition of one equivalent of water is unstable with respect to tautomerization to the carbonyl structure. Thus addition of water to an alkyne yields either acetaldehyde from acetylene, or a ketone from a substituted alkyne.

$$HC\equiv CH \xrightarrow[\substack{H^+ \\ (Hg^{+2})}]{H_2O} \left[\underset{H_2C=\overset{|}{CH}}{\overset{OH}{}}\right] \underset{\longleftarrow}{\overset{Rearr.}{\longrightarrow}} CH_3\overset{O}{\overset{||}{CH}}$$

$$RC\equiv CH \longrightarrow \left[\underset{RC=CH_2}{\overset{OH}{|}}\right] \underset{\longleftarrow}{\overset{Rearr.}{\longrightarrow}} R-\overset{O}{\overset{||}{C}}-CH_3$$

## Friedel–Crafts Acylation

Aromatic ketones are readily prepared by reaction of an acyl halide (RCOX) with an aromatic ring in the presence of $AlCl_3$. This process is an electrophilic attack on the aromatic ring by the polarized acyl group.

Example:

INDUSTRIAL PREPARATION
OF COMMERCIALLY IMPORTANT
CARBONYL COMPOUNDS

Many aldehydes are produced by the Oxo process, in which carbon monoxide reacts with an alkene in the presence of hydrogen and an appropriate catalyst while heating under pressure. The process forming the major product is equivalent to an anti-

$$RCH=CH_2 \xrightarrow[\substack{[Co(CO)_4]_2 \\ \text{catalyst} \\ \Delta, P}]{CO+H_2} R-\underset{\underset{CHO}{|}}{CH}-CH_3 + RCH_2CH_2CHO$$

Minor         Major
product      product

Markownikoff addition of formaldehyde to the double bond, although the actual mechanism is more complicated than that.

$$\underset{R-\overset{.}{C}H=\overset{.}{C}H_2}{\overset{O}{\overset{\|}{H-C-H}}} \longrightarrow \underset{R-CH-CH_2}{\overset{H}{\underset{|}{\overset{|}{H}}\;\overset{|}{C}=O}}$$

*Formaldehyde* is produced commercially by the vapor-phase oxidation of methanol over a metal catalyst. Formaldehyde produced in this fashion is usually sold as *formalin*, a 37% solution in water. Formalin is familiar to the biologist or medical student as a preservative, and it may also be used as a disinfectant. Formaldehyde is also important in the manufacture of Bakelite and other phenolic resins (Chapter I-9).

$$CH_3OH + O_2 \xrightarrow[250°]{Ag} \overset{O}{\overset{\|}{H-C-H}}$$

*Acetaldehyde* (ethanal) is normally produced by the hydration of acetylene, or by vapor-phase oxidation of ethanol. It is an important industrial intermediate in the manufacture of acetic acid, ethyl acetate, and vinyl acetate, all of which are extremely important industrial chemicals whose uses will be described in other chapters. Acetaldehyde, like formaldehyde, is also difficult to handle in pure form and is often sold as an aqueous solution or as a liquid trimer called *paraldehyde*. Monomeric acetaldehyde may be obtained from paraldehyde by simple heating in the presence of an acid catalyst.

    *Acetone* is the simplest ketone and is water soluble. It is produced by oxidizing isopropyl alcohol, as a byproduct of the cumene hydroperoxide process in the

187

manufacture of phenol, or by bacterial fermentation. Acetone is an extremely good solvent and is utilized as such in lacquers, paint removers, polymerization processes, and various organic syntheses. It has also been used by generations of organic chemistry students to "dry" and "clean" glassware in the laboratory.

## REACTIONS OF ALDEHYDES AND KETONES

### Addition to the Carbonyl

The carbonyl group is polarized, which explains much of the group's addition chemistry. Electrophilic reagents are attracted to the negative oxygen, while nucleophiles invariably attack the carbonyl carbon.

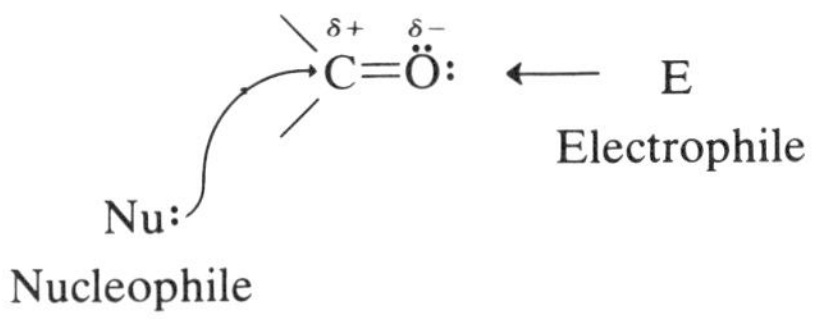

Thus the addition of an unsymmetrical reagent may be predicted with a fair degree of certainty: The positive group will bond with the oxygen while the negative group will bond at the carbon. The reactions of several typical reagents are shown below to underline this principle. Many of these additions are reversible, and subject to acid catalysis.

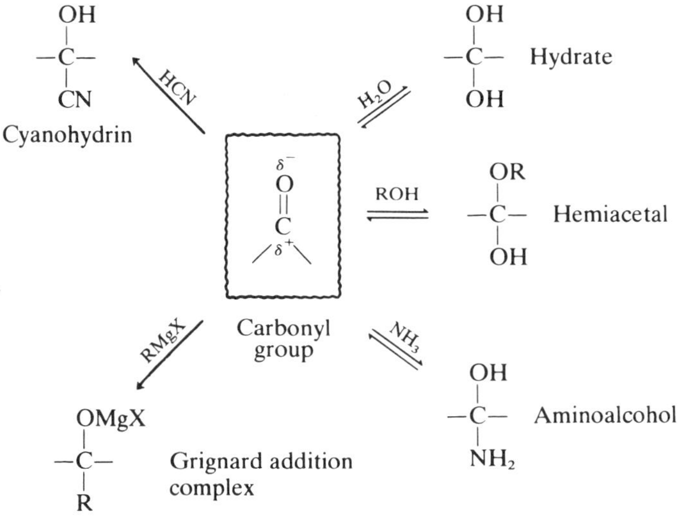

These various additions and their importance in organic syntheses are discussed below.

**Hydrates, hemiacetals, and acetals.**    Hydrates of aldehydes and ketones are formed when the carbonyl compounds are dissolved in water.  The percentage of hydrate in the equilibrium varies according to the nature of the groups attached to the carbonyl group.  Formaldehyde and acetaldehyde are almost completely converted to the hydrate form in solution, while aliphatic ketones are not.  One notable example is chloral hydrate, the infamous "Mickey Finn," or knockout drops, administered to unsavory characters in bars in low-budget movies.

$$
\underset{\substack{\text{Trichloro-}\\\text{acetaldehyde}\\\text{(chloral)}}}{Cl_3C-\overset{\displaystyle O}{\overset{\|}{C}}-H} + H_2O \longrightarrow \underset{\substack{\text{Chloral hydrate}}}{Cl_3C-\overset{\displaystyle OH}{\underset{\displaystyle OH}{C}}-H}
$$

Hemiacetals are formed in alcohol solutions, usually by acid catalysis.  Like hydrates, hemiacetals are extremely difficult to isolate.  Hemiacetals may be converted to *acetals* in excess alcohol, although this conversion is usually much slower than the first stage formation of the hemiacetal.  Acetals are similar to ethers in their

$$
R-\overset{\displaystyle O}{\overset{\|}{C}}-H \;\overset{H^+}{\rightleftharpoons}\; \left[ R-\overset{\displaystyle \overset{+}{O}H}{\overset{\|}{C}}-H \longleftrightarrow R-\overset{\displaystyle OH}{\underset{+}{C}}-H \right]
$$

$$\downarrow CH_3OH$$

$$
R-\overset{\displaystyle OH}{\underset{\displaystyle \underset{H \diagup \overset{+}{\diagdown} CH_3}{O}}{C}}-H \;\overset{-H^+}{\rightleftharpoons}\; \underset{\substack{\text{An aldehyde}\\\text{hemiacetal}}}{R-\overset{\displaystyle OH}{\underset{\displaystyle OCH_3}{C}}-H}
$$

reactivity and may be isolated from solution.  They are stable toward base, but in acid solution they are hydrolyzed back to the aldehyde form.  Because of this ease of reversion, aldehyde functional groups may be "protected" by conversion to the

$$
R-\overset{\displaystyle OH}{\underset{\displaystyle OCH_3}{C}}-H \;\overset{H^+}{\rightleftharpoons}\; \left[ R-\overset{\displaystyle \overset{+}{O}H_2}{\underset{\displaystyle OCH_3}{C}}-H \right] \;\overset{-H_2O}{\rightleftharpoons}\; \left[ R-\overset{+}{\underset{\displaystyle OCH_3}{C}}-H \right]
$$

$$\Big\updownarrow CH_3OH$$

$$
\underset{\substack{\text{Aldehyde dimethyl acetal}}}{R-\overset{\displaystyle OCH_3}{\underset{\displaystyle OCH_3}{C}}-H} \;\overset{-H^+}{\rightleftharpoons}\; \left[ R-\overset{\displaystyle \overset{+}{H}OCH_3}{\underset{\displaystyle OCH_3}{C}}-H \right]
$$

acetal form when one wishes to perform a reaction in another portion of the molecule. After completion of the reaction, the aldehyde functional group may be regenerated by acid hydrolysis.

Both hemiacetals and acetals are extremely important in carbohydrate chemistry, in terms of both their overall structure and their reactions. We will return to this topic in the next chapter.

**Ammonia and amine derivatives.** Ammonia, like water, adds to carbonyl compounds in solution to yield compounds that are relatively unstable. However,

$$
\underset{\substack{\|\\ O}}{R-\overset{\displaystyle O}{\overset{\|}{C}}-H} + :NH_3 \;\rightleftharpoons\; R-\overset{\displaystyle OH}{\underset{\displaystyle NH_2}{\overset{|}{C}}}-H \quad \text{Unstable—cannot be isolated}
$$

various derivatives of ammonia form stable addition compounds that are important synthetically and also as a means of identifying aldehydes and ketones. These

$$
R-\overset{\displaystyle O}{\overset{\|}{C}}-H + :NH_2Z \;\longrightarrow\; \left[ R-\overset{\displaystyle OH}{\underset{\displaystyle NHZ}{\overset{|}{C}}}-H \right] \xrightarrow{-H_2O} R-CH=NZ
$$

$$
R-\overset{\displaystyle O}{\overset{\|}{C}}-R + :NH_2Z \;\longrightarrow\; \left[ R-\overset{\displaystyle OH}{\underset{\displaystyle NHZ}{\overset{|}{C}}}-R \right] \xrightarrow{-H_2O} R-\underset{\displaystyle R}{\overset{|}{C}}=NZ
$$

derivatives, formed by spontaneous dehydration of the intermediary addition product, are often crystalline solids which are easily purified. The more important classes of these derivatives are illustrated below for a typical aldehyde. Ketones form similar derivatives.

$$
\underset{\text{Hydrazine}}{CH_3CH_2CH_2\overset{\displaystyle O}{\overset{\|}{C}}H + H_2NNH_2} \xrightarrow{-H_2O} \underset{\text{Butanal hydrazone}}{CH_3CH_2CH_2CH=NNH_2}
$$

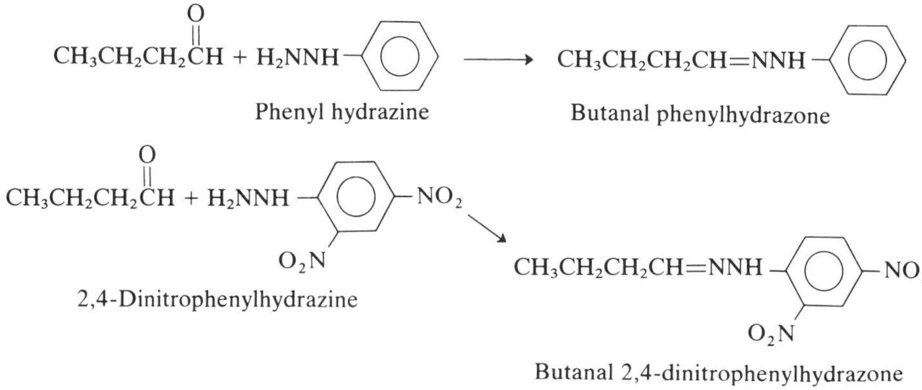

$$
CH_3CH_2CH_2\overset{\displaystyle O}{\overset{\|}{C}}H + H_2NNH\text{—}\langle\bigcirc\rangle \;\longrightarrow\; CH_3CH_2CH_2CH=NNH\text{—}\langle\bigcirc\rangle
$$
Phenyl hydrazine                    Butanal phenylhydrazone

$$
CH_3CH_2CH_2\overset{\displaystyle O}{\overset{\|}{C}}H + H_2NNH\text{—}\langle\bigcirc\rangle\text{—}NO_2
$$
2,4-Dinitrophenylhydrazine ($O_2N$)

$$
CH_3CH_2CH_2CH=NNH\text{—}\langle\bigcirc\rangle\text{—}NO_2
$$
($O_2N$)

Butanal 2,4-dinitrophenylhydrazone

$$CH_3CH_2CH_2\overset{\overset{\displaystyle O}{\|}}{C}H + H_2NNH\overset{\overset{\displaystyle O}{\|}}{C}NH_2 \longrightarrow CH_3CH_2CH_2CH=NNH\overset{\overset{\displaystyle O}{\|}}{C}NH_2$$

Semicarbazide · · · · · · · · · · · · · · · · · · · · · Butanal semicarbazone

$$CH_3CH_2CH_2\overset{\overset{\displaystyle O}{\|}}{C}H + H_2NOH \longrightarrow CH_3CH_2CH_2CH=NOH$$

Hydroxylamine · · · · · · · · · · · · · · · Butanal oxime

**Cyanohydrins.**   Addition of HCN to a carbonyl compound produces a cyanohydrin. These bifunctional compounds are important in carbohydrate synthesis because they

$$R-\overset{\overset{\displaystyle O}{\|}}{C}-H + HCN \longrightarrow R-\overset{\overset{\displaystyle OH}{|}}{C}HC\equiv N \quad \text{Aldehyde cyanohydrin}$$

$$R-\overset{\overset{\displaystyle O}{\|}}{C}-R + HCN \longrightarrow R-\underset{\underset{\displaystyle R}{|}}{\overset{\overset{\displaystyle OH}{|}}{C}}-C\equiv N \quad \text{Ketone cyanohydrin}$$

yield α-hydroxy carboxylic acids upon hydrolysis in either acid or base. We will return to this method of building carbohydrate molecules in the next chapter.

$$R-\overset{\overset{\displaystyle OH}{|}}{C}HC\equiv N \xrightarrow{H_3O^+} R-\overset{\overset{\displaystyle OH}{|}}{C}HCOOH$$

α-Hydroxycarboxylic acid

**Grignard addition—formation of 1°, 2°, and 3° alcohols.**   In Chapter I-8 we outlined the Grignard synthesis of alcohols from aldehydes and ketones: 1° and 2° alcohols may be formed from aldehydes, while 3° alcohols result from Grignard addition to ketones.

$$RMgX + CH_2=O \longrightarrow [RCH_2OMgX] \xrightarrow{H_3O^+} RCH_2OH$$

1° Alcohol

$$RMgX + R'CHO \longrightarrow \left[\underset{\underset{\displaystyle R'}{|}}{RCHOMgX}\right] \xrightarrow{H_3O^+} \underset{\underset{\displaystyle R'}{|}}{RCHOH}$$

2° Alcohol

$$RMgX + R'\overset{\overset{\displaystyle O}{\|}}{C}R'' \longrightarrow \left[R-\underset{\underset{\displaystyle R'}{|}}{\overset{\overset{\displaystyle OMgX}{|}}{C}}-R''\right] \xrightarrow{H_3O^+} R-\underset{\underset{\displaystyle R'}{|}}{\overset{\overset{\displaystyle OH}{|}}{C}}-R''$$

3° Alcohol

**Exercise 10-3:**   Give structures for the products derived from the reaction of benzaldehyde and 3-pentanone with the following reagents:

(a) $CH_3CH_2MgBr$

(b) ⟨◯⟩—$NHNH_2$

(c) $CH_3CH_2OH/H^+$
(Excess)

## Reduction of Carbonyl Compounds

Aldehydes and ketones may be reduced either chemically or by catalytic hydrogenation. The various methods available to the synthetic chemist all have their advantages and disadvantages, and in some cases, differing products. These are outlined below for typical compounds.

**Catalytic hydrogenation**

$$CH_3CH_2CH_2\overset{\overset{O}{\|}}{C}H \xrightarrow[\text{Pressure}]{H_2/Pd} CH_3CH_2CH_2CH_2OH$$

$$CH_3CH_2\overset{\overset{O}{\|}}{C}CH_2CH_3 \xrightarrow[\Delta, \text{Pressure}]{H_2/Ni} CH_3CH_2\overset{\overset{OH}{|}}{C}HCH_2CH_3$$

Aldehydes and ketones both yield alcohols upon catalytic reduction in the presence of a finely divided metal catalyst.

*Advantages:*   Reaction usually proceeds to 100% completion; product is easy to isolate; reaction may be followed quantitatively by monitoring hydrogen pressure.

*Disadvantages:*   Other functional groups, such as C=C and C≡C, will also be reduced.

**Metal hydride reduction.**   Several metal hydrides, such as $NaBH_4$ and $LiAlH_4$, are sources of hydride ion ($H:^-$), a powerful reducing agent. Reductions are usually carried out in ether solution and the free alcohol obtained by acid hydrolysis of the intermediary metal complex.

$$CH_3CH_2CH_2\overset{\overset{O}{\|}}{C}-H \xrightarrow[\text{Et}_2O]{LiAlH_4} [CH_3CH_2CH_2CH_2O]_4AlLi$$

$$\downarrow H_3O^+$$

$$CH_3CH_2CH_2CH_2OH$$

*Advantages:*   Reactions are clean and easy to run; yields are high; C=C and
C≡C are not reduced by LAH.

$$CH_3CH_2-CH=CH\overset{O}{\overset{\|}{C}}H \xrightarrow{\text{LAH}} \xrightarrow{\text{H}_3\text{O}^+} CH_3-CH=CHCH_2OH$$

*Disadvantages:*   Reagent is expensive and reacts violently with active sources
such as moisture in the air, causing fires.

**Dissolving-metal reduction (Clemmensen reduction).**  Aldehydes and ketone
carbonyl groups may be reduced to methylene groups by certain dissolving metals in
mineral acid solution.  By far the most famous of these reactions is the Clemmensen
reduction, which employs amalgamated zinc in aqueous HCl.  Aromatic rings and
isolated double bonds are not affected.

## Oxidation of Carbonyl Compounds
## (Tollens', Fehling's, and Benedict's Tests)

As stated at the beginning of this chapter, aldehydes are easily oxidized to
carboxylic acids.  Ketones, on the other hand, are quite resistant.  Even relatively
weak oxidizing agents such as silver ion ($Ag^+$) effect aldehyde oxidation.  This is the
basis of the *Tollens' test* for aldehydes.  If the aldehyde is treated with ammoniacal
silver nitrate in a *clean* test tube or flask, a silver mirror is formed; the formation of

the mirror is regarded as a positive test for the presence of $-\overset{O}{\overset{\|}{C}}-H$ in the unknown
compound.  Ketones do not react with the test solution and so give a negative test.

$$CH_3CH_2CH_2\overset{O}{\overset{\|}{C}}-H + Ag(NH_3)_2^+NO_3^- \longrightarrow Ag^0\downarrow + CH_3CH_2CH_2\overset{O}{\overset{\|}{C}}O^-$$

Tollens' solution  (clear)           Silver mirror

$$CH_3CH_2\overset{O}{\overset{\|}{C}}CH_2CH_3 + Ag(NH_3)_2^+NO_3^- \longrightarrow \text{No reaction}$$
(Solution remains clear)

Similar behavior with $Cu^{+2}$ complexes gives rise to two other well-known tests,
Fehling's and Benedict's.  In both these tests, the blue color of the copper complex is

destroyed and a precipitate of $Cu_2O$ (cuprous oxide) is obtained.

$$R-\overset{\overset{\displaystyle O}{\|}}{C}-H + Cu^{+2} \text{ (tartrate) complex} \longrightarrow R-\overset{\overset{\displaystyle O}{\|}}{C}-OH + Cu_2O\downarrow$$

Fehling's solution

$$Red$$

or

$Cu^{+2}$ (citrate) complex $\diagdown$ Blue

Benedict's solution

$$R-\overset{\overset{\displaystyle O}{\|}}{C}-R + \text{Fehling's or Benedict's solution} \longrightarrow \text{No reaction}$$

## Sodium Bisulfite Addition Compounds

An unusual reaction of aldehydes and a few special classes of ketones involves the addition of sodium bisulfite to the carbonyl group. The reaction may be reversed by the addition of dilute acid and the aldehyde recovered. This behavior enables aldehydes to be separated from crude reaction mixtures and thus purified. Apparently the addition is affected by the size of groups attached to the carbonyl, because unhindered ketones also give addition complexes. Two such ketone classes are methyl ketones and some cyclic ketones.

$$R-\overset{\overset{\displaystyle O}{\|}}{C}-H + Na^+\overset{-}{S}O_3H \longrightarrow R-\overset{\overset{\displaystyle OH}{|}}{\underset{\underset{\displaystyle SO_3^-Na^+}{|}}{C}}-H \qquad \text{White crystalline solid}$$

$$R-\overset{\overset{\displaystyle O}{\|}}{C}-CH_3 \quad \text{and} \quad (CH_2)_n\ \ C{=}O \quad \text{Also give addition products}$$

---

**Exercise 10-4:** The labels have been misplaced for bottles of the following compounds. By using the simple chemical tests described in this chapter, how would you label each bottle correctly;

(a)

$$CH_3CH_2CH_2CH_2\overset{\overset{\displaystyle O}{\|}}{C}H$$

(b)

$$CH_3CH_2CH_2\overset{\overset{\displaystyle O}{\|}}{C}CH_3$$

(c)

$$CH_3CH_2\overset{\overset{\displaystyle O}{\|}}{C}CH_2CH_3$$

(d)

$$CH_3CH_2\overset{\overset{\displaystyle OH}{|}}{C}HCH_2CH_3$$

---

## Reactions of Carbonyl Enolate Anions—
## Acidity of the $C_\alpha$—H Bond

The polarization of the carbonyl group not only controls addition but also affects the acidity of hydrogens on the carbons adjacent to partially positive carbon.

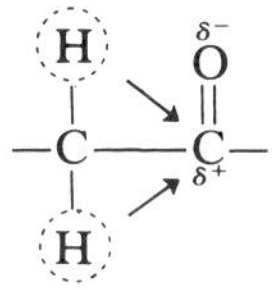

Carbonyl carbon ($\delta^+$) attracts electron density from adjacent $C_\alpha$—H bonds, causing these hydrogens to be more acidic.

$\alpha$-Hydrogens

These hydrogens may be removed by a strong base ($OH^-$, $OR^-$, etc.), forming a resonance stabilized *enolate anion*.

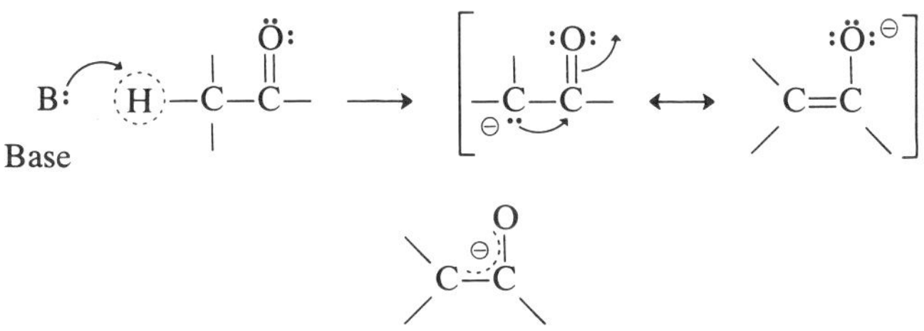

Enolate anion of an aldehyde or ketone

---

**Exercise 10-5:** For each of the following compounds, identify the enolizable hydrogens in basic solution. Draw structures for the enolate anion.

(a)
$$CH_3CH_2CH_2\overset{\displaystyle O}{\overset{\|}{C}}H$$

(b)
$$CH_3CH_2\overset{\displaystyle O}{\overset{\|}{C}}CH_3$$

(c)

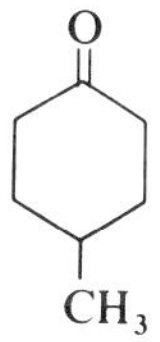

---

## Reactions of Enolate Anions

**Aldol condensation.**  If an aldehyde or ketone is treated with a basic catalyst, forming a small quantity of enolate anion (Step a below), a *condensation reaction* occurs.  The enolate anion is a powerful nucleophile and attacks the carbonyl group of a second molecule (Step b).  In essence we have "condensed" two molecules into one whose carbon chain is now doubled.  The condensation product is not resonance stabilized and abstracts a proton from the solvent, thus regenerating the basic catalyst, B, and starting the process all over again. The condensation process is referred to as an aldol condensation because the product obtained from acetaldehyde bears the trivial name *aldol*. Condensation reactions are extremely important to synthetic organic chemistry because of the ease by which long and complicated carbon chains may be constructed.

(a)

$$CH_3\overset{\overset{\displaystyle O}{||}}{C}-H \ \underset{}{\overset{B:}{\rightleftharpoons}} \ \left[ :\overset{\ominus}{C}H_2-\overset{\overset{\displaystyle O}{||}}{C}-H \ \longleftrightarrow \ CH_2{=}\overset{\overset{\displaystyle O^{\ominus}}{|}}{C}-H \right]$$

Acetaldehyde

(b)

$$H-\overset{\overset{\displaystyle O}{||}}{C}-\overset{\ominus}{\overset{..}{C}}H_2 \ + \ \overset{\overset{\displaystyle CH_3}{|}}{\underset{\underset{\displaystyle H}{|}}{C}}{=}O \ \rightleftharpoons \ H-\overset{\overset{\displaystyle O}{||}}{C}-CH_2-\overset{\overset{\displaystyle CH_3}{|}}{\underset{\underset{\displaystyle H}{|}}{C}}-O^{\ominus}$$

Condensation step                     Not resonance stabilized

(c)

$$H-\overset{\overset{\displaystyle O}{||}}{C}-CH_2-\overset{\overset{\displaystyle O^{\ominus}}{|}}{C}H-CH_3 \ \xrightarrow[\text{(Solvent)}]{BH} \ H-\overset{\overset{\displaystyle O}{||}}{C}CH_2\overset{\overset{\displaystyle OH}{|}}{C}HCH_3 + B:$$

Aldol condensation product

Aldol condensation products are $\beta$-hydroxy aldehydes or ketones. In the presence of even small quantities of acid, or when heated, they tend to dehydrate readily, forming $\alpha,\beta$-unsaturated carbonyl compounds.

$$R-\overset{\overset{\displaystyle OH}{|}}{C}H-\overset{\overset{\displaystyle}{}}{\underset{\underset{\displaystyle R}{|}}{C}}H-CHO \ \xrightarrow[\substack{\text{or } \Delta \\ (-H_2O)}]{H^+} \ R-\overset{\beta}{C}H{=}\overset{\alpha}{\underset{\underset{\displaystyle R}{|}}{C}}-\overset{\overset{\displaystyle O}{||}}{C}-H$$

$\beta$-Hydroxyaldehyde                     $\alpha,\beta$-Unsaturated
                                                      aldehyde

---

**Exercise 10-6:**   (a) Write products for each of the following condensations:

(1)

$$CH_3CH_2CH_2\overset{\overset{\displaystyle O}{||}}{C}H \ \xrightarrow[\text{EtOH}]{NaOEt}$$

(2)

$$CH_3\overset{\overset{\displaystyle O}{||}}{C}CH_3 \ \xrightarrow{KOH}$$

(b) Give structures for the products obtained from each of the above condensation products after warming in dilute acid solution.

Aldehydes containing no $\alpha$-hydrogens cannot undergo condensation. However, in the presence of another aldehyde which does have $\alpha$-hydrogens they

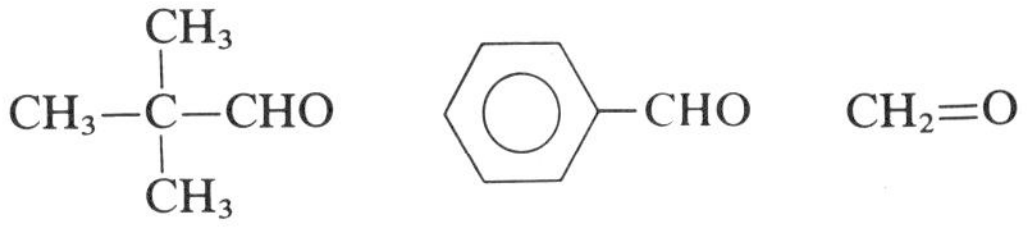

$$CH_3-\underset{\underset{CH_3}{|}}{\overset{\overset{CH_3}{|}}{C}}-CHO \qquad \langle O \rangle-CHO \qquad CH_2{=}O$$

These aldehydes have no $\alpha$-hydrogens

participate in a *mixed* or *crossed aldol condensation*. In this process, the enolate anion generated from the aldehyde containing $\alpha$-hydrogens attacks the carbonyl of the other aldehyde. Several examples are shown below.

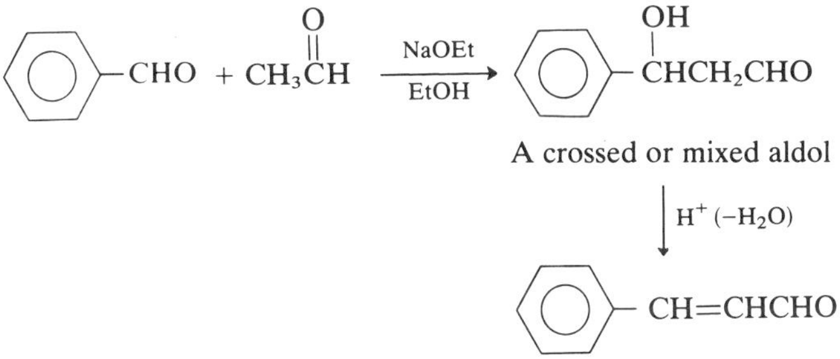

$$\langle O \rangle-CHO \;+\; CH_3\overset{\overset{O}{\|}}{CH} \;\xrightarrow[\text{EtOH}]{\text{NaOEt}}\; \langle O \rangle-\underset{\underset{OH}{|}}{CH}CH_2CHO$$

A crossed or mixed aldol

$$\Big\downarrow \text{H}^+ \, (-\text{H}_2\text{O})$$

$$\langle O \rangle - CH{=}CHCHO$$

Cinnamaldehyde

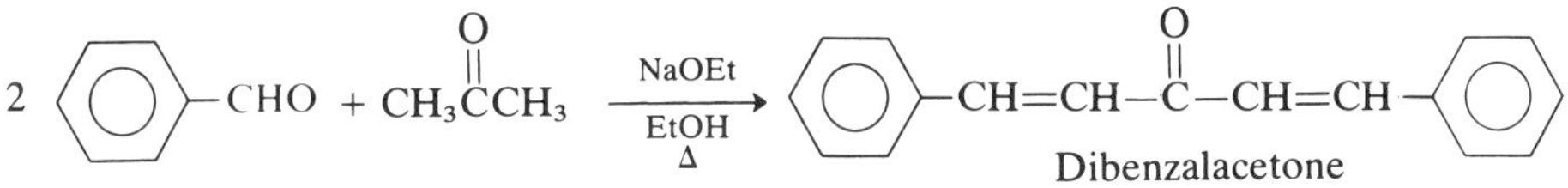

$$2\,\langle O \rangle-CHO \;+\; CH_3\overset{\overset{O}{\|}}{C}CH_3 \;\xrightarrow[\substack{\text{EtOH}\\\Delta}]{\text{NaOEt}}\; \langle O \rangle-CH{=}CH-\overset{\overset{O}{\|}}{C}-CH{=}CH-\langle O \rangle$$

Dibenzalacetone

$$(CH_3)_3C\overset{\overset{O}{\|}}{C}H \;+\; CH_3CHO \;\xrightarrow[\text{MeOH}]{\text{KOH}}\; (CH_3)_3C\underset{\underset{OH}{|}}{C}HCH_2CHO$$

$$\Big\downarrow \text{H}^+, \Delta$$

$$(CH_3)_3CCH{=}CHCHO$$

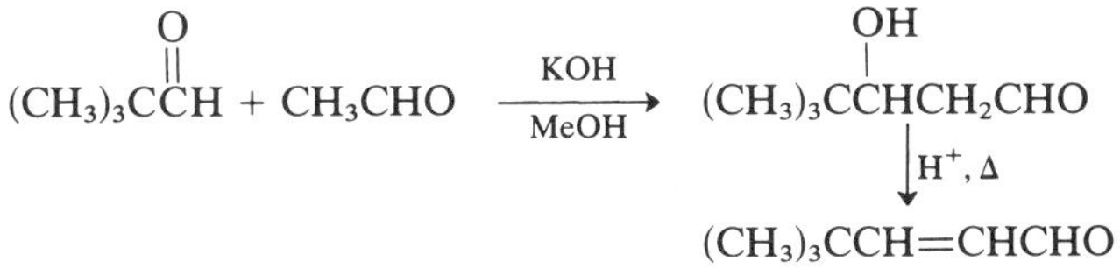

**Halogenation and the haloform reaction.**    Aldehydes and ketones both react with halogens such as $Cl_2$ and $Br_2$ to yield $\alpha$-halogenated carbonyl compounds. The reaction proceeds best in basic solution. If the compound contains more than one

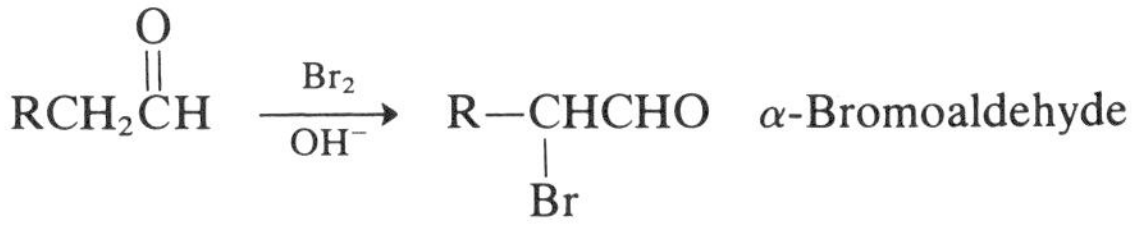

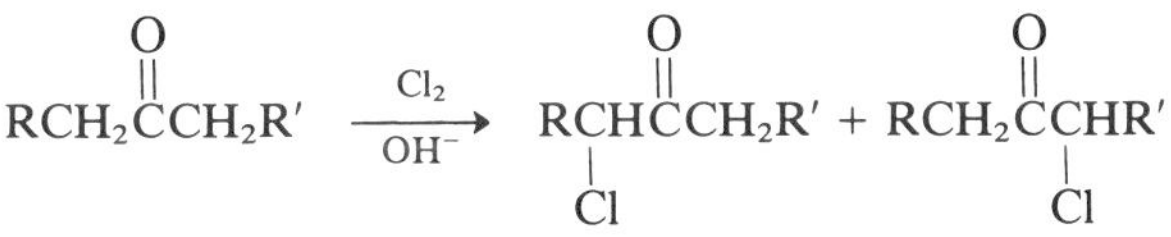

$$RCH_2\overset{\overset{O}{\|}}{CH} \;\xrightarrow[\text{OH}^-]{\text{Br}_2}\; R-\underset{\underset{Br}{|}}{CH}CHO \quad \alpha\text{-Bromoaldehyde}$$

$$RCH_2\overset{\overset{O}{\|}}{C}CH_2R' \;\xrightarrow[\text{OH}^-]{\text{Cl}_2}\; R\underset{\underset{Cl}{|}}{CH}\overset{\overset{O}{\|}}{C}CH_2R' \;+\; RCH_2\overset{\overset{O}{\|}}{C}\underset{\underset{Cl}{|}}{CH}R'$$

type of $\alpha$-hydrogen, a mixture usually results. An interesting exception is methyl ketones. The methyl hydrogens are especially reactive and give rise to a method by which they can be readily recognized. In the presence of a basic iodine solution, methyl ketones yield iodoform, $CHI_3$, a yellow solid which precipitates from solution. No other type of ketone yields iodoform in this reaction. Briefly, the mechanism can be envisioned as follows:

$$CH_3\overset{\overset{O}{\|}}{C}R \xrightarrow[OH^-]{X_2} CHX_3 + RCOO^- + X^- + H_2O$$
$$\text{Haloform}$$

$$CH_3\overset{\overset{O}{\|}}{C}R \xrightarrow[\substack{(OH^-) \\ \text{via enolate} \\ \text{anion}}]{I_2} \underset{I}{CH_2}\overset{\overset{O}{\|}}{C}-R \xrightarrow[(OH^-)]{I_2} \underset{I}{\overset{I}{\diagdown}}CH-\overset{\overset{O}{\|}}{C}-R \xrightarrow{I_2(OH^-)}$$

$$:\overset{\ominus}{C}I_3 + RCOOH \longleftarrow \underset{OH}{\overset{O^\ominus}{\underset{|}{C}I_3-C-R}} \longleftarrow \overset{O}{CI_3\overset{\|}{C}-R} \quad HO^\ominus$$

$$CHI_3 + RCOO^\ominus \quad \text{(In basic solution)}$$
$$\text{Iodoform}$$

---

**Exercise 10-7:** The iodoform test is an excellent method for the identification of methyl ketones, or rather the $-\overset{\overset{O}{\|}}{C}-CH_3$ grouping. However, ethanol, acetaldehyde, and all compounds having the structure $RCH(OH)CH_3$ also give positive iodoform tests. How would you explain this result?

*Hint:* $X_2 + 2\,OH^- \rightleftharpoons X^- + XO^- + H_2O$

$X = Cl, Br, I$

Good oxidizing agent

---

## CARBONYL COMPOUNDS IN NATURE

Volatile aldehydes and ketones are important constituents of many plant and animal odors. Propanol is one of the odor constituents of apples, while both acetaldehyde and 2-pentanone are present in the odor of fresh pineapples. 2-Hexenal is often referred to as "leaf aldehyde" because of its dominance in the odor of freshly crushed leaves. It is quite probable that the role of these substances is to attract insects and animals to act as pollinators and seed-dispersers. Several additional aldehydes and ketones responsible for familiar odors are shown below. One should remember that natural odors are actually produced by mixing many different chemical odors. It is just this complexity that makes the production of artificial flavors and odors so difficult.

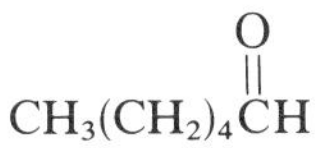

$$CH_3(CH_2)_4\overset{\overset{\displaystyle O}{\|}}{C}H$$

Eucalyptus

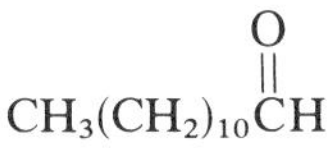

$$CH_3(CH_2)_{10}\overset{\overset{\displaystyle O}{\|}}{C}H$$

Citrus fruits

$$CH_3\overset{\overset{\displaystyle O}{\|}}{C}\overset{\overset{\displaystyle O}{\|}}{C}CH_3$$

Raspberries

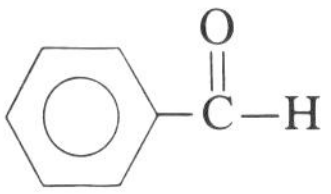

Oil of almonds

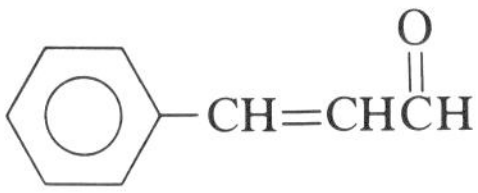

Oil of cinnamon

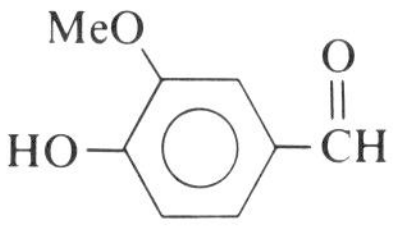

Vanillin

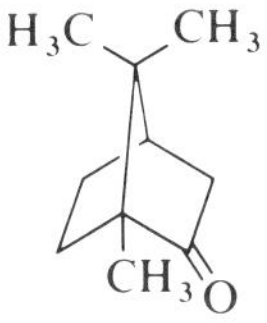

Camphor (mothballs)

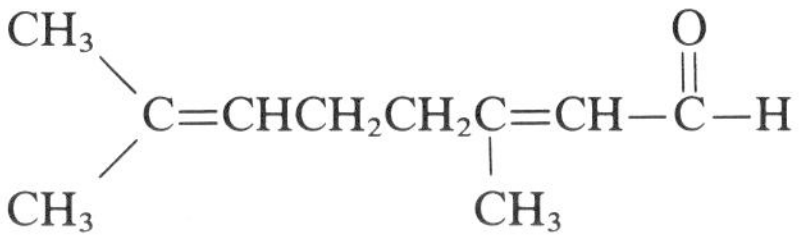

Citral (oil of bitter almonds)

Recently much research has been directed to the study of *pheromones*, chemicals that are secreted by organisms to provoke a specific response in another member of the same species. There are two different types of pheromones—*primer pheromones*, which produce prolonged physiological changes, and *releaser pheromones*, which cause rapid, reversible changes. Aldehydes and ketones are extremely important functional groups found in many pheromones, as are ester (—COOR), hydroxy (OH), and double-bond functionalities. Many pheromones are sex attractants, designed primarily to attract members of the opposite sex of the same species. For example, macrocyclic ketones are found in the scent glands of musk ox (muscone) and civet cat (civetone). However, these are often not pure compounds, but rather mixtures of several closely related compounds.

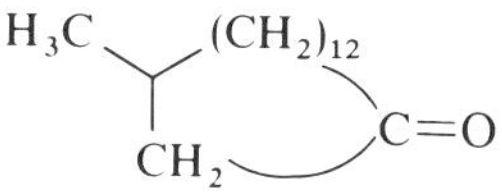

Muscone
(musk ox)

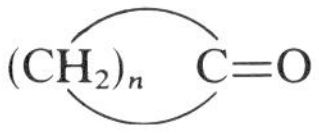

$n = 15, 16$

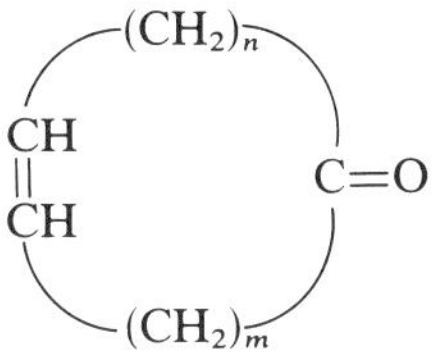

$m = 4, n = 10$
$m = 7, n = 7$
$m = 7, n = 9$

Civetone mixture
(civet cat)

Even simple carbonyl compounds can function as pheromones. For example, 4-methyl-3-heptanone is known to function as an *alarm pheromone* in harvester ants, as are 2-heptanone and 2-nonanone.

$$\underset{\substack{|\\CH_3}}{CH_3CH_2\overset{\overset{O}{\|}}{C}CHCH_2CH_3} \qquad CH_3\overset{\overset{O}{\|}}{C}(CH_2)_nCH_3 \quad n = 4, 6$$

Alarm pheromones in ants

Alarm pheromones not only signal danger to other members of the species, but they can also direct these members to attack. For example, honeybees release iso-amylacetate when they attack an intruder, which leads other bees to attack the *same* intruder.

$$(CH_3)_2CHCH_2CH_2O\overset{\overset{O}{\|}}{C}CH_3 \qquad \begin{array}{l}\text{Isoamyl acetate—honeybee}\\ \text{alarm and attack pheromone}\end{array}$$

Many *sex attractant pheromones* have been utilized in insect control. The most amazing discovery of this research is the infinitesimal quantities of pheromones that can produce response. It has been estimated that a female gypsy moth has only 0.1 microgram of sex attractant at her disposal, and that male gypsy moths can detect a concentration of only a *few hundred molecules* per cubic centimeter in the air! Research chemists at the U.S. Department of Agriculture have been able to isolate and identify many of these sex attractants. Several of these attractants have been synthesized and used to bait traps for undesirable species. Since many insect species have developed high tolerances to chemical insecticides, this method holds great promise as a natural means of controlling insect populations. Several examples of these insect sex attractants are shown below.

$$CH_3(CH_2)_7CH{=}CH(CH_2)_{12}CH_3 \quad \text{Common house fly}$$

$$(CH_3)_2CH(CH_2)_4\overset{\overset{O}{\diagup\;\diagdown}}{CH{-}CH}(CH_2)_9CH_3 \quad \text{Gypsy moth}$$

$$\underset{\substack{|\\H\;\;H\;\;\;\;H}}{CH_3(CH_2)_7C{=}C{-}\overset{\overset{H}{|}}{C}{=}CCH_2COOH} \quad \text{Black carpet beetle}$$

Many insect attractants are combinations of several closely related compounds; for example, the boll weevil, scourge of Southern cotton growers, has a sex attractant composed of four related substances:

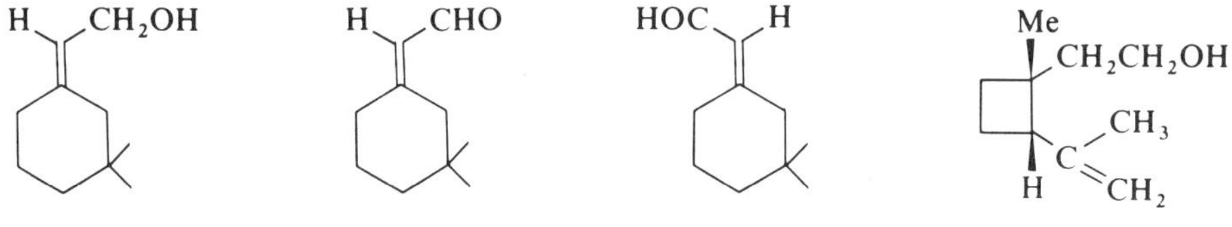

The field of pheromone chemistry is still in its infancy. If insects and animals excrete alarm and attractant pheromones, the next obvious question is whether humans do as well. We cannot answer that question at the present time, although the many possibilities are intriguing if we do!

**Exercise 10-8:**   Devise a synthesis of the ant alarm pheromone 4-methyl-3-heptanone from compounds containing 3 carbons or fewer.


## REVIEW OF NEW TERMS AND CONCEPTS

Define each term and, if possible, give an example of each:

Carbonyl compound                        Aldehyde
Ketone                                   Catalytic dehydrogenation
Friedel–Crafts acylation                 Oxo process
Carbonyl addition                        Hemiacetals and acetals
Cyanohydrin                              Hydrazone
Semicarbazone                            Oxime
Metal hydride reduction                  Clemmensen reduction
Tollens', Fehling's, and Benedict's tests   Sodium bisulfite addition compound
Enolate anion                            Aldol condensation
Mixed or crossed aldol condensation      Haloform reaction
Pheromone


## PROBLEMS: CHAPTER I-10

1. Name the following compounds according to the IUPAC system. Where possible, also give common names.

(a)
$$CH_3CH_2CH_2CH_2\overset{\overset{\displaystyle O}{\|}}{C}H$$

(b)
$$CH_3CH_2\overset{\overset{\displaystyle O}{\|}}{C}CH_2CH_2CH_3$$

(c)

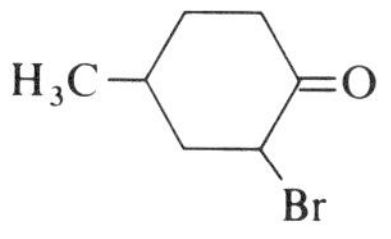

(d)

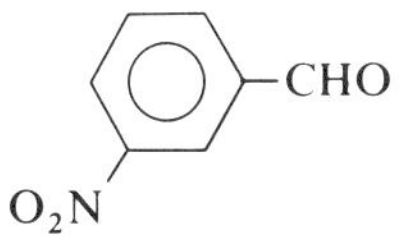

(e) $CH_2{=}\underset{\underset{\displaystyle Cl}{|}}{C}{-}CHO$

(f)

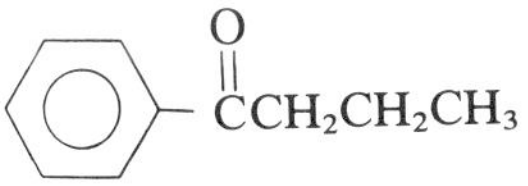

(g)
$$CH_3CH_2\overset{\overset{\displaystyle O}{\|}}{C}CH{=}CH_2$$

2. Complete the following reactions, showing only organic products:

(a) $(CH_3)_2CHCH_2CHO \xrightarrow[\text{(Hot, conc.)}]{KMnO_4}$

(b) $CH_3CH_2CH_2OH \xrightarrow[250°]{Cu}$

(c)

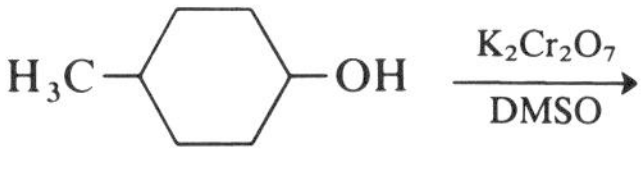

$H_3C-\!\!\bigcirc\!\!-OH \xrightarrow[DMSO]{K_2Cr_2O_7}$

(d)

$(CH_3)_2CHCH_2\overset{\overset{\displaystyle O}{\|}}{C}Cl \xrightarrow{AlCl_3}$

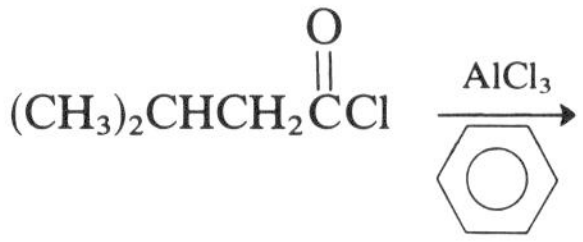

(e)

$CH_3CH_2CH_2\overset{\overset{\displaystyle O}{\|}}{C}CH_3 \xrightarrow[Pd]{H_2}$

(f)

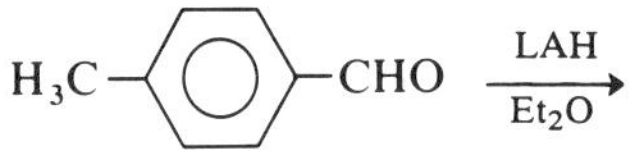

$H_3C-\!\!\bigcirc\!\!-CHO \xrightarrow[Et_2O]{LAH}$

(g)

$CH_3CH_2CH_2\overset{\overset{\displaystyle O}{\|}}{C}CH_2CH_3 \xrightarrow[\text{2. } H_3O^+]{\text{1. } CH_3MgBr}$

(h)

$(CH_3)_3C\overset{\overset{\displaystyle O}{\|}}{C}CH_2CH_3 \xrightarrow[(H^+)]{H_2NOH}$

(i)

$CH_3CH_2\overset{\displaystyle CH}{\underset{\displaystyle CH_3}{|}}\overset{\overset{\displaystyle O}{\|}}{C}CH_3 \xrightarrow[H^\oplus]{H_2NNH\overset{\overset{\displaystyle O}{\|}}{C}NH_2}$

(j)

$\bigcirc\!\!-CHO \xrightarrow{HCN}$

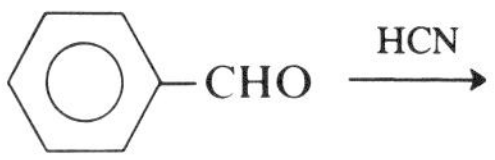

(k) $CH_3CH_2CH_2CH_2CHO \xrightarrow[EtOH]{NaOEt}$

(l)

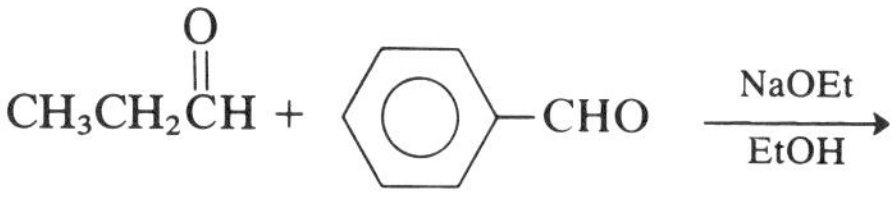

$CH_3CH_2\overset{\overset{\displaystyle O}{\|}}{C}H + \bigcirc\!\!-CHO \xrightarrow[EtOH]{NaOEt}$

(m)

$H_3C-\!\!\bigcirc\!\!=O \xrightarrow[(OH^-)]{\overset{Br_2}{DMSO}}$

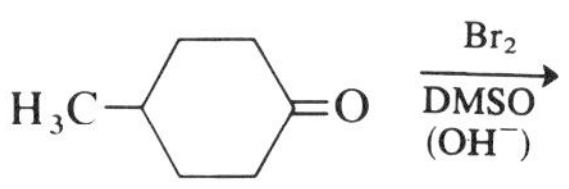

(n)

$(CH_3)_3C\overset{\overset{\displaystyle O}{\|}}{C}CH_3 \xrightarrow[NaOH]{\text{Excess } Br_2}$

**3.** Give reaction products for the reaction, if any, of each of the following reagents with benzaldehyde and 2-pentanone:

(a) Zn/HCl

(b)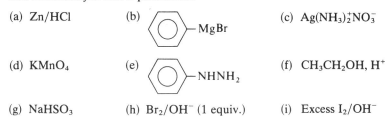

(c) $Ag(NH_3)_2^+NO_3^-$

(d) $KMnO_4$

(e) 

(f) $CH_3CH_2OH, H^+$

(g) $NaHSO_3$

(h) $Br_2/OH^-$ (1 equiv.)

(i) Excess $I_2/OH^-$

**4.** Distinguish between the following pairs of compounds by using a simple chemical test with a visual result:

(a) $CH_3CH_2CH_2CH_2CHO$   and   $CH_3CH_2CH_2CH_2OH$

(b)
$$\underset{\displaystyle CH_3CH_2CH_2\overset{\textstyle O}{\overset{\|}{C}}CH_3}{} \quad and \quad \underset{\displaystyle CH_3CH_2\overset{\textstyle O}{\overset{\|}{C}}CH_2CH_3}{}$$

(c)

(d)

$and$   $CH_3CH_2\overset{\textstyle O}{\overset{\|}{C}}CH_2CH_2CH_3$

(e) $CH_3CH_2CH_2CH_2CH_2CHO$   and   $CH_3CH_2CH_2\overset{\textstyle O}{\overset{\|}{C}}CH_2CH_3$

**5.** Prepare the following compounds from benzene, any organic compounds containing 3 carbons or fewer, and any necessary inorganic reagents:

(a)
$$\overset{\displaystyle OH}{\underset{}{CHCH_2CH_3}}$$

(b)
$$CH_3CH_2CH_2\overset{\displaystyle OH}{\underset{}{CHCN}}$$

(c)
$$CH_3CH_2CH_2\overset{\textstyle O}{\overset{\|}{C}}CH_2CH_3$$

(d)
$$CH_3CH_2CH_2CH_2-$$

(e) $CH_3CH_2CH=CHCHCHO$
     $\overset{|}{CH_2CH_3}$

(f) $CH=CCHO$
     $\overset{|}{CH_2CH_3}$

**6.** Carry out the following conversions using any necessary organic reagents:

(a)
$$H_3C-\!\!\!\bigcirc\!\!\!-OH \longrightarrow H_3C-\!\!\!\bigcirc\!\!\!=O$$

(b)
$$CH_3CH_2CH_2\overset{\textstyle O}{\overset{\|}{CH}} \longrightarrow CH_3CH_2CH_2\overset{\displaystyle OH}{\underset{}{CHCOOH}}$$

(c)

$$CH_3\overset{\displaystyle O}{\overset{\displaystyle \|}{C}}CH_3 \longrightarrow \langle\hspace{-0.3em}\bigcirc\hspace{-0.3em}\rangle\text{-}CH_2CH_2\overset{\displaystyle O}{\overset{\displaystyle \|}{C}}CH_3$$

(d) $CH_3CH_2CH_2OH \longrightarrow CH_3CH_2CH_2COOH$

7. An unknown compound, A ($C_8H_{16}$), absorbs one mole of hydrogen in the presence of palladium to give B ($C_8H_{18}$). When A is treated with ozone, an intermediary ozonide, C ($C_8H_{16}O_3$), is obtained which yields two products, D ($C_3H_6O$), and E ($C_5H_{10}O$), when reduced with zinc dust. D and E yielded the following test data:

D or E + 2,4-dinitrophenylhydrazine $\longrightarrow$ Ppt

D or E + Tollens' reagent $\longrightarrow$ No reaction

D + NaHSO₃ $\longrightarrow$ Ppt

E + NaHSO₃ $\longrightarrow$ No reaction

Give reasonable structures for A, B, C, D, and E and write equations for all reactions.

8. Give structure for all intermediates in the following syntheses:

(a) $CH_3CH_2CH_2OH \xrightarrow{PBr_3} A \xrightarrow[Et_2O]{Mg} B \xrightarrow{\phi CHO} C$

$$C \xrightarrow{H_2SO_4} D \text{ (Major product)}$$

(b)

$$\xrightarrow[H^+]{\text{Excess } CH_3OH} A \xrightarrow[\Delta]{KMnO_4} B \xrightarrow[\Delta]{H_3O^+} C$$

(c)

$$CH_3CH_2\overset{\displaystyle O}{\overset{\displaystyle \|}{C}}CH_2CH_3 \xrightarrow[\text{2. } H_3O^+]{\text{1. LAH}} A \xrightarrow{PBr_3} B$$

$$B \xrightarrow{Mg/Et_2O} C$$

$$E \xleftarrow[DMSO]{K_2Cr_2O_7} D \xleftarrow[\text{2. } H_3O^+]{\text{1. } CH_2=O} C$$

9. When cyclohexanone ($C_6H_{10}O$) is treated with a solution of NaOD in $D_2O$, recovered, and analyzed, the molecular formula is found to be $C_6H_6D_4O$. If this product is reduced with lithium aluminum hydride, the resulting alcohol is found to

have the formula $C_6H_8D_4O$. Acid-catalyzed dehydration yields an olefin, $C_6H_7D_3$. Write equations for all of the above reactions which account for the observed deuterium distributions.

10. An unsaturated ketone, $C_{17}H_{28}O$, present in the scent glands of the muskrat, yields two products (A and B) when submitted to ozonolysis:

$$A = \overset{\displaystyle O}{\overset{\|}{HC}}(CH_2)_3\overset{\displaystyle O}{\overset{\|}{C}}(CH_2)_5\overset{\displaystyle O}{\overset{\|}{CH}}$$

$$B = C_6H_{10}O_2$$

B gives a positive Tollens' test, but a negative iodoform test. Clemmensen reduction of B yields *n*-hexane. Write structures for B and the original unsaturated ketone.

# 11

# Acyl Compounds:
## Carboxylic Acids and Their Derivatives

Carboxylic acids, RCOOH, are the end product of chemical oxidation in organic systems. They are found in nature as free acids or as various *acyl derivatives*, RCOX, such as esters (X=OR) or amides (X=$NH_2$, NHR, $NR_2$). In this chapter we shall learn how carboxylic acids and their derivatives may be prepared and how these various classes of compounds react with other molecules. We shall also learn that carboxylic acids are extremely important in nature as prime ingredients of flavor and odor constituents as well as fats and oils.

## STRUCTURE AND NOMENCLATURE OF CARBOXYLIC ACIDS

The carboxyl group is the end product of the oxidation of primary alcohols discussed briefly in Chapter I-10. The group may be envisioned as being a combination of

$$RCH_2OH \xrightarrow{[O]} \underset{\text{Aldehyde}}{R-\overset{\overset{\displaystyle O}{\|}}{C}-H} \xrightarrow{[O]} \underset{\substack{\text{Carboxylic} \\ \text{acid}}}{R-\overset{\overset{\displaystyle O}{\|}}{C}-OH} \qquad\qquad \underset{\text{Carboxyl group}}{-\overset{\overset{\displaystyle O}{\|}}{C}-OH}$$

1° alcohol

carbonyl (C=O) and hydroxyl (—OH) and much of the group's chemistry may be interpreted from this viewpoint.

In the IUPAC system carboxylic acids are named by replacing the -e of the longest hydrocarbon chain containing the COOH with -oic acid. The —COOH carbon is then designated C-1 in the chain.

Examples:

$\overset{5}{C}H_3\overset{4}{C}H\overset{3}{C}H_2\overset{2}{C}H_2\overset{1}{C}OOH$     Longest chain: pentan*e*
|                                  IUPAC name: 4-methylpentan*oic acid*
$CH_3$

$\overset{4}{C}H_3\overset{3}{C}H=\overset{2}{C}H\overset{1}{C}OOH$     Longest chain: 2-butene
IUPAC name: 2-butenoic acid

Aromatic acids are named as derivatives of the parent benzoic acid, while acids in which the carboxylic group is attached to a saturated ring are named as illustrated below.

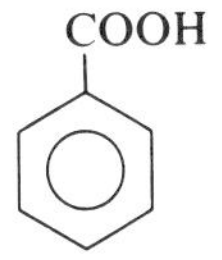

COOH

COOH

COOH

Br

CH$_3$

CH$_3$

Benzoic acid 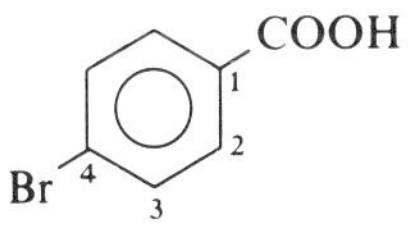          4-Bromobenzoic          3-Methylcyclohexane
                      acid                    carboxylic acid

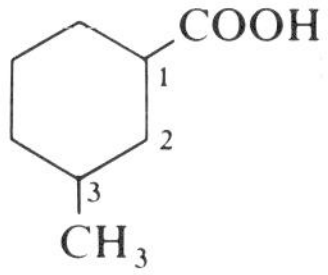

Many carboxylic acids have been known for a long time, some since antiquity, frequently because of their strong odors. Several common names are actually derived from early Latin names identifying the source of these pungent liquids. In Table 11.1, common names for several carboxylic acids are listed with their derivation. IUPAC names are also listed for comparison.

Dicarboxylic acids are also prevalent in nature. These difunctional compounds are extremely useful in organic synthesis. They are named in the IUPAC system by adding the ending -*dioic acid*.

HOOCCH$_2$CH$_2$COOH     Parent chain: butane
IUPAC name: butandioic acid

Table 11-1.   Common and IUPAC Names for Simple Carboxylic Acids

| Structure | Common name | Derivation | IUPAC name |
| --- | --- | --- | --- |
| HCOOH | Formic acid | L. formica (ant) | Methanoic acid |
| CH$_3$COOH | Acetic acid | L. acetum (vinegar) | Ethanoic acid |
| CH$_3$CH$_2$CH$_2$COOH | Butyric acid | L. butyrum (butter) | Butanoic acid |
| CH$_3$(CH$_2$)$_4$COOH | Caproic acid | L. caper (goat) | Hexanoic acid |

However, dicarboxylic acids are almost always referred to by their common names, some of which are listed in Table 11-2 as a representative sampling of the more important diacids.

Table 11-2.  Common and IUPAC Names for Some
Representative Dicarboxylic Acids

| Structure | Common name | IUPAC name |
|---|---|---|
| HOOCCOOH | Oxalic acid | Ethandioic acid |
| HOOCCH$_2$COOH | Malonic acid | Propandioic acid |
| HOOC(CH$_2$)$_2$COOH | Succinic acid | Butandioic acid |
| HOOC(CH$_2$)$_3$COOH | Glutaric acid | Pentandioic acid |
| HOOC(CH$_2$)$_4$COOH | Adipic acid | Hexandioic acid |

H   COOH
  ‖              Maleic acid        $Z$-Butendioic acid
H   COOH

H   COOH
  ‖              Fumaric acid       $E$-Butendioic acid
HOOC   H

COOH
       Phthalic acid
COOH

============================================================

**Exercise 11-1:**   Name the following compounds by the IUPAC system

(a)          CH$_3$
          |
      CH$_3$CHCHCOOH
          |
          Cl

(b)     COOH
          NO$_2$

          Br

(c)           CH$_3$
              |
      HOOCCH$_2$CCOOH
              |
              CH$_3$

============================================================

## ACIDITY OF CARBOXYLIC ACIDS
## AND THE CARBOXYLATE ANION

As their name implies, carboxylic acids are proton donors in aqueous solution.  You
may recall from Chapter I-8 that alcohols are not proton donors.  In fact, the
difference in acidity is reflected in their respective $K_a$'s.  Carboxylic acids are better

$$RCOOH + H_2O \rightleftharpoons RCOO^{\ominus} + H_3O^{\oplus} \qquad K_a \sim 10^{-5}$$

$$RCH_2OH + H_2O \rightleftharpoons RCH_2O^{\ominus} + H_3O^{\oplus} \qquad K_a \sim 10^{-15} - 10^{-20}$$

proton donors by a factor of $10^{10}$–$10^{15}$. How may we account for this extremely large acidity difference?

The answer lies in the difference in stability of the anions. The charge on the alkoxide ion is localized on the oxygen, but the carboxylic anion is resonance stabilized. This explanation is dramatically confirmed by examining the x-ray

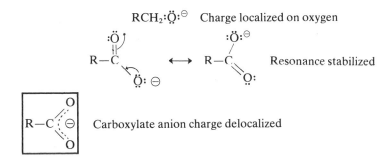

diffraction patterns of alkali metal salts of carboxylic acids. The x-ray patterns clearly show both C—O bond lengths to be exactly equivalent, as required for complete delocalization. In the free, undissociated acid the C=O and C—O have quite different bond lengths.


## SALTS OF CARBOXYLIC ACIDS

Carboxylic acids can be completely neutralized by reaction with an equivalent amount of strong base. From solution these salts can be isolated as stable crystalline

$$RCOOH + NaOH \longrightarrow RCOO^{\ominus}Na^{\oplus} + H_2O$$

Carboxylic acid            Sodium carboxylate

solids by evaporation of the solvent, usually water. Acids of unknown structure can actually be titrated with standard sodium hydroxide solution to obtain a

Examples:

$$\langle\bigcirc\rangle\text{—COOH} \xrightarrow{\text{Aq. NaOH}} \langle\bigcirc\rangle\text{—COO}^{\ominus}\text{Na}^{\oplus}$$

Sodium benzoate

$$CH_3CH_2COOH \xrightarrow{\text{Aq. KOH}} CH_3CH_2COO^{\ominus}K^{\oplus}$$

Potassium
propanoate

*neutralization equivalent.* The neutralization equivalent, or N.E., is directly related to the molecular weight of the unknown acid, as shown below.

$$N.E. = \text{Equivalent weight of the acid}$$

$$= \frac{\text{Molecular weight}}{n}$$

$$n = \text{Number of acid groups (COOH) present in molecule}$$

Example:   0.122 g of benzoic acid is exactly neutralized by 100 ml of 0.01 $N$ sodium hydroxide solution. What is the N.E.? At the equivalence point, the number of equivalents acid equals the number of equivalents base ($E_A = E_B$). Where $V_B$ equals the volume of base, and $N_B$ equals the normality,

$$\frac{0.122}{N.E.} = V_B \times N_B = (0.10 \; \ell) \times (0.01) = 0.001$$

$$N.E. = \frac{0.122}{0.001} = 122$$

Benzoic acid has a neutralization equivalent of 122. Since there is only one COOH group in the molecule, this is also equivalent to the molecular weight.

---

**Exercise 11-2:**   An organic chemist has isolated an unknown acid from plant material, and he believes the structure to be either A or B.

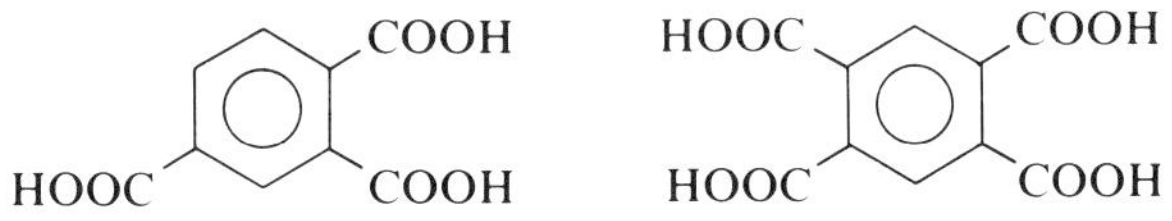

A 0.150 g sample of the unknown requires 17 ml of 0.1 $N$ sodium hydroxide solution for neutralization. What is the N.E. of the unknown acid? Can you distinguish between the two possible formulas for the unknown acid using this information?

---

## SUBSTITUENT EFFECTS ON ACIDITY

We have previously seen that substituents can affect acidity in phenols (Chap. I-9). The effect is even more dramatic in carboxylic acid acidity.

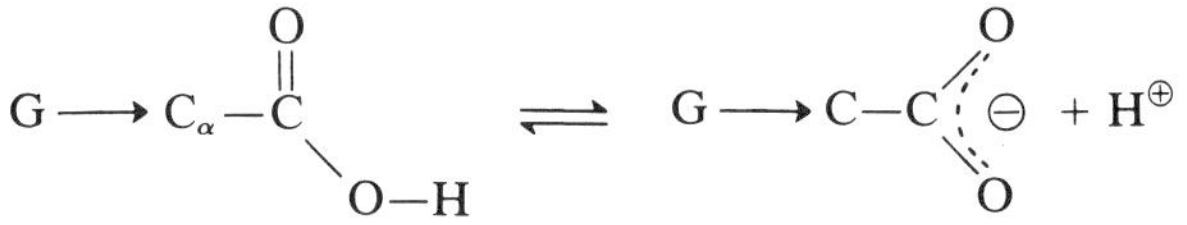

G donates electrons, $K_a$ decreases

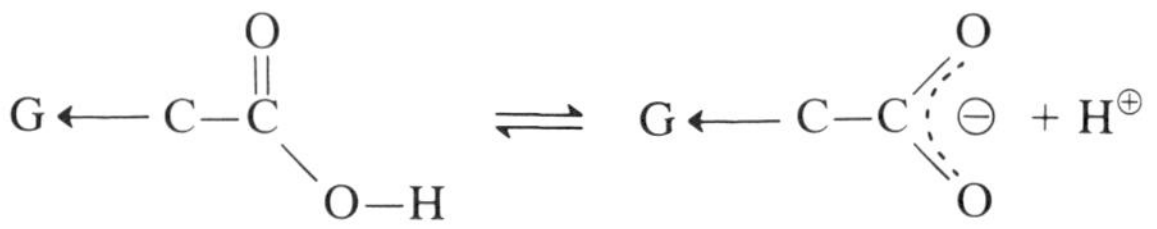

$$G \longleftarrow C - C \overset{O}{\underset{O-H}{\parallel}} \rightleftharpoons G \longleftarrow C - C \overset{O}{\underset{O}{\diagup}} \ominus + H^{\oplus}$$

G withdraws electrons, $K_a$ increases

If a group, G, attached to the carbon chain in the $\alpha$-position can donate electron density to the carboxyl group, the net effect is to increase the strength of the O—H bond. Thus ionization becomes more difficult. On the other hand, if G withdraws electron density, the O—H is weakened and ionization becomes more facile. These effects are clearly reflected in the $K_a$'s for various carboxylic acids, shown below in Table 11-3. The acidity constants for phenol and ethanol are included for comparison.

Table 11-3.   Acidities of Various Carboxylic Acids

| Compound | G | Effect on electron density | $K_a$ |
|---|---|---|---|
| $CH_3COOH$ | H | none | $1.8 \times 10^{-5}$ |
| $CH_3CH_2COOH$ | $CH_3$ | + | $1.4 \times 10^{-5}$ |
| $ClCH_2COOH$ | Cl | – | $1.5 \times 10^{-3}$ |
| $Cl_2CHCOOH$ | 2 Cl | – | $5.0 \times 10^{-2}$ |
| $Cl_3CCOOH$ | 3 Cl | – | $9.0 \times 10^{-1}$ |
| 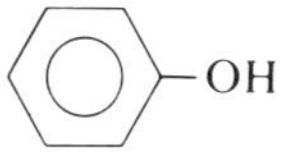 ⬡—OH | | | $1.0 \times 10^{-10}$ |
| $CH_3CH_2OH$ | | | $1.0 \times 10^{-20}$ |

This electron withdrawing or donating effect is commonly referred to as an *inductive effect*. The transmission of this effect is thought to occur through the $\sigma$-bonding network of the carbon chain and should not be confused with more common resonance effects which transmit electronic effects through the $\pi$-electron network.

Table 11-4.   Inductive Effect as a Function of Chain Position

| Compound | Distance from reaction site ( # of $\sigma$ bonds) | $K_a$ |
|---|---|---|
| $CH_3CH_2CH_2COOH$ | — | $1.5 \times 10^{-5}$ |
| $CH_3CH_2CHCOOH$<br>$\quad\vert$<br>$\quad Cl$ | 2 | $1.4 \times 10^{-3}$ |
| $CH_3CHCH_2COOH$<br>$\quad\vert$<br>$\quad Cl$ | 3 | $8.9 \times 10^{-5}$ |
| $ClCH_2CH_2CH_2COOH$ | 4 | $3.0 \times 10^{-5}$ |

Inductive effects fall off quickly as chain length increases and are usually only effective through a maximum of 3–4 bonds. This "falling off" of inductive influence can clearly be seen in a series of carboxylic acids in which the substituent is farther and farther from the reaction site (Table 11-4).

---

**Exercise 11-3:**  Arrange the following compounds in order of increasing acidity.

(a) $CH_3COOH$

(b) $CH_3CH_2CH_2OH$

(c)

(d)

(e) $F_3CCOOH$

---

## PHYSICAL PROPERTIES OF CARBOXYLIC ACIDS

Carboxylic acids are quite polar, and physical behavior may be thought of as a sum of both carbonyl and alcohol properties. Acids through $C_4$ are soluble in water

$$R-\overset{\overset{\textstyle O^{\delta-}}{\|}}{\underset{\delta+}{C}}-\underset{\delta-}{O}-\underset{\delta+}{H}$$

and have boiling points higher than alcohols of similar carbon content or molecular weight as illustrated in Table 11-5.

Table 11-5.  Boiling Points of Some Typical Carboxylic Acids and Alcohols

| Acid | B.P. | M.W. | Alcohol | B.P. | M.W. |
|---|---|---|---|---|---|
|  |  |  | $CH_3CH_2OH$ | 78 | 46 |
| $CH_3COOH$ | 118 | 60 | $CH_3CH_2CH_2OH$ | 97 | 60 |
| $CH_3CH_2COOH$ | 141 | 74 | $CH_3CH_2CH_2CH_2OH$ | 118 | 74 |
| $CH_3CH_2CH_2COOH$ | 164 | 88 | $CH_3(CH_2)_4OH$ | 138 | 88 |
| $CH_3CH_2CH_2CH_2COOH$ | 187 | 102 | $CH_3(CH_2)_5OH$ | 157 | 102 |
| $CH_3(CH_2)_4COOH$ | 202 | 116 | $CH_3(CH_2)_6OH$ | 176 | 116 |

In protic solvents, carboxylic acids will be hydrogen-bonded to solvent molecules. However, in nonpolar solvents, acid molecules tend to be attracted to one another, forming mostly dimers. The existence of intermolecular hydrogen bonds explains, to a large extent, the high boiling points of acids with respect to other polar organic molecules of similar molecular weight.

In polar solvents, acid molecules bond to solvent molecules.

In nonpolar solvent or in neat liquid form, acids tend to form cyclic dimers held together by intermolecular hydrogen bonds.

# PREPARATION OF MONO- AND DICARBOXYLIC ACIDS

## Chemical Oxidation of 1° Alcohols and Aldehydes

Carboxylic acids represent the end stage of oxidation processes in straight-chain aliphatic compounds. We have previously indicated this in our discussion of both alcohol and aldehyde oxidation. Under normal conditions, oxidation ceases at

$$RCH_2OH \xrightarrow{[O]} R-\overset{\overset{\displaystyle O}{\|}}{C}-H \xrightarrow{[O]} R-\overset{\overset{\displaystyle O}{\|}}{C}-OH$$

| 1° alcohol | Aldehyde | Carboxylic acid |

$$[O] = KMnO_4 \text{ or } K_2Cr_2O_7$$

the carboxylic acid stage. You should also recall that aldehydes are more easily oxidized than primary alcohols.

$$CH_3CH_2CH_2CH_2CH_2OH \xrightarrow[\Delta]{Alk.\ KMnO_4} [CH_3CH_2CH_2CH_2\overset{\overset{\displaystyle O}{\|}}{C}-H]$$

Difficult to isolate in good yield

Alk. KMnO₄
Δ, conc

Alk. KMnO₄
Δ

(Drastic conditions)

$$CH_3CH_2CH_2CH_2COOH$$

$$RCH_2CH_2CH_2\overset{\overset{\displaystyle O}{\|}}{C}-H \xrightarrow[\substack{Tollens'\ reagent \\ (Mild\ conditions)}]{Ag(NH_3)_2^+} RCH_2CH_2CH_2\overset{\overset{\displaystyle O}{\|}}{C}OH + Ag\downarrow$$

Olefins may also be oxidized to yield, among other things, carboxylic acids. If chemical oxidation is carried out under strenuous conditions, olefins having at least one $R—CH=$ unit will yield carboxylic acids as products. This reaction is particularly useful for symmetrical olefins or cyclic olefins, yielding either a single monocarboxylic acid or a dicarboxylic acid in the latter case.

$$\boxed{RCH} = \boxed{CH_2} \xrightarrow{KMnO_4} RCOOH + CO_2\uparrow$$

$$CH_3CH_2CH_2CH=CH_2 \xrightarrow[\Delta]{KMnO_4} CH_3CH_2CH_2\overset{\displaystyle O}{\overset{\displaystyle \|}{C}}OH + CO_2\uparrow$$

$$RCH=CHR \xrightarrow[\Delta]{KMnO_4} 2RCOOH$$

Symmetrical
alkene

$$\langle\ \rangle-CH=CH-\langle\ \rangle \xrightarrow[\Delta]{KMnO_4} 2\ \langle\ \rangle-COOH$$

$$(CH_2)_n \overset{CH}{\underset{CH}{\|}} \xrightarrow[\Delta]{KMnO_4} HOOC(CH_2)_nCOOH$$

Cycloalkene

$$\text{(cyclopentene)} \xrightarrow[\Delta]{KMnO_4} HOOC(CH_2)_3COOH$$

## Oxidation of Alkylbenzenes

Alkyl group side chains may also be oxidized to carboxyl groups as long as there is at least one $\alpha$-hydrogen (see Chapter I-6). The length or complexity of the alkyl chain has no effect on the overall course of the oxidation.

$$\langle\ \rangle{-}CH_2CH_2CH_3 \xrightarrow[\Delta]{KMnO_4} \langle\ \rangle{-}COOH \xleftarrow[\Delta]{KMnO_4} \langle\ \rangle{-}CH_3$$

$$\uparrow \begin{array}{c} KMnO_4 \\ \Delta \end{array}$$

$$\langle\ \rangle{-}CH(CH_2)_{10}\underset{\underset{\displaystyle CH_3}{|}}{\overset{\overset{\displaystyle CH_3}{|}}{C}}HCH_3$$

## Grignard Addition to $CO_2$

Grignard reagents will undergo ordinary carbonyl addition with carbon dioxide, yielding a magnesium salt of a carboxylic acid. Subsequent acid hydrolysis

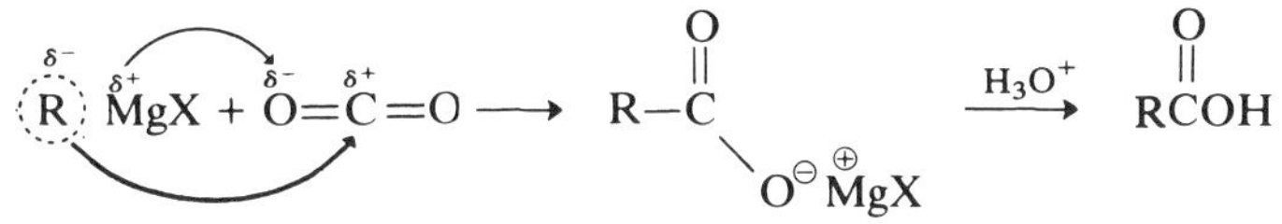

releases the free acid. Both aromatic and aliphatic Grignard reagents give good yields of corresponding acids.

$$\underset{\displaystyle CH_3}{CH_3CH_2\overset{|}{C}HCH_2MgBr} \xrightarrow[2.\ H_3O^+]{1.\ CO_2} \underset{\displaystyle CH_3}{CH_3CH_2\overset{|}{C}HCH_2COOH}$$

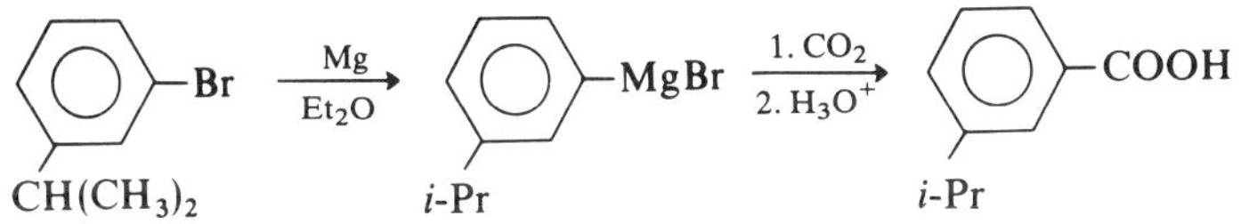

## Hydrolysis of Nitriles

Nitriles (organic cyanides) may be hydrolyzed in either acidic or basic solutions to carboxylic acids. You may recall that nitriles are normally produced by nucleophilic substitution of cyanide for halide in either 1° or 2° halides. Thus this sequence actually performs the same net synthetic result described above for the Grignard: a halide is converted into an acid with a net increase of one carbon in the chain.

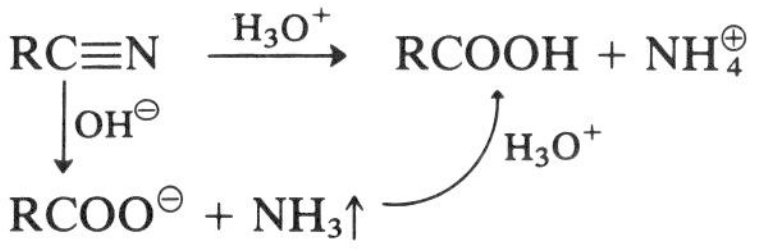

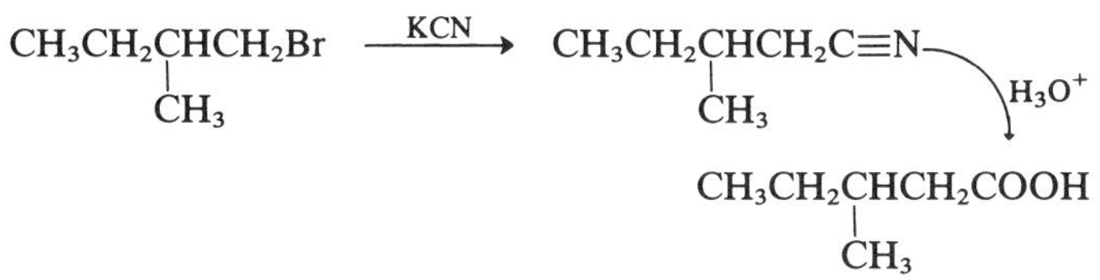

These reaction sequences are also quite useful for building up dicarboxylic acid chains. The synthetic sequences are illustrated below for the building up of both odd and even numbered chains.

$$
\begin{array}{c}
\text{CH}_2\text{Br} \\
| \\
\text{CH}_2\text{Br}
\end{array}
\xrightarrow{\text{KCN}}
\begin{array}{c}
\text{CH}_2\text{CN} \\
| \\
\text{CH}_2\text{CN}
\end{array}
\xrightarrow{\text{H}_3\text{O}^+}
\begin{array}{c}
\text{CH}_2\text{COOH} \\
| \\
\text{CH}_2\text{COOH}
\end{array}
\xrightarrow[\text{2. H}_3\text{O}^+]{\text{1. LAH}}
\begin{array}{c}
\text{CH}_2\text{CH}_2\text{OH} \\
| \\
\text{CH}_2\text{CH}_2\text{OH}
\end{array}
$$

$$
\begin{array}{c}
\text{CH}_2\text{CH}_2\text{COOH} \\
| \\
\text{CH}_2\text{CH}_2\text{COOH}
\end{array}
\xleftarrow{\text{H}_3\text{O}^+}
\begin{array}{c}
\text{CH}_2\text{CH}_2\text{CN} \\
| \\
\text{CH}_2\text{CH}_2\text{CN}
\end{array}
\xleftarrow{\text{KCN}}
\begin{array}{c}
\text{CH}_2\text{CH}_2\text{Br} \\
| \\
\text{CH}_2\text{CH}_2\text{Br}
\end{array}
\Big\rangle \text{PBr}_3
$$

↓ etc.

↓ Mg/Et$_2$O

$$
\xleftarrow{\text{etc.}}
\begin{array}{c}
\text{CH}_2\text{CH}_2\text{COOH} \\
| \\
\text{CH}_2\text{CH}_2\text{COOH}
\end{array}
\xleftarrow[\text{2. H}_3\text{O}^+]{\text{1. CO}_2}
\begin{array}{c}
\text{CH}_2\text{CH}_2\text{MgBr} \\
| \\
\text{CH}_2\text{CH}_2\text{MgBr}
\end{array}
$$

$$
\text{CH}_2
\begin{array}{c}
\diagup \text{CH}_2\text{Br} \\
\diagdown \text{CH}_2\text{Br}
\end{array}
\xrightarrow{\text{KCN}}
\text{CH}_2
\begin{array}{c}
\diagup \text{CH}_2\text{CN} \\
\diagdown \text{CH}_2\text{CN}
\end{array}
\xrightarrow{\text{H}_3\text{O}^+}
\text{CH}_2
\begin{array}{c}
\diagup \text{CH}_2\text{COOH} \\
\diagdown \text{CH}_2\text{COOH}
\end{array}
\xrightarrow{\text{etc.}}
$$

---

**Exercise 11-4:**   In the previous reaction sequences, 1,3-dibromopropane is utilized as a starting material for the synthesis of dicarboxylic acids with odd-numbered chains. Prepare 1,3-dibromopropane from any mono-functional 3-carbon compound and then use it to prepare HOOC(CH$_2$)$_7$COOH.

---

**Exercise 11-5:**   Carry out the following multistep conversion:

$$\longrightarrow \ \longrightarrow \ \longrightarrow \ \text{HOOC(CH}_2)_6\text{COOH}$$

---

### Hydrolysis of Benzonitriles

We have already seen how an aromatic methyl group may be vigorously oxidized to a carbonyl group. However, this method is severely limited if other functional groups on the ring cannot survive this treatment. Aromatic methyl groups may also be halogenated under the influence of UV radiation. The resulting benzotrihalide may then be hydrolyzed under mildly basic conditions to yield the desired carboxylic acid.

Benzotrichloride

## Industrial Production of Important Acids

*Formic acid* is produced industrially from carbon monoxide and sodium hydroxide in solution under the influence of both heat and pressure, followed by acid hydrolysis of the crude sodium formate.

$$CO + NaOH \xrightarrow[\text{Pressure}]{200°} \underset{\text{Sodium formate}}{H-\overset{\overset{\textstyle O}{\|}}{C}-O^{\ominus}Na^{\oplus}} \xrightarrow{H_3O^+} \underset{\text{Formic acid}}{HCOOH}$$

*Acetic acid* is normally produced by air oxidation of acetaldehyde. You will recall that acetaldehyde is produced from the hydration of acetylene.

$$HC\equiv CH \xrightarrow[\text{(Hg}^{+2})]{H_3O^+} \underset{}{CH_3\overset{\overset{\textstyle O}{\|}}{C}-H} \xrightarrow[\text{air (O}_2)]{\text{Warm}} CH_3\overset{\overset{\textstyle O}{\|}}{C}OH$$

*Phthalic acid* is obtained from naphthalene by air oxidation at elevated temperature in the presence of vanadium pentoxide.

Phthalic acid

## REACTIONS OF CARBOXYLIC ACIDS

Carboxylic acid derivatives are formed by reactions in which the $-OH$ group is replaced by some other functionality. The $R-\overset{\overset{\textstyle O}{\|}}{C}-$ portion is usually referred to as

$$\underset{\substack{\text{Carboxylic} \\ \text{acid}}}{R-\overset{\overset{\textstyle O}{\|}}{C}-OH} \longrightarrow \underset{\substack{\text{Carboxylic acid derivative} \\ \text{or ``acyl'' derivative}}}{R-\overset{\overset{\textstyle O}{\|}}{C}-Y}$$

an *acyl group*, while the Y group determines the nature and chemical properties of the derivative. In this section we will explore the preparation of these derivatives.

Acyl Halides:   $R-\overset{\overset{\displaystyle O}{\|}}{C}-X$

Acyl halides, or as they sometimes called, acid halides, are formed by reaction of a carboxylic acid with an inorganic acid halide, such as $PCl_5$ or $SOCl_2$. Acyl halides are easy to prepare and purify and are extremely reactive. This reactivity explains why acyl chlorides are utilized by organic chemists in preference to the other halides in organic synthesis: they are much cheaper to prepare and you gain little advantage by using the more expensive bromides.

$$R-\overset{\overset{\displaystyle O}{\|}}{C}-OH \xrightarrow[\substack{or\ PCl_3 \\ or\ PCl_5}]{SOCl_2} R-\overset{\overset{\displaystyle O}{\|}}{C}-Cl + SO_2\uparrow + HCl\uparrow$$

Acyl            or
chloride        $H_3PO_3$
                or
                $POCl_3$

Examples:

$$(CH_3)_2CHCH_2COOH \xrightarrow{SOCl_2} (CH_3)_2CHCH_2\overset{\overset{\displaystyle O}{\|}}{C}Cl$$

3-Methylbutanoic acid          3-Methylbutanoyl chloride

3-Nitrobenzoic acid            3-Nitrobenzoyl chloride

Esters:   $R-\overset{\overset{\displaystyle O}{\|}}{C}-OR'$

Esters are normally formed from acid catalyzed reaction of an alcohol with the carboxylic acid. The reaction is an equilibrium process which is shifted to favor the ester by using large excesses of the alcohol, usually as the solvent. Esters, as pointed out previously in this and other chapters, are extremely important flavor and odor constituents. They are important industrially as solvents and as primary ingredients in the manufacture of soaps and synthetic fibers.

$$\overset{O}{\underset{\|}{R\overset{\|}{C}OH}} + R'OH \underset{}{\overset{H^{\oplus}}{\rightleftharpoons}} \overset{O}{\underset{\|}{R-\overset{\|}{C}-OR'}} + H_2O$$
Excess

$$(CH_3)_2CHCH_2COOH \overset{EtOH/H^+}{\rightleftharpoons} (CH_3)_2CHCH_2\overset{O}{\overset{\|}{C}}OCH_2CH_3$$
3-Methylbutanoic acid                Ethyl 3-methylbutanoate

$$H_3C-\langle\bigcirc\rangle-COOH \overset{i\text{-PrOH/H}^+}{\rightleftharpoons} H_3C-\langle\bigcirc\rangle-\overset{O}{\overset{\|}{C}}OCH(CH_3)_2$$
4-Methylbenzoic acid                i-Propyl 4-methylbenzoate

Methyl esters may be prepared by the above procedure, or by reaction of the free acid with diazomethane. This method is extremely useful when working with small quantities of acid, but great care must be exercised when using diazomethane because it is both explosive and poisonous.

$$R-\overset{O}{\overset{\|}{C}}-OH \overset{CH_2N_2}{\underset{Et_2O}{\longrightarrow}} R-\overset{O}{\overset{\|}{C}}-OCH_3 + N_2\uparrow$$

Recently, solutions of $BCl_3$ or $BF_3$, Lewis acid catalysts, in various alcohols have been developed commercially as esterification reagents. Addition of the acid to these mixtures usually results in complete esterification within 5–10 minutes.

$$\overset{CH_3}{\underset{CH_3}{\diagdown}}C=CHCOOH \overset{BF_3\cdot MeOH}{\longrightarrow} \overset{CH_3}{\underset{CH_3}{\diagdown}}C=CHCOOCH_3$$

Amides:   $R-\overset{O}{\overset{\|}{C}}-NH_2$

Amides can be prepared directly by heating the ammonium salts of carboxylic acids, eliminating water. It is often more convenient, however, to prepare amides

$$R-\overset{O}{\overset{\|}{C}}-O^{\ominus}NH_4^{\oplus} \overset{\Delta}{\longrightarrow} R-\overset{O}{\overset{\|}{C}}-NH_2 + H_2O$$
Ammonium salt            Carboxylic
of carboxylic acid       acid amide

indirectly by using the acyl chlorides. The indirect method may be carried out under much milder conditions and often gives a higher yield of the desired amide.

$$R-\overset{\overset{\displaystyle O}{\|}}{C}OH \xrightarrow{SOCl_2} R-\overset{\overset{\displaystyle O}{\|}}{C}-Cl \xrightarrow{NH_3} R-\overset{\overset{\displaystyle O}{\|}}{C}-NH_2$$

Examples:

$$CH_3CH_2CH_2\overset{\overset{\displaystyle O}{\|}}{C}O^{\ominus}NH_4^{\oplus} \xrightarrow{\Delta} CH_3CH_2CH_2\overset{\overset{\displaystyle O}{\|}}{C}NH_2$$

Ammonium butanoate                    Butanamide

$$H_3C-\left\langle\!\!\bigcirc\!\!\right\rangle-COCl \xrightarrow{NH_3} H_3C-\left\langle\!\!\bigcirc\!\!\right\rangle-\overset{\overset{\displaystyle O}{\|}}{C}-NH_2$$

4-Methylbenzoyl          4-Methylbenzamide
chloride

Anhydrides:   $R-\overset{\overset{\displaystyle O}{\|}}{C}-O-\overset{\overset{\displaystyle O}{\|}}{C}-R$

Anhydrides may be prepared by reaction of an acyl chloride with a carboxylate salt. This process is a nucleophilic substitution reminiscent of the Williamson ether synthesis. Both symmetrical and mixed anhydrides can be prepared by this method.

$$R-\overset{\overset{\displaystyle O}{\|}}{C}-Cl + M^{\oplus}\overset{\ominus}{O}-\overset{\overset{\displaystyle O}{\|}}{C}-R \longrightarrow R-\overset{\overset{\displaystyle O}{\|}}{C}-O-\overset{\overset{\displaystyle O}{\|}}{C}-R + M^{\oplus}Cl^{\ominus}$$

$$M = Na, Ag, etc.$$

Two important industrial anhydrides, phthalic and acetic, are prepared by specialized procedures. Phthalic anhydride, widely used as a plasticizer and in polyester resins, can be prepared directly from phthalic acid by simple heating. Acetic anhydride is prepared from *ketene*, a highly reactive intermediate generated

Phthalic acid          Phthalic anhydride

from high temperature pyrolysis of acetone. Ketene reacts rapidly and completely with any compound containing an active hydrogen. Passage of ketene vapor into pure acetic acid (glacial) produces acetic anhydride.

$$CH_3\overset{\overset{\displaystyle O}{\|}}{C}CH_3 \xrightarrow{700°} CH_2=C=O + CH_4$$

Ketene

$$H-Z + CH_2=C=O \longrightarrow CH_3-\overset{\overset{\displaystyle O}{\|}}{C}-Z \quad \text{General ketene addition}$$

$$H-O-\overset{\overset{\displaystyle O}{\|}}{C}-CH_3 + CH_2=C=O \longrightarrow CH_3-\overset{\overset{\displaystyle O}{\|}}{C}-O-\overset{\overset{\displaystyle O}{\|}}{C}-CH_3$$

Acetic acid          Ketene                          Acetic anhydride

Dicarboxylic acids containing four and five carbon atoms form *cyclic anhydrides* when heated. Longer chain dicarboxylic acids form linear anhydrides in polymeric chains of varying lengths.

Succinic acid          Succinic anhydride

Glutaric acid          Glutaric anhydride

$$HO\overset{\overset{\displaystyle O}{\|}}{C}(CH_2)_n\overset{\overset{\displaystyle O}{\|}}{C}OH \xrightarrow{\Delta} \left[O\overset{\overset{\displaystyle O}{\|}}{C}(CH_2)_n-\overset{\overset{\displaystyle O}{\|}}{C}\right]_z$$

$n > 3$               Linear anhydride
                      polymer

**Exercise 11-6:**   Give products and write complete reactions for the reaction of ketene with each of the following reagents:

(a) $H_2O$     (b) $CH_3CH_2OH$     (c) $(CH_3)_2NH$     (d) $CH_3CH_2COOH$

## Reduction of Carboxylic Acids

The direct reduction of carboxylic acids was, for many years, an impossible task. Normal chemical or catalytic reducing agents such as $Zn/HCl$ and $Pd/H_2$ have no effect on the carboxyl group. However, in recent years, more powerful reducing agents have become available. Either lithium aluminum hydride ($LiAlH_4$) or diborane ($B_2H_6$) may now be used to effect reduction, usually in ether or THF. The reduction occurs smoothly at either room temperature or 0°, and the primary alcohol product is obtained by careful neutralization with dilute acid. LAH will not reduce

$$RCOOH \xrightarrow[Et_2O]{LiAlH_4} (RCH_2O)_4LiAl \xrightarrow{H_3O^+} RCH_2OH$$

$$RCOOH \xrightarrow[\substack{THF \\ 0°}]{B_2H_6} RCH_2OH$$

double bonds also present in the carboxylic acid, while $B_2H_6$ will not affect reducible groups such as $NO_2$, CN, or carbonyl groups.

Examples:

$$\substack{CH_3 \\ \diagdown \\ \diagup \\ CH_3}C=CHCOOH \xrightarrow[2.\ H_3O^+]{1.\ LAH} \substack{CH_3 \\ \diagdown \\ \diagup \\ CH_3}C=CHCH_2OH$$

$$O_2N-\langle\bigcirc\rangle-COOH \xrightarrow[\substack{THF \\ 0°}]{B_2H_6} O_2N-\langle\bigcirc\rangle-CH_2OH$$

## NOMENCLATURE OF ACYL DERIVATIVES

*Acyl halides:* Drop -ic acid ending of IUPAC acid name, add -yl and the name of the *halide.*

$$\underset{\displaystyle CH_3CH_2\overset{\textstyle O}{\overset{\textstyle \|}{C}}-Cl}{} \qquad \text{acid name:} \quad \text{propano}\cancel{\text{ic acid}}$$

$$\downarrow$$

propanoyl chloride

*Esters*:   Drop *-ic acid* ending of IUPAC name, add *-ate*. Precede name with alkyl group of the alcohol portion.

$$\underbrace{CH_3CH_2\overset{\overset{\textstyle O}{\|}}{C}}_{\text{Acid portion}}\underbrace{OCH_2CH_3}_{\text{Alcohol}}$$   *Ethyl* alcohol + propano~~ic~~ ~~acid~~

Ethyl propanoate

*Amides*:   Drop *-oic acid* ending of IUPAC name, add *-amide*.

$$CH_3CH_2\overset{\overset{\textstyle O}{\|}}{C}NH_2$$   propano~~ic~~ ~~acid~~

propanamide

If the amine portion is substituted, the substituents are identified by adding the prefix N- and the name of the substituent.

$$CH_3CH_2\overset{\overset{\textstyle O}{\|}}{C}NHCH_2CH_3$$   N-ethylpropanamide

↑

Substituent

*Anhydrides*:   The acid residues are identified by their IUPAC names followed by the word *anhydride*.

$$CH_3CH_2\overset{\overset{\textstyle O}{\|}}{C}O\overset{\overset{\textstyle O}{\|}}{C}CH_2CH_3$$   Propanoic anhydride

$$CH_3\overset{\overset{\textstyle O}{\|}}{C}O\overset{\overset{\textstyle O}{\|}}{C}CH_2CH_3$$   Ethanoic propanoic anhydride

---

**Exercise 11-7:**   Name the following compounds by the IUPAC system as described above:

(a) $(CH_3)_2C{=}CHCOOCH_2CH_3$

(b) 
$$(CH_3)_2CHCH_2\overset{\overset{\textstyle O}{\|}}{C}NHCH_3$$

(c)

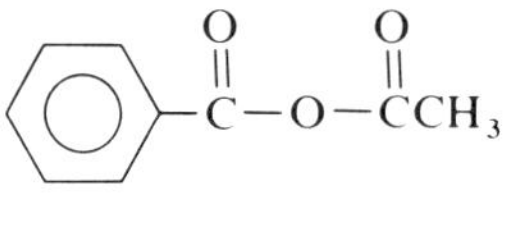

(d)

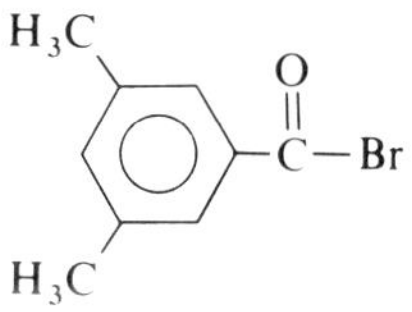

# REACTIONS OF ACYL DERIVATIVES

Acyl derivatives of carboxylic acids undergo nucleophilic substitution reactions with great facility. At first glance, it might appear that simple $S_N1$ or $S_N2$ mechanisms are operative. However, displacement actually occurs via a two-step addition-elimination sequence as shown below. In general, strong nucleophiles will displace weak

$$R-\overset{\overset{\displaystyle O}{\|}}{C}-X + :Z \longrightarrow R-\overset{\overset{\displaystyle O}{\|}}{C}-Z + :X$$

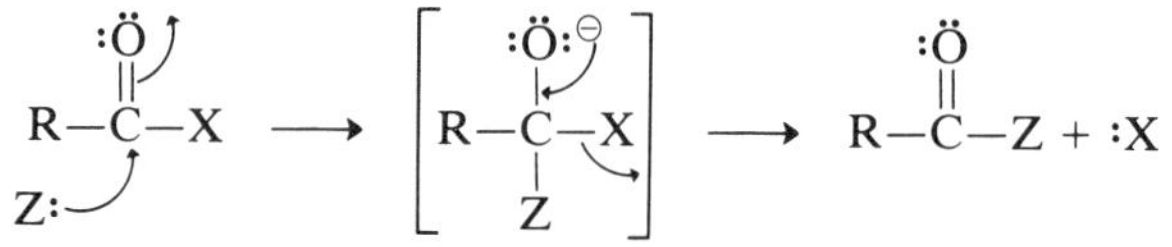

$$R-\overset{\overset{\displaystyle :\ddot{O}}{\|}}{\underset{\overset{\displaystyle |}{Z:}}{C}}-X \longrightarrow \left[ R-\overset{\overset{\displaystyle :\ddot{O}:^{\ominus}}{|}}{\underset{\overset{\displaystyle |}{Z}}{C}}-X \right] \longrightarrow R-\overset{\overset{\displaystyle :\ddot{O}}{\|}}{C}-Z + :X$$

General displacement mechanism for acyl derivatives
of carboxylic acids

nucleophiles. Thus alkoxide can displace chloride, but not the reverse.

$$R-\overset{\overset{\displaystyle O}{\|}}{C}-Cl + {}^{\ominus}OR' \longrightarrow R-\overset{\overset{\displaystyle O}{\|}}{C}-OR' + Cl^{\ominus}$$

Reverse reaction does not occur

Examples:

1. Esters from acyl chlorides

$$CH_3CH_2\overset{\overset{\displaystyle O}{\|}}{C}Cl + Na^{\oplus}\overset{\ominus}{O}-CH_2CH_3 \longrightarrow CH_3CH_2\overset{\overset{\displaystyle O}{\|}}{C}OCH_2CH_3$$

2. Amides from acyl chlorides

$$CH_3CH_2\overset{\overset{\displaystyle O}{\|}}{C}Cl + H\ddot{N}(CH_3)_2 \longrightarrow CH_3CH_2\overset{\overset{\displaystyle O}{\|}}{C}N(CH_3)_2$$

3. Anhydrides from acyl chlorides

$$CH_3CH_2\overset{\overset{\displaystyle O}{\|}}{C}Cl + Na^{\oplus}\overset{\ominus}{O}-\overset{\overset{\displaystyle O}{\|}}{C}CH_3 \longrightarrow CH_3CH_2\overset{\overset{\displaystyle O}{\|}}{C}-O-\overset{\overset{\displaystyle O}{\|}}{C}CH_3$$

4. Amides from esters

$$CH_3CH_2\overset{\overset{\displaystyle O}{\|}}{C}OCH_3 + H_2\ddot{N}CH_3 \longrightarrow CH_3CH_2\overset{\overset{\displaystyle O}{\|}}{C}NHCH_3 + CH_3OH$$

224

5. Solvolysis (protic solvents)

$$CH_3CH_2\overset{\overset{\displaystyle O}{\|}}{C}Cl \xrightarrow[\Delta]{H_2O} CH_3CH_2\overset{\overset{\displaystyle O}{\|}}{C}OH + HCl\,(aqueous)$$

$$\xrightarrow{CH_3CH_2OH} CH_3CH_2\overset{\overset{\displaystyle O}{\|}}{C}OCH_2CH_3 + HCl\,(soln.)$$

All acyl derivatives can be hydrolyzed back to the original carboxylic acid in either strong acid or base. The latter method is usually employed because the carboxylate anion is soluble in water and thus easily isolated. Also, the stability of the anion essentially makes the hydrolysis irreversible and drives the reaction to completion.

$$CH_3CH_2\overset{\overset{\displaystyle O}{\|}}{C}Cl \xrightarrow[H_2O]{Na^+OH^-} CH_3CH_2\overset{\overset{\displaystyle O}{\|}}{C}O^{\ominus}Na^{\oplus} + Na^{\oplus}Cl^{\ominus}$$

$$CH_3CH_2\overset{\overset{\displaystyle O}{\|}}{C}OCH_3 \xrightarrow[H_2O]{Na^+OH^-} CH_3CH_2\overset{\overset{\displaystyle O}{\|}}{C}O^{\ominus}Na^{\oplus} + CH_3OH$$

$$\langle\bigcirc\rangle-\overset{\overset{\displaystyle O}{\|}}{C}-NH_2 \xrightarrow[H_2O]{Na^+OH^-} \langle\bigcirc\rangle-\overset{\overset{\displaystyle O}{\|}}{C}-O^{\ominus}Na^{\oplus} + NH_3\uparrow$$

$$(CH_3)_2CH\overset{\overset{\displaystyle O}{\|}}{C}O\overset{\overset{\displaystyle O}{\|}}{C}CH_2CH_2CH_3 \xrightarrow[H_2O]{Na^+OH^-} (CH_3)_2CH\overset{\overset{\displaystyle O}{\|}}{C}O^{\ominus}Na^{\oplus}$$

$$+$$

$$CH_3CH_2CH_2\overset{\overset{\displaystyle O}{\|}}{C}O^{\ominus}Na^{\oplus}$$

## OTHER REACTIONS OF ACYL DERIVATIVES

### Friedel–Crafts Acylation

Aromatic ketones may be prepared readily from either acyl halides or anhydrides. The reaction is similar to Friedel–Crafts alkylation (see Chapter I-6), and proceeds by an electrophilic substitution pathway.

$$R-\overset{\overset{\displaystyle O}{\|}}{C}-Cl + \langle\bigcirc\rangle \xrightarrow{AlCl_3} \langle\bigcirc\rangle-\overset{\overset{\displaystyle O}{\|}}{C}\diagdown R + HCl$$

or

$$R-\overset{\overset{\displaystyle O}{\|}}{C}-O-\overset{\overset{\displaystyle O}{\|}}{C}-R$$

or

RCOOH

Examples:

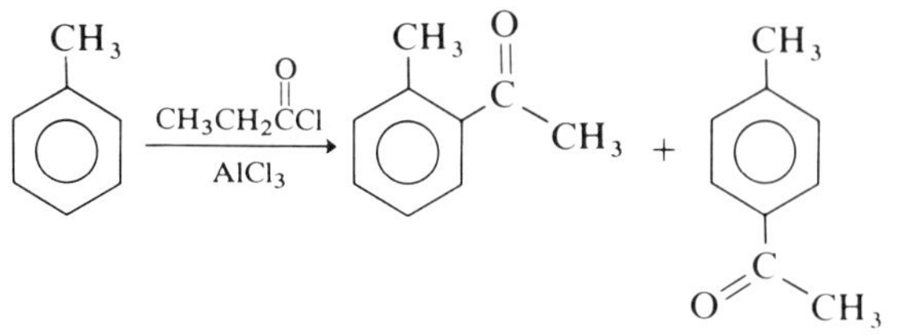

Minor product    Major product

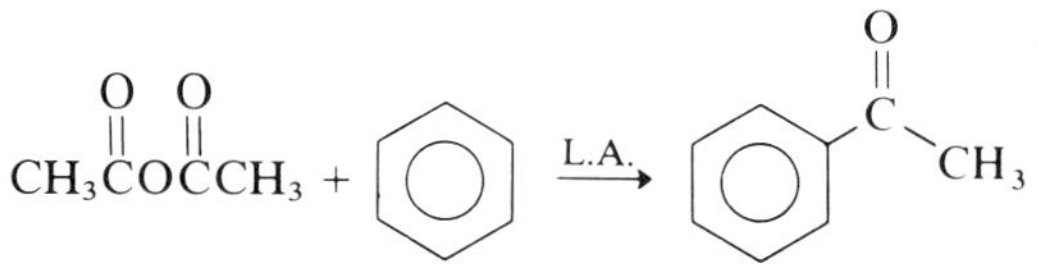

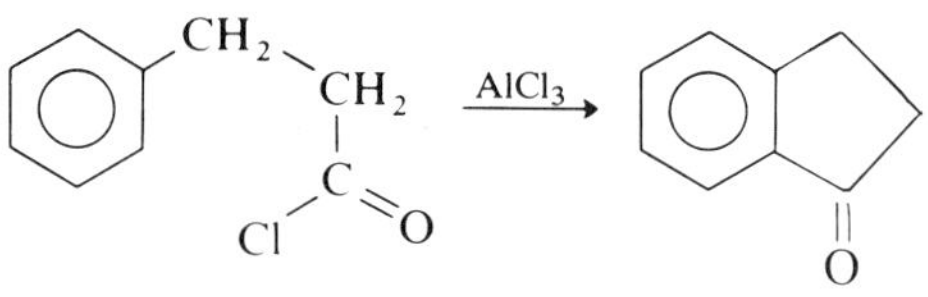

## Dehydration of Amides

Amides may be dehydrated, forming nitriles, in the presence of strong dehydrating agents.

$$R-\overset{\overset{\displaystyle O}{\|}}{C}-NH_2 \xrightarrow{-H_2O} R-C\equiv N$$

This route is particularly important in the synthesis of cyanides that cannot be obtained by direct nucleophilic substitution.

Example:

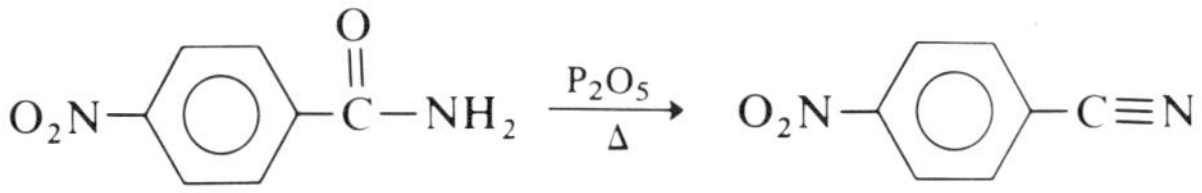

## Partial Reduction of Acyl Halides

Acyl halides are reduced to primary alcohols by most reducing agents, but, under certain conditions, the reduction may be halted at the aldehyde stage. This may be accomplished by using either a poisoned catalyst (containing some sulfur) or a sterically hindered reducing agent. Two examples of this behavior are illustrated below:

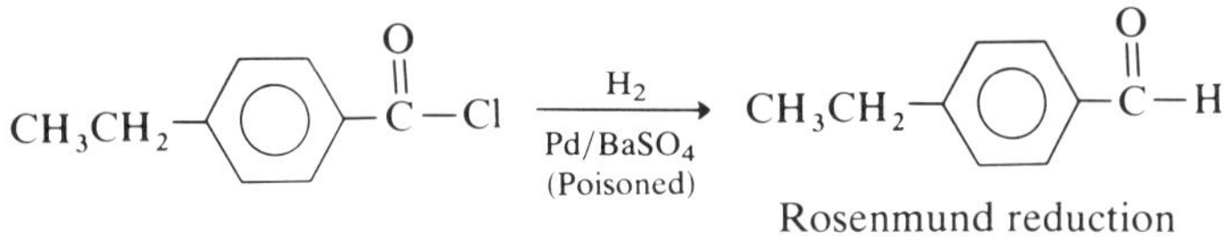


Rosenmund reduction

$$CH_3CH_2CH{=}CH\overset{O}{\overset{\|}{C}}Cl \xrightarrow[\text{(Hindered reducing agent)}]{LiAl(tBuO)_3H} CH_3CH_2CH{=}CH\overset{O}{\overset{\|}{C}}H$$

## CARBOXYLIC ACIDS AND ACYL DERIVATIVES IN NATURE

Straight-chain carboxylic acids and their esters are widely distributed in nature. Even-numbered chains dominate, and unsaturated acids are common. Table 11-6 lists some of the more common acids. These acids are often referred to as "fatty acids" and are generally insoluble in water. Oleic acid is an extremely common unsaturated fatty acid occurring in nature and is found in virtually all plant and animal sources.

Table 11-6. Common Fatty Acids Occurring in Nature

| Name | Formula |
|---|---|
| Lauric | $CH_3(CH_2)_{10}COOH$ |
| Myristic | $CH_3(CH_2)_{12}COOH$ |
| Palmitic | $CH_3(CH_2)_{14}COOH$ |
| Stearic | $CH_3(CH_2)_{16}COOH$ |
| Oleic | $CH_3(CH_2)_7CH{=}CH(CH_2)_7COOH$ |
| Linoleic | $CH_3(CH_2)_4CH{=}CHCH_2CH{=}CH(CH_2)_7COOH$ |
| Linolenic | $CH_3CH_2CH{=}CHCH_2CH{=}CHCH_2CH{=}CH(CH_2)_7COOH$ |

Fatty acids and their esters are members of a complex class of compounds known as *lipids.* Lipids are normally regarded as compounds from living systems that are insoluble in water, but soluble in organic solvents such as chloroform, methanol, benzene, and acetone. Some of the many classes of lipids that exist include fats, oils, waxes, phospholipids, sphingolipids, and steroids. We will discuss many of these classes in Chapter II-6, but we will limit ourselves in this chapter to the fatty acid esters.

Fatty acid esters normally occur as triglycerides—esters of three fatty acids and glycerol. Triglycerides can be broken down into their component parts by hydrolysis in aqueous base, a process known as *saponification*. We will discuss this process in more detail in Chapter II-6. Triglycerides function as food storage molecules, but ordinary fatty acid esters seem to serve as protective coatings in plants (leaves, stems, etc.).

$$
\begin{array}{l}
\underset{\displaystyle \text{CH}_2\text{O}\overset{\textstyle O}{\overset{\|}{\text{C}}}-\text{R}^1} {} \\[2ex]
\underset{\displaystyle \text{CHO}\overset{\textstyle O}{\overset{\|}{\text{C}}}-\text{R}^2} {} \xrightarrow[\Delta]{\text{Aq.NaOH}} \\[2ex]
\underset{\displaystyle \text{CH}_2\text{O}\overset{\textstyle O}{\overset{\|}{\text{C}}}-\text{R}^3} {}
\end{array}
$$

$$
\begin{array}{l}
\text{CH}_2\text{OH} \quad {}^{\ominus}\text{O}-\overset{\textstyle O}{\overset{\|}{\text{C}}}-\text{R}^1 \\[2ex]
\text{CHOH} + {}^{\ominus}\text{O}-\overset{\textstyle O}{\overset{\|}{\text{C}}}-\text{R}^2 \\[2ex]
\text{CH}_2\text{OH} \quad {}^{\ominus}\text{O}-\overset{\textstyle O}{\overset{\|}{\text{C}}}-\text{R}^3
\end{array}
$$

Triglyceride                    Glycerol

Saponification of a triglyceride

Low molecular weight esters serve as natural flavors and fragances, usually in complicated mixtures. For example, the volatile flavor constituents associated with apples include at least eight such esters:

$$\overset{\text{O}}{\overset{\|}{\text{CH}_3\text{C}}}\text{OCH}_2\text{CH}_3, \quad \overset{\text{O}}{\overset{\|}{\text{CH}_3\text{CH}_2\text{C}}}\text{OCH}_2\text{CH}_3, \quad \overset{\text{O}}{\overset{\|}{\text{CH}_3\text{CH}_2\text{CH}_2\text{C}}}\text{OCH}_2\text{CH}_3$$

$$\overset{\text{O}}{\overset{\|}{\text{CH}_3(\text{CH}_2)_3\text{C}}}\text{OCH}_2\text{CH}_3, \quad \overset{\text{O}}{\overset{\|}{\text{CH}_3\text{C}}}\text{OCH}_2\text{CH}_2\text{CH}_3, \quad \overset{\text{O}}{\overset{\|}{\text{CH}_3\text{CH}_2\text{C}}}\text{OCH}_2\text{CH}_2\text{CH}_3$$

$$\overset{\text{O}}{\overset{\|}{\text{CH}_3(\text{CH}_2)_3\text{C}}}\text{OCH}_2\text{CH}_2\text{CH}_3, \quad \overset{\text{O}}{\overset{\|}{\text{CH}_3\text{CH}_2\text{CH}_2\text{C}}}\text{OCH}_2\text{CH}_2\text{CH}_3$$

To some extent, chemists can duplicate natural fragrances with simple esters, but artificial flavors are more difficult. Some of the flavors and fragrances and the chemicals responsible for them are shown in Table 11-7. It can be readily appreciated that minor changes in the chain structure produce dramatic changes in our perception of odor.

Table 11-7. Chemicals Responsible for Common
Flavors and Fragrances

| Compound | Flavor and/or Fragrance |
|---|---|
| $CH_3\overset{O}{\overset{\|}{C}}OCH_2CH_2CH_2CH_2CH_3$ | Banana |
| $CH_3CH_2CH_2\overset{O}{\overset{\|}{C}}OCH_2CH_3$ | Pineapple |
| $CH_3\overset{O}{\overset{\|}{C}}OCH_2CH_2CH(CH_3)_2$ | Pear |
| $CH_3CH_2\overset{O}{\overset{\|}{C}}OCH_2CH(CH_3)_2$ | Rum |
| $CH_3CH_2CH_2\overset{O}{\overset{\|}{C}}OCH_2CH_2CH_2CH_2CH_3$ | Apricot |
| $H-\overset{O}{\overset{\|}{C}}-OCH_2CH(CH_3)_2$ | Raspberry |
| $CH_3\overset{O}{\overset{\|}{C}}O(CH_2)_7CH_3$ | Orange |
| $CH_3\overset{O}{\overset{\|}{C}}OCH_2-\langle\bigcirc\rangle$ | Jasmine |
| $\langle\bigcirc\rangle-COOCH_3$ with OH | Wintergreen |

# DERIVATIVES OF CARBONIC ACID

Carbonic acid is the hypothetical product of the addition of water to carbon dioxide. Although carbonic acid is unstable, as are the monoaddition products involving

$$H_2O + CO_2 \rightleftharpoons [HO-\overset{O}{\overset{\|}{C}}-OH] \quad \text{Carbonic acid}$$

$$ROH + CO_2 \rightleftharpoons [RO-\overset{O}{\overset{\|}{C}}-OH] \quad \text{Alkyl hydrogen carbonate (“half ester”)}$$

$$NH_3 + CO_2 \rightleftharpoons [H_2N-\overset{O}{\overset{\|}{C}}-OH] \quad \text{Carbamic acid (“half amide”)}$$

alcohols, ammonia, and amines, the derivatives of these unstable structures *are* stable. The diamide of carbonic acid is urea. You may recall that urea was originally prepared by Wohler from ammonium cyanate, the first example of an organic

$$NH_4^{\oplus} \overset{\ominus}{O}CN \xrightarrow{\Delta} H_2N-\overset{\overset{\displaystyle O}{\|}}{C}-NH_2$$

Urea–diamide
of carbonic acid

synthesis from inorganic material. It is the end product of all protein metabolism in animal systems. It is also widely used as a fertilizer to add nitrogen to the soil.

Phosgene, a deadly poison used in World War I trench warfare, is the diacyl chloride of carbonic acid and is prepared by direct combination of carbon monoxide and chlorine at 200°.

$$CO + Cl_2 \xrightarrow[\text{active C}]{200°} Cl-\overset{\overset{\displaystyle O}{\|}}{C}-Cl$$   Diacyl chloride

Phosgene        of carbonic acid

Phosgene is an important starting material for the synthesis of other carbonic acid derivatives.

$$Cl-\overset{\overset{\displaystyle O}{\|}}{C}-Cl$$

$$\xrightarrow{NH_3} H_2N-\overset{\overset{\displaystyle O}{\|}}{C}-NH_2 \quad \text{Urea}$$

$$\xrightarrow{ROH} RO-\overset{\overset{\displaystyle O}{\|}}{C}-OR \quad \text{Dialkyl carbonate}$$

Hydrolysis of phosgene yields carbon dioxide and HCl.

## REVIEW OF NEW CONCEPTS AND TERMS

Define and explain each term and, if possible, give a specific example.

| | |
|---|---|
| Carboxylic acid | Acyl derivative |
| Carboxylate anion | Neutralization equivalent |
| Inductive effect | Acyl halide |
| Carboxylic acid ester | Carboxylic acid amide |
| Carboxylic acid anhydride | Ketene |
| Acyl displacement reaction | Friedel-Crafts acylation |
| Fatty acid | Lipid |
| Triglyceride | Saponification |

**1.** Name the following compounds by the IUPAC system:

(a) 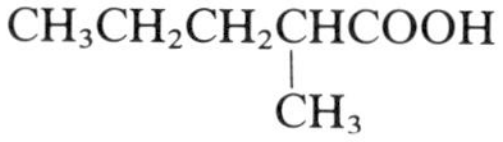 $CH_3CH_2CH_2CHCOOH$
$\phantom{CH_3CH_2CH_2CH}|$
$\phantom{CH_3CH_2CH_2CH}CH_3$

(b) 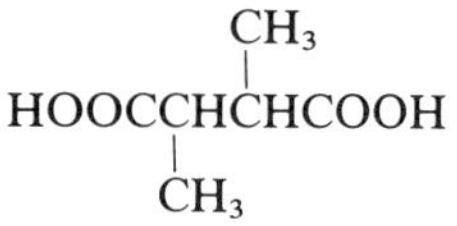 $\phantom{HOOCCHCH}CH_3$
$\phantom{HOOCCH}|$
$HOOCCHCHCOOH$
$\phantom{HOOCCHCH}|$
$\phantom{HOOCCHCH}CH_3$

(c)

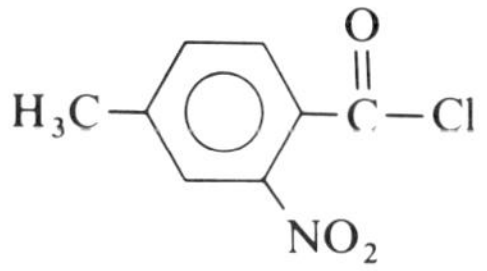

(d) $(CH_3)_2C{=}CHCH_2COOMe$

(e)

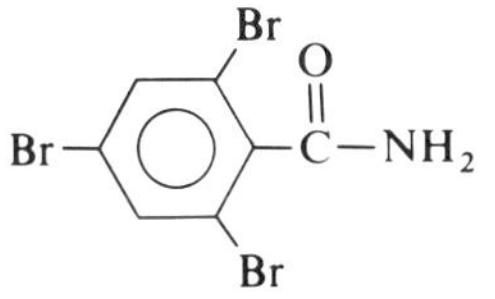

(f)

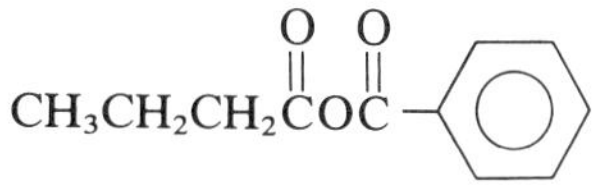

(g) $CH_3CH{=}CHCH{=}CHCOOH$

(h) $(CH_2)_2[COOCH(CH_3)_2]_2$

**2.** Complete the following reactions showing only organic products:

(a) $CH_3CH_2CH_2COOH \xrightarrow[H^{\oplus}]{(CH_3)_2CHOH}$

(b) $CH_3CH_2CH{=}CHCH_2CH_3 \xrightarrow[\Delta]{KMnO_4}$

(c)

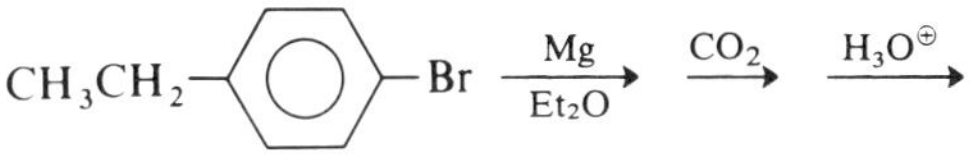

$\xrightarrow[Et_2O]{Mg} \xrightarrow{CO_2} \xrightarrow{H_3O^{\oplus}}$

(d) $CH_3CH{=}CHCH_2CH_2COOH \xrightarrow{SOCl_2}$

(e)

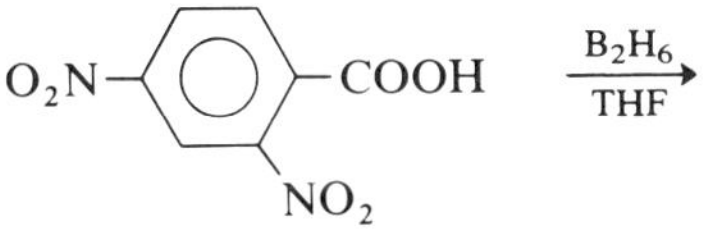

$\xrightarrow[THF]{B_2H_6}$

(f)
$\overset{\displaystyle O}{\overset{\displaystyle \|}{CH_3CH_2CH_2C}}{-}Br + (CH_3CH_2)_2NH \longrightarrow$

(g) 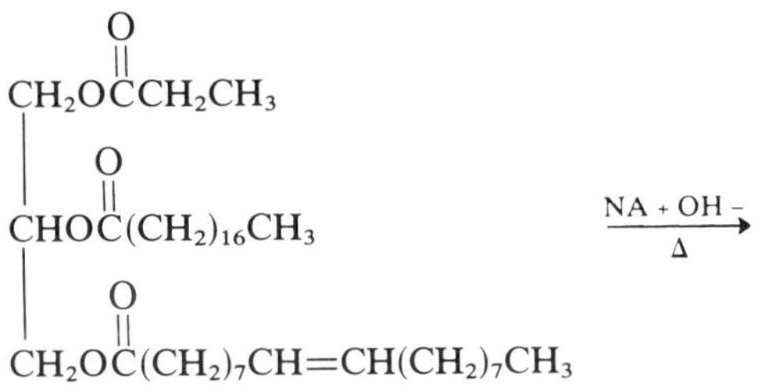
$\xrightarrow[\Delta]{NA + OH -}$

(h)

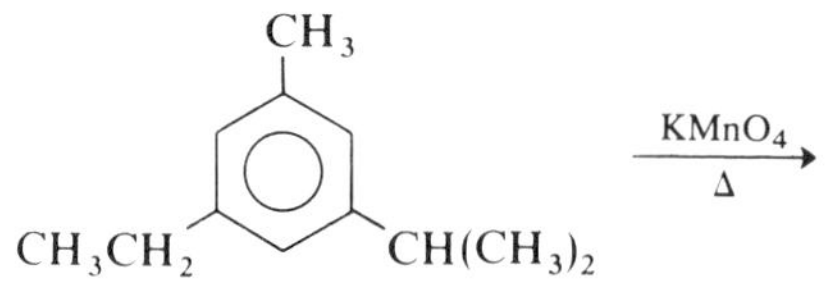

$$\xrightarrow[\Delta]{KMnO_4}$$

(i)  $(CH_3)_2C=CHCH_2C\equiv N \xrightarrow{H_3O^{\oplus}}$

(j)

$$\xrightarrow[Et_2O]{CH_2N_2}$$

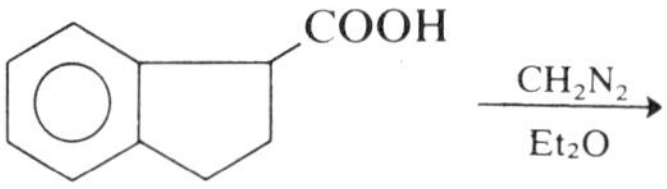

(k)

$$CH_3CH_2CH_2CH_2\overset{\overset{\displaystyle O}{\|}}{C}OH \xrightarrow{NH_3} A \xrightarrow{\Delta} B \xrightarrow{P_2O_5} C$$

(l)

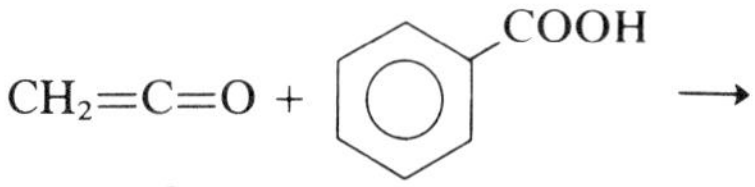

$$CH_2=C=O + $$ $$\longrightarrow$$

(m)

$$\xrightarrow{CH_3OH}$$

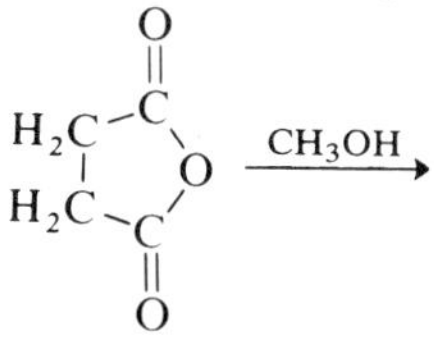

**3.** Carry out the following conversions utilizing any necessary organic or inorganic reagents

(a)

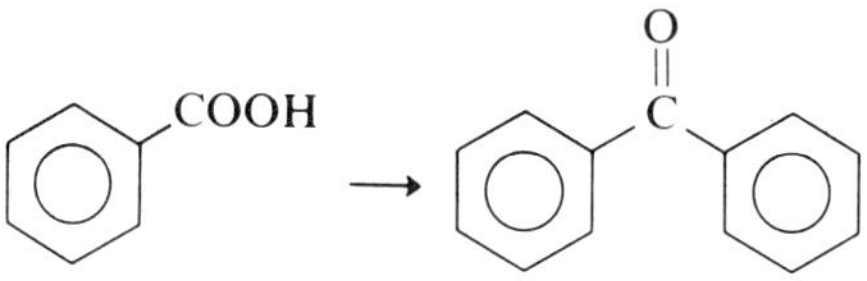

(b)

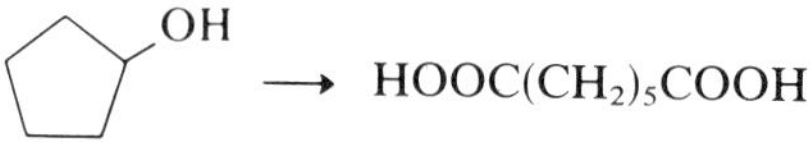

$\longrightarrow HOOC(CH_2)_5COOH$

(c)

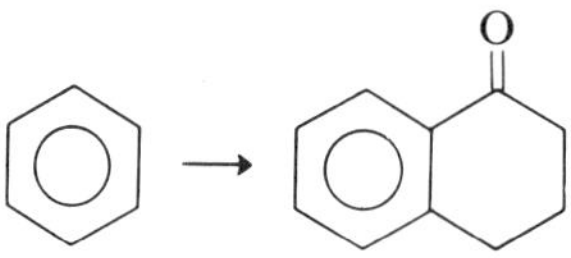

**4.** Arrange the following acids in order of increasing acid strengths, and justify your ranking:

$Cl_3CCOOH$,    $FCH_2COOH$,    $ICH_2COOH$,    $CF_3COOH$,

$BrCH_2COOH$,    $ClCH_2COOH$

**5.** Starting with benzene, toluene, and/or any carbon compounds containing four or fewer carbon atoms, prepare each of the following:

(a)

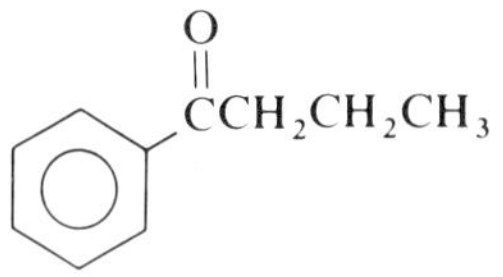

(b)

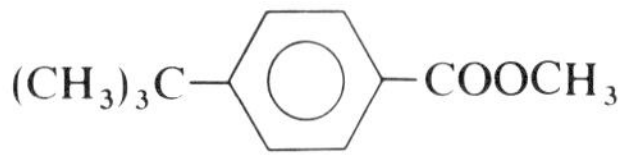

$(CH_3)_3C$—⟨benzene⟩—$COOCH_3$

(c) $EtOOC(CH_2)_8COOEt$

(d)

$CH_3$
$\phantom{CH_3}\backslash$
$\phantom{xxx}C=CHCH_2O\overset{\displaystyle O}{\overset{\displaystyle \|}{C}}$—⟨benzene⟩
$\phantom{CH_3}/$
$CH_3$

6. Again, starting with the same materials used in problem 5, show how you would prepare all the flavors and fragrances in Table 11-7.

7. An undergraduate student reacted phenyl magnesium halide with ethyl acetate and obtained the tertiary alcohol shown below. How could you explain this result? (Hint: Look at the structure of the initial addition product.)

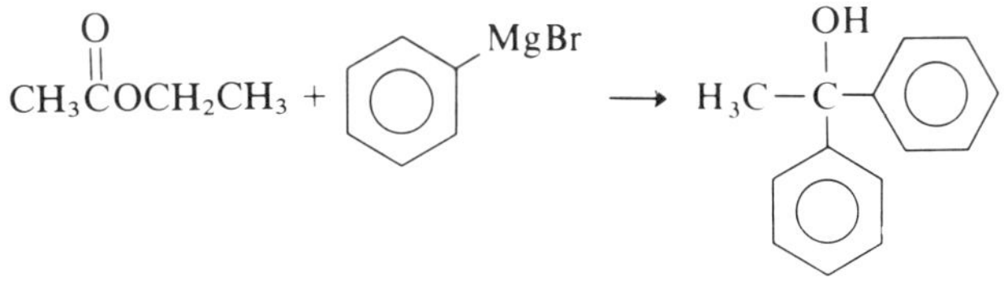

$$CH_3\overset{\displaystyle O}{\overset{\displaystyle \|}{C}}OCH_2CH_3 + ⟨benzene⟩MgBr \longrightarrow H_3C-\overset{\displaystyle OH}{\underset{\displaystyle}{C}}⟨\text{two benzene rings}⟩$$

8. Before lithium aluminum hydride and diborane were available to the organic chemist, carboxylic acids were "reduced" by a two-step procedure in which the acid was first converted to a derivative before reduction. Can you suggest a possible method?

$$RCH_2COOH \xrightarrow{?} A \xrightarrow{?} RCH_2CH_2OH$$

9. How would you distinguish between the following pairs of compounds?

(a) $CH_3CH=CHCOOH$   and   $CH_3CH_2CH_2COOH$

(b)

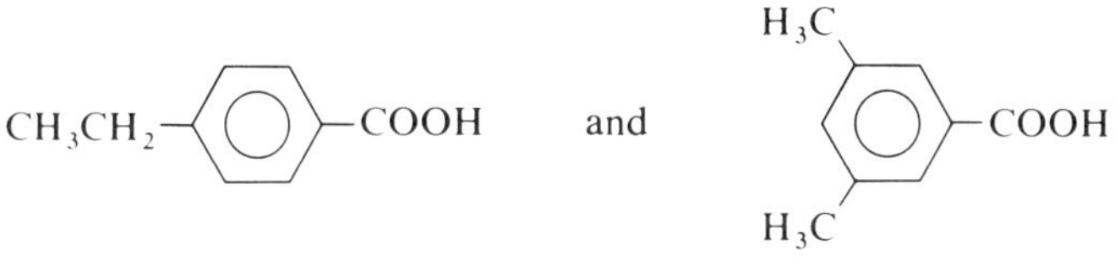

$CH_3CH_2$—⟨benzene⟩—$COOH$   and   

$H_3C$—⟨benzene with $H_3C$ substituents⟩—$COOH$
$H_3C$

(c) $CH_3(CH_2)_8COOH$   and   $HOOC(CH_2)_6COOH$

10. An unknown compound A, $C_{10}H_{14}O_2$, was oxidized with hot $KMnO_4$ solution, yielding B, $C_9H_{10}O_4$, whose N.E. was $182 \pm 1$. B was synthesized independently according to the following scheme:

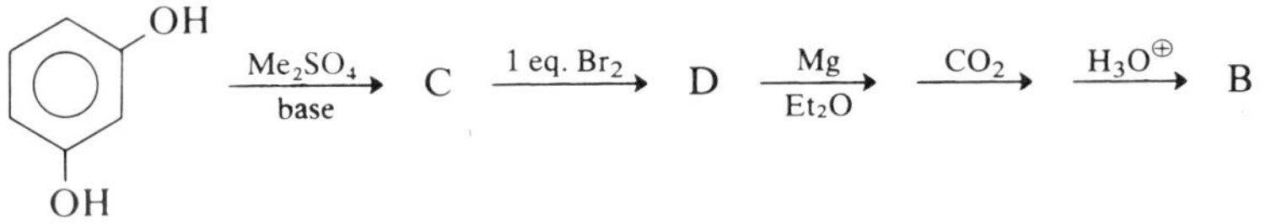

⟨benzene with two OH groups⟩ $\xrightarrow[\text{base}]{Me_2SO_4}$ C $\xrightarrow{\text{1 eq. } Br_2}$ D $\xrightarrow[\text{Et}_2\text{O}]{Mg} \xrightarrow{CO_2} \xrightarrow{H_3O^{\oplus}}$ B

Suggest reasonable structures for compounds A, B, C, and D.

# 12

# Carbohydrates

Carbohydrates are composed of carbon, hydrogen, and oxygen and their general formula may be expressed as $C_n(H_2O)_n$. The chemistry of carbohydrates is determined by their carbonyl and alcohol functional groups, and is extremely important to the maintenance of both plant and animal life. Sugars, starch, and cellulose are common examples of carbohydrates produced in plants by photosynthesis. These, in turn, are converted into glycogen by animals, which is used as an energy reservoir in animal tissue. In this respect, all animal life is dependent on plant life. Carbohydrates not only provide us with food but also clothing (cotton), shelter (wood) and enlightenment (newspapers and books). In this chapter we shall learn how carbohydrates are produced and how their structure determines their chemical properties. We shall also learn why the stereochemistry of carbohydrates is as important as their structure.

## STRUCTURE OF SIMPLE CARBOHYDRATES

Simple carbohydrates are either polyhydroxy ketones or polyhydroxy aldehydes. Complex carbohydrates, on the other hand, yield polyhydroxy carbonyl compounds upon hydrolysis, usually from hemiacetal or acetal structures. In Chapter I-4, glyceraldehyde was discussed in terms of its stereochemistry; it is also one of the simplest of carbohydrates. Glyceraldehyde is an extremely important family

CHO
|
CHOH        Dihydroxyaldehyde
|           $C_3H_6O_3 \equiv C_3(H_2O)_3$
CH$_2$OH    Simple carbohydrate
Glyceraldehyde

member in that all stereochemical assignments for carbohydrates are related to chiral glyceraldehyde. It is also a convenient starting material for the synthesis of more complex carbohydrates. We shall return to both of these topics later in this chapter. Simple polyhydroxy aldehydes are referred to as *alsoses*, and simple polyhydroxy ketones are *ketoses*.

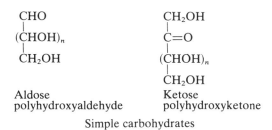

Aldose
polyhydroxyaldehyde

Ketose
polyhydroxyketone

Simple carbohydrates

---

**Exercise 12-1:**  Keeping in mind that simple carbohydrates are either aldoses or ketoses, draw structures for all possible carbohydrates corresponding to the following empirical formulas. Ignore stereoisomerism at this time.

(a) $C_4(H_2O)_4$     (b) $C_5(H_2O)_5$

---

## CLASSIFICATION OF CARBOHYDRATES

Carbohydrates may be classified according to several structural features:

1. the number of simple carbohydrate units, either aldoses or ketoses, produced upon hydrolysis;
2. the total number of carbons in each structural unit; and
3. the stereochemical configuration of each aldose or ketose.

In terms of gross structure, carbohydrates may be divided into three important categories. We have already discussed simple aldose and ketose structures. Molecules having these structures are referred to as *monosaccharides*. Many important carbohydrates are composed of two or more simple monosaccharides bonded together by hemiacetal or acetal links. If only two of these monosaccharide units are bonded together in this fashion, the molecule is referred to as a *disaccharide*, although a more general name for this class is *oligosaccharides* (oligo = few, 3–10 units are common in this class). Complex carbohydrates that yield many monosaccharides upon acid hydrolysis are named *polysaccharides*. The relationships between these various classes can be outlined as follows:

$$\text{Polysaccharides} \xrightarrow{\text{H}_3\text{O}^{\oplus}} \text{Oligosaccharides} \xrightarrow{\text{H}_3\text{O}^{\oplus}} \text{Disaccharides}$$

$$(>10 \text{ monosaccharide} \qquad\qquad (3\text{–}10 \text{ units}) \qquad\qquad (2 \text{ units})$$

$$\text{units}) \qquad\qquad\qquad\qquad\qquad\qquad\qquad\qquad \Big\downarrow \text{H}_3\text{O}^{\oplus}$$

$$\text{monosaccharides}$$

Typical monosaccharides:

$$
\begin{array}{ccc}
 & & \text{CH}_2\text{OH} \\
 & & | \\
\text{CHO} & \text{CHO} & \text{C=O} \\
| & | & | \\
(\text{CHOH})_4 & (\text{CHOH})_3 & (\text{CHOH})_3 \\
| & | & | \\
\text{CH}_2\text{OH} & \text{CH}_2\text{OH} & \text{CH}_2\text{OH} \\
\text{Glucose} & \text{Ribose} & \text{Fructose}
\end{array}
$$

Typical disaccharides:

$$\text{Sucrose} \xrightarrow{\text{H}_3\text{O}^{\oplus}} 2 \text{ glucose units}$$

$$\text{Lactose} \xrightarrow{\text{H}_3\text{O}^{\oplus}} 1 \text{ glucose} + 1 \text{ galactose unit}$$

Typical polysaccharides:

$$\text{Starch} \xrightarrow{\text{H}_3\text{O}^{\oplus}} \text{glucose units}$$

$$\text{Cellulose} \xrightarrow{\text{H}_3\text{O}^{\oplus}} \text{glucose units}$$

It is convenient, when discussing carbohydrate units, not only to refer to a unit as being an aldose or a ketose but also to include the total number of carbons in the structure. Thus glucose may be referred to as an *aldohexose*, while fructose is then a *ketohexose*.

$$
\begin{array}{l}
\text{CHO} \longleftarrow (\text{Aldo}) \, hexose \\
| \\
(\text{CHOH})_4 \\
| \qquad\qquad 6 \text{ carbons in total structure} \\
\text{CH}_2\text{OH}
\end{array}
$$

$$
\begin{array}{l}
\text{CH}_2\text{OH} \\
| \\
\text{C=O} \longleftarrow (\text{Keto}) \, hexose \\
| \\
(\text{CHOH})_3 \qquad 6 \text{ carbons} \\
| \\
\text{CH}_2\text{OH}
\end{array}
$$

CHO  $\overleftarrow{\qquad}$  (Aldo) *pentose*
|
(CHOH)$_3$
|
CH$_2$OH          5 carbons

## CARBOHYDRATE STEREOCHEMISTRY

Carbohydrates isolated from nature are invariably chiral. Early research in carbohydrate chemistry was severely hindered until the stereochemical relationships
between various isomeric monosaccharides were understood. The key to understanding carbohydrate chirality is in the absolute configurations of the simple
building block aldoses and ketoses. Let us consider the simplest aldose containing a
chiral center, glyceraldehyde. It exists in $R$ and $S$ forms, as previously discussed in
Chapter I-4. The $R$ and $S$ system, however is a rather recent convention for

$$\begin{array}{ccc} \text{CHO} & \text{CHO} & \text{CHO} \\ \text{H}\!-\!\!\!-\!\text{OH} \equiv \text{H}\!\!-\!\!\!\!\!\!-\!\!\!\blacktriangleleft\!\text{OH} \equiv \text{HOCH}_2\cdots\!\!\!\blacktriangleleft\!\!\!\!\!\!\!\!\!\!\! \\ \text{CH}_2\text{OH} & \text{CH}_2\text{OH} & \overset{|}{\text{H}} \;\; \text{OH} \end{array}$$

$R(+)$ Glyceraldehyde—a chiral aldohexose

$$\begin{array}{ccc} \text{CHO} & \text{CHO} & \text{CHO} \\ \text{HO}\!-\!\!\!-\!\text{H} \equiv \text{HO}\!\!-\!\!\!\!\!\!-\!\!\!\blacktriangleleft\!\text{H} \equiv \text{H}\cdots\!\!\!\!\!\!\!\!\!\! \\ \text{CH}_2\text{OH} & \text{CH}_2\text{OH} & \text{CH}_2\text{OH}\;\text{OH} \end{array}$$

$S$-(−)-Glyceraldehyde

describing absolute configuration. Older carbohydrate literature uses a system in
which the absolute configuration of glyceraldehyde was arbitrarily defined. In other
words, the (+) isomer of glyceraldehyde was assigned the "D" configuration. The
chemists who originally suggested this convention had a 50-50 chance of being
correct and, as it turned out, they made the correct guess! The other enantiomer was
assigned the opposite or "L" configuration.

CHO
|
(+) isomer assigned the    H $-\!\!\!-$ OH configuration
|
CH$_2$OH

$$\begin{array}{cc} \text{CHO} & \text{CHO} \\ \text{H}\!-\!\!\!-\!\text{OH} & \text{HO}\!-\!\!\!-\!\text{H} \\ \text{CH}_2\text{OH} & \text{CH}_2\text{OH} \end{array}$$
D-(+)-Glyceraldehyde    L-(−)-Glyceraldehyde

Certain advantages accrue from the defining of D and L configurations. If other, more complex, carbohydrates could be related to D- or L-glyceraldehyde, they could be referred to as members of the D series or L series. In determining to which series a monosaccharide belongs, we always look at the chirality of the carbon *farthest* from the carbonyl group.

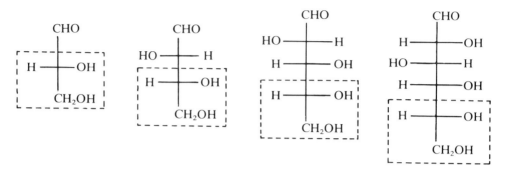

All the above carbohydrates belong to the D series.

Thus we can identify a stereochemical family of carbohydrates by comparing configurations at a specific chiral center. The chirality of the remaining centers in, say, an aldohexose, will determine which particular monosaccharide we are dealing with. Most naturally occurring monosaccharides belong to the D family. Figure 12-1 shows the common aldoses related to D-glyceraldehyde containing up to 6 carbons.

**Exercise 12-2:**   In the following reaction sequence two stereoisomers are produced. Are they both members of the same stereochemical series (D or L)? Explain.

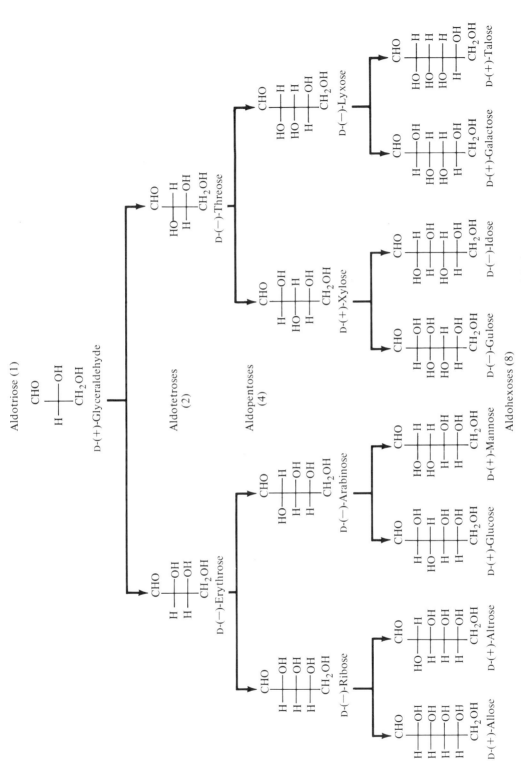

**Figure 12-1.** Simple aldoses derived from D-glyceraldehyde.

## PHYSICAL PROPERTIES
## OF SIMPLE CARBOHYDRATES

Carbohydrates contain large numbers of polar groups (—OH, C=O, etc.). Thus they dissolve well in polar liquids such as water. They have extremely high boiling points, even at reduced pressure. Most carbohydrates, in fact, cannot be distilled without significant thermal decomposition. Many carbohydrates are also difficult to crystallize, because they have a tendency to form glass-like syrups. Many of these "glasses" have taken *years* to crystallize in the laboratory.

## THE STRUCTURES
## OF SIMPLE MONOSACCHARIDES

Simple carbohydrates, such as glucose, can exist in either cyclic or acyclic structures. You will recall from our discussions of carbonyl chemistry that aldehydes or ketones react with alcohols to form hemiacetals. The "alcohol" is already built into a carbohydrate chain, and thus *intramolecular hemiacetal* formation becomes feasible, as shown below for glucose.

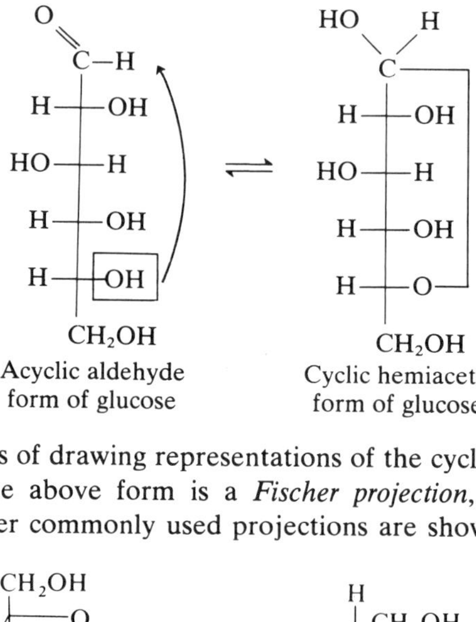

Acyclic aldehyde
form of glucose

Cyclic hemiacetal
form of glucose

Several methods of drawing representations of the cyclic form of glucose are commonly used. The above form is a *Fischer projection*, similar to those for glyceraldehyde. Other commonly used projections are shown below for glucose.

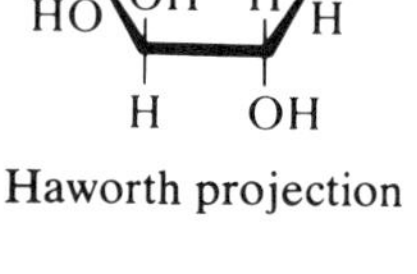

Haworth projection

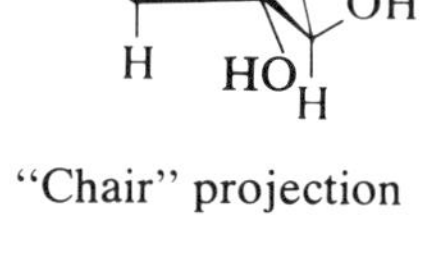

"Chair" projection

When we draw projection formulas for the cyclic hemiacetals, it becomes readily apparent that two different hemiacetals may be formed when ring closure occurs. These isomers are referred to as either $\alpha$ or $\beta$, and these designations must be included in the nomenclature of the carbohydrate. Carbohydrates form either five-

$\beta$-Form

$\alpha$-Form

$\alpha$-D-Glucose
($\alpha$-D-glucopyranose)

$\beta$-D-Glucose
($\beta$-D-glucopyranose)

or six-membered hemiacetal rings. This fact may also be incorporated in the name. The heterocyclic ring formed by the carbohydrate may be related in structure to either *furan* or *pyran*. Thus we can refer to glucose as a *pyranose*, while ribose may be

Furan—5-membered
heterocycle containing O

Pyran—6-membered
heterocycle containing O

referred to as a *furanose*, since it forms a five-membered ring in the hemiacetal form.

$\beta$-D-Ribose
$\beta$-D-Ribofuranose (hemiacetal)

---

**Exercise 12-3:**   Glucose preferentially forms a hemiacetal by reacting with the OH on carbon 5 in the chain. Why this OH rather than any other OH in the molecule? Draw the hemiacetals formed by reaction with the OH groups on carbon atoms 2, 3, 4 and 6 before you answer the question.

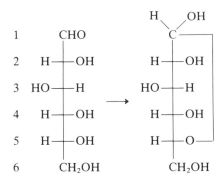

# MUTAROTATION

The $\alpha$ and $\beta$ forms of any particular carbohydrate are diastereomers—they differ in absolute configuration only at the hemiacetal carbon. Thus these diastereomers have different physical and optical properties, and often both can be obtained in pure form. For example, pure $\alpha$ and $\beta$ forms of glucose can be obtained by recrystallization from different solvents.

$$\text{Pure } \beta \text{ form} \xleftarrow[\text{from HOAc}]{\text{Recrystallize}} \text{D-glucose} \xrightarrow[\text{from } CH_3OH]{\text{Recrystallize}} \text{Pure } \alpha \text{ form}$$
$$[\alpha] = +19° \qquad\qquad\qquad\qquad\qquad\qquad [\alpha] = +113°$$

When either of these pure forms are dissolved in water, however, the rotations gradually change and eventually reach equilibrium at +52°. This change is referred to as *mutarotation*.

$$\text{Pure } \alpha \xrightarrow[\text{in } H_2O]{\text{Dissolve}} \text{Rotation gradually changes to } +52°$$

$$\text{Pure } \beta \xrightarrow[\text{in } H_2O]{\text{Dissolve}} \text{Rotation gradually changes to } +52°$$

In solution, the hemiacetal is in equilibrium with a small quantity of open-chain aldehyde. Eventually, an equilibrium is established between $\alpha$ and $\beta$ forms since they both are in equilibrium with a common intermediate. The carbon atom undergoing the change from $\alpha$ to $\beta$ configuration is referred to as the *anomeric carbon*, whose chirality is continuously being created and destroyed.

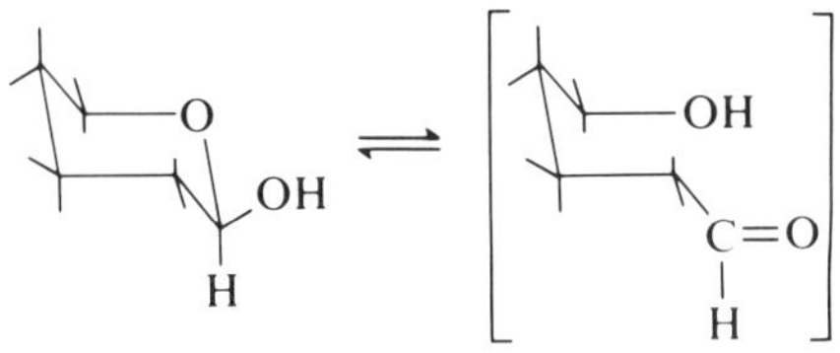

## FORMATION OF FURANOSES

Ribose (an aldopentose) and fructose (a ketohexose) both form five-membered *furanose* structures. Ribose is similar to the aldohexoses previously discussed in that hemiacetal formation occurs; however, fructose forms a *hemiketal*.

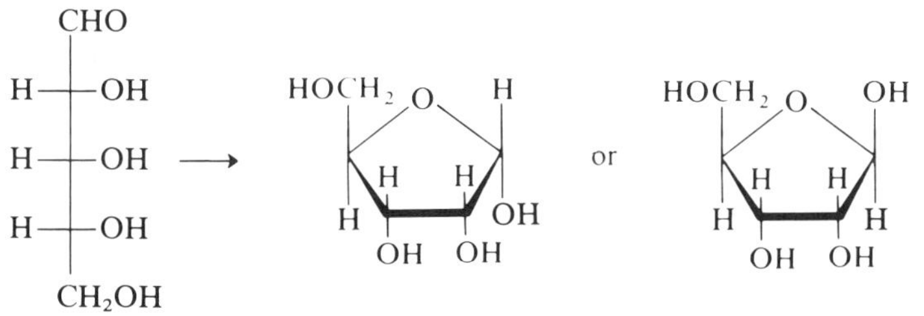

D-Ribose            α-D-Ribofuranose            β-D-Ribofuranose

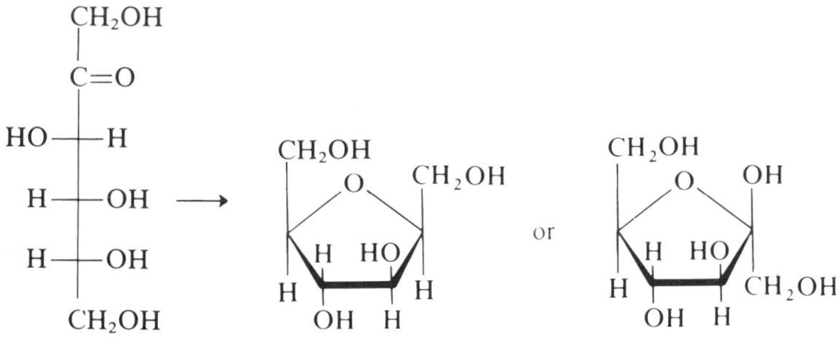

D-Fructose           α-D-Fructofuranose          β-D-Fructofuranose

---

**Exercise 12-4:** Fructose can also exist in a pyranose form. Draw Haworth and chair forms of α- and β-D-fructopyranose.

---

## SIMPLE REACTIONS OF MONOSACCHARIDES

Carbohydrates contain alcohol and either aldehyde or keto functional groups. We have previously studied the reactions of these functionalities separately, and we should expect that monosaccharide chemistry will be a combination of alcohol and carbonyl chemistry. To a large extent this is true, although a few unusual reactions do result from this assemblage of reactive functional groups. It might be a good idea at this time to review the reactions of both alcohol and carbonyl groups before reading this section.

### Formation of Carbonyl Derivatives and Osazones

$$>C{=}O + H_2NZ \longrightarrow >C{=}N{-}Z + H_2O$$

$$Z = OH,\ NH_2,\ \text{etc.}$$

The open chain forms of monosaccharides react with carbonyl derivatization reagents to form oximes, hydrazones, and so forth. Although most of these reactions are identical to those of simple aldehydes and ketones (see Chapter I-10), phenylhydrazine undergoes a peculiar reaction known as *osazone formation* in which *three*

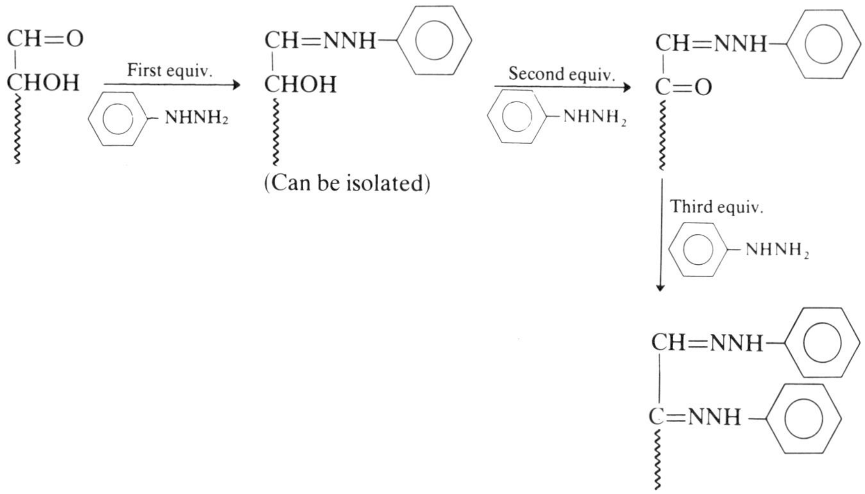


Osazone formation

equivalents of phenylhydrazine react with the carbohydrate. This reaction is useful in the determination of carbohydrate stereochemistry in that one chiral center is destroyed in the reaction. Thus, if two monosaccharides differ only in the stereochemistry of the 2-position, they will form the same osazone. If they differ at another position, the osazones will be different, as illustrated below.

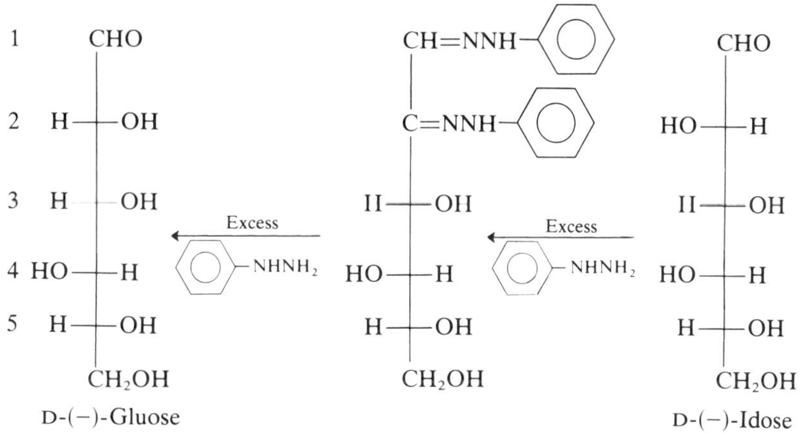

```
   1    CHO              CH=NNH-⬡                CHO
                            |
   2   H——OH             C=NNH-⬡              HO——H
                            |
   3   H——OH        Excess  H——OH      Excess  H——OH
              ⟵————————              ⟵————————
   4  HO——H     ⬡-NHNH₂  HO——H   ⬡-NHNH₂  HO——H
                            |
   5   H——OH             H——OH                H——OH
                            |
       CH₂OH             CH₂OH                CH₂OH
    D-(−)-Gluose                           D-(−)-Idose
```

D-(−)-Gluose and D-(−)-Idose form the same osazone because they differ in stereochemistry only at C-2.

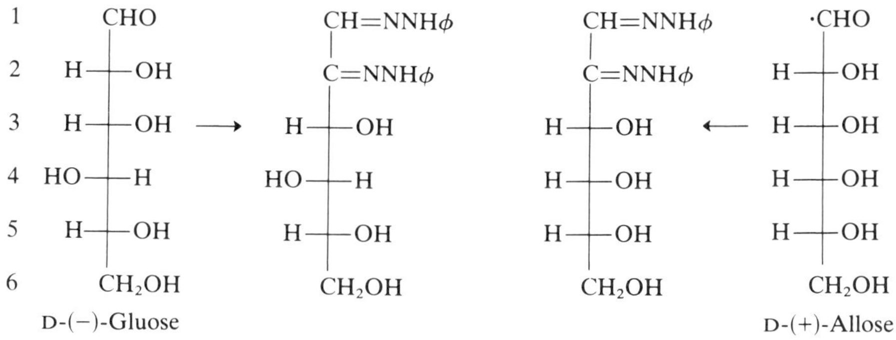

```
   1    CHO         CH=NNHφ        CH=NNHφ        ·CHO
                       |              |
   2   H——OH         C=NNHφ         C=NNHφ       H——OH
                       |              |
   3   H——OH  ⟶    H——OH         H——OH   ⟵   H——OH
   4  HO——H        HO——H         H——OH        H——OH
   5   H——OH        H——OH         H——OH        H——OH
   6   CH₂OH         CH₂OH         CH₂OH        CH₂OH
    D-(−)-Gluose                               D-(+)-Allose
```

D-(−)-Gluose and D-(+)-Allose form different osazones because they differ in stereochemistry at C-4.

Osazones are crystalline compounds that are also useful for identifying monosaccharides, which are difficult to crystallize in pure form.

## Formation of Cyanohydrins

Cyanohydrin formation is a useful method of building carbohydrate chains. Although the overall synthetic sequence is beyond the scope of this text, the creation of chiral centers in the HCN addition is the key. In the initial addition, diastereomers are formed which can be separated by such physical means as recrystallization or chromatography. They can then both be converted by chemical means to other monosaccharides, in this case, D-(−)-ribose and D-(−)-arabinose. Note that we are

building a D series of carbohydrates by this method, since the chiral center responsible for the D designation is not involved in the reaction.

## Glycoside Formation

The hemiacetal and hemiketal forms of single carbohydrates may be converted to acetals or ketals respectively, by reaction with alcohols under acid catalysis.

Hemiacetal form          Acetal form

Carbohydrates in the acetal or ketal form are referred to as glycosides. The reaction may be reversed by dilute aqueous acid hydrolysis, or by various enzymes. Enzymatic hydrolysis of O-glycosides can be quite specific in that the $\alpha$ or $\beta$ form will be hydrolyzed to the total exclusion of the other.

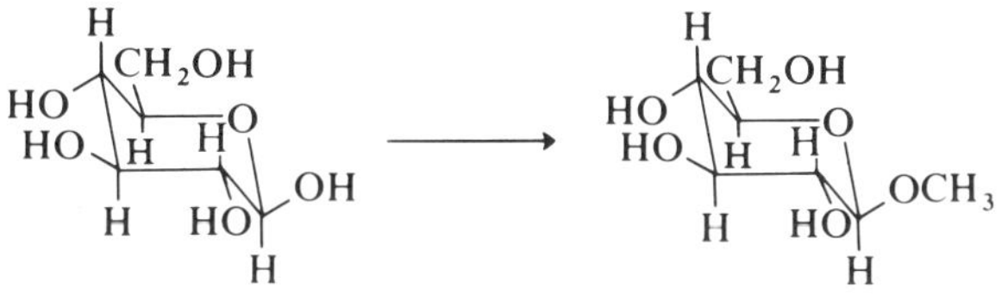

β-D-Glucose                    Methyl β-D-glucoside

The glycosidic bond is perhaps the most important aspect of carbohydrate chemistry. It is the basic linkage holding the individual monosaccharides together in polysaccharides, and also bonds sugars to other biologically important molecules such as steroids. N-glycosides, in which an amino group reacts with a hemiacetal, are extremely important classes of chemicals that include ribonucleosides and deoxyribonucleosides, N-glycosyl derivatives of heterocyclic bases. These linkages are extremely important in protein replication as the binding force linking nitrogen bases to sugars in both DNA and RNA (see Chapter I-14).

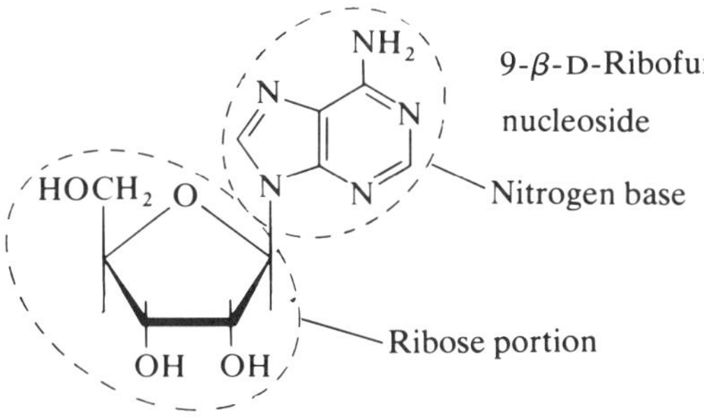

9-β-D-Ribofuranosyladenine—a typical nucleoside

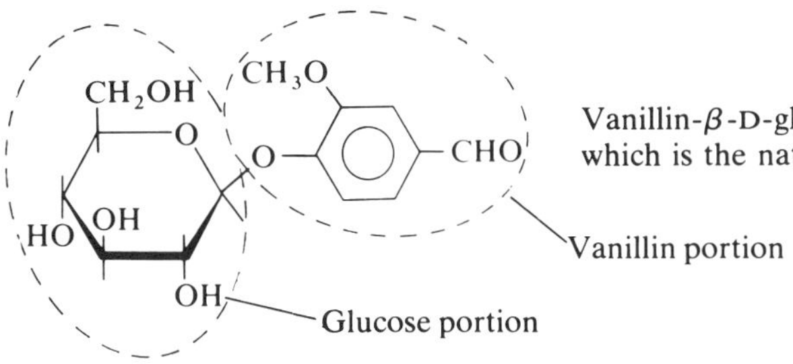

Nitrogen base

Ribose portion

Vanillin-β-D-glucoside—a typical glycoside which is the natural source of vanilla flavor

Vanillin portion

Glucose portion

## Reducing Behavior

Aldehydes reduce Fehling's, Benedict's, or Tollens' solutions, and this test is useful in classifying carbohydrates as either *reducing* or *non-reducing* sugars.

$$CHO + Ag(NH_3)_2^+ \longrightarrow COOH + Ag\downarrow (\text{silver mirror})$$
$$\text{Tollens}$$

$$CHO + Cu^{+2} (\text{citrate complex}) \longrightarrow COOH + Cu_2O\downarrow$$
$$\text{or} \qquad \text{red ppt.}$$
$$Cu^{+2} (\text{tartrate complex})$$

Functional groups which give (+) test:

$$\begin{matrix} | \\ -C-CHO \\ | \\ OH \end{matrix}$$  $\alpha$-Hydroxyaldehyde (aldoses)

$$\begin{matrix} & O \\ | & || \\ -C-C- \\ | \\ OH \end{matrix}$$  $\alpha$-Hydroxyketones (ketoses)

$$\begin{matrix} & OH \\ & / \\ C \\ & \backslash \\ & OR \end{matrix}$$  Hemiacetals

Acetals or ketals do not react, and thus glycosides fall into the category of non-reducing sugars. This observation is often useful in determining the structures of di- and polysaccharides.

---

**Exercise 12-5:** Give reaction products for the reaction of D-(+)-mannose with each of the following:

(a) $Ag(NH_3)_2^+NO_3^-$      (b) $Cu^{+2}$ (Citrate complex)

(c)  ⬡—NHNH$_2$      (d) HCN      (e) $H_2$/Pd

(Excess)

---

## THE STRUCTURE AND CHEMISTRY OF DISACCHARIDES

Disaccharides are composed of two monosaccharide units joined together by a glycoside linkage. Thus an alcohol group of one monosaccharide must react with the carbonyl of the second unit to form the hemiacetal glycoside. Lactose and sucrose are two of the more common disaccharides, and their formulas are illustrated below:

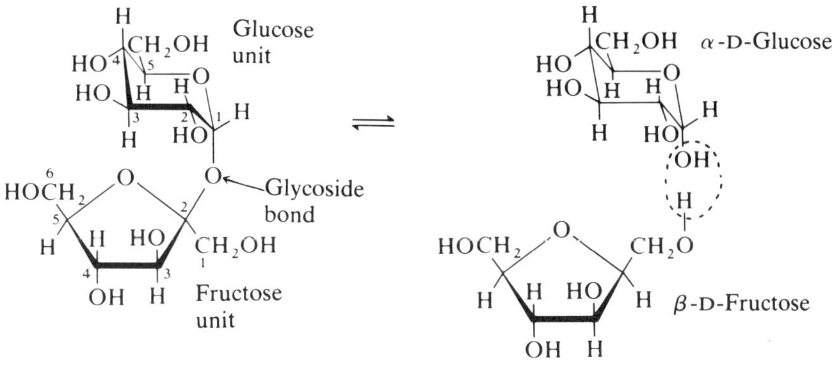

Sucrose

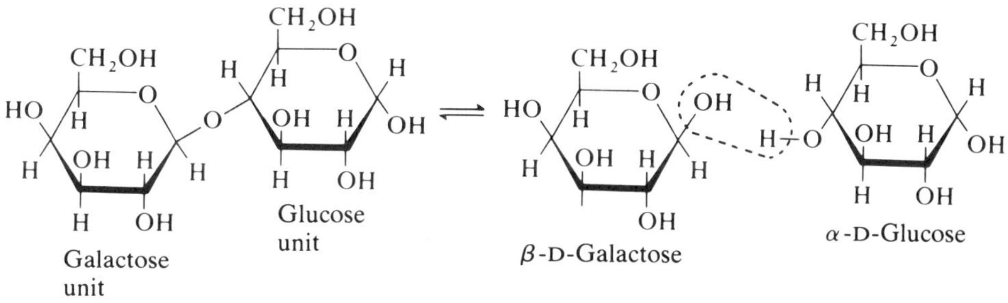

Lactose

Disaccharides may be hydrolyzed by dilute acid catalysis or enzymatically. Several questions, however, remain concerning the disaccharide structure even though we know the two participating monosaccharide units. First, it is necessary that we know which monosaccharide is acting as the hemiacetal and which is acting as the alcohol in forming the glycoside link. Second, we must also know which of the alcohol groups is being utilized to form the glycoside; and last, we must also know the $\alpha$ or $\beta$ configurations of *each* of the participating monosaccharide units. In sucrose, for example, the glucose unit acts as the hemiacetal and is in the $\alpha$ form. The OH group on the number 2 carbon in the $\beta$ form of fructose acts as the alcohol. Thus we can specify the structure of sucrose as

$\alpha$-D-glucose-(1,2)-$\beta$-D-fructose
or
$\beta$-D-fructose-(2,1)-$\alpha$-D-glucose

Sucrose is ordinary table sugar and is ubiquitous in our everyday lives. It is produced photosynthetically in plants and functions as an energy source. Lactose is the primary sugar found in milk, and thus is sometimes referred to as "milk sugar." Both lactose and sucrose produce glucose, which is essential to human metabolism, and is transported through the blood stream. Sucrose is a nonreducing sugar, while lactose does reduce Fehling's or Benedict's solutions. These facts indicate that sucrose must have a glycosidic link involving the glucose portion acting as the hemiacetal. In lactose, it doesn't matter which unit is acting as the hemiacetal—it would still behave as a reducing sugar. Osazone formation and mutarotation behavior follow similar logic.

| Lactose | Sucrose |
|---|---|
| Reduces Fehling's, Benedict's and Tollens' solutions | Does not reduce Fehling's, Benedict's or Tollens' solutions |
| Undergoes mutarotation | Does not mutarotate |
| Does form osazones | Does not form osazones |

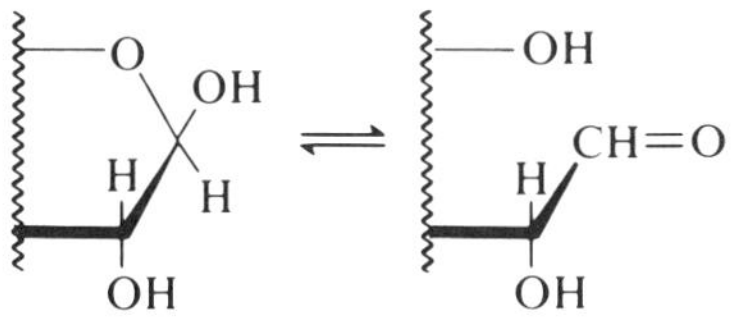

Glucose unit at end of lactose responsible for reducing behavior, osazone formation, and mutarotation.

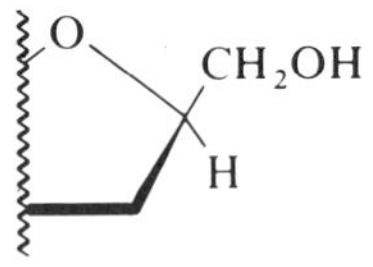

Fructose unit at end of sucrose responsible for nonreducing behavior, lack of mutarotation of osazone formation.

Several other disaccharides are important in nature. *Maltose* is not normally found in the free state but may be produced by the action of the enzyme maltase on starch. It yields only glucose upon hydrolysis. *Cellobiose* also produces only glucose when hydrolyzed and is an isomer of maltose. It is not affected by maltase but is hydrolyzed by emulsin enzyme. Maltose and cellobiose differ in the configuration of the glycoside bridge, the former being formed from $\alpha$-D-glucose, while the latter is formed from $\beta$-D-glucose, as shown below.

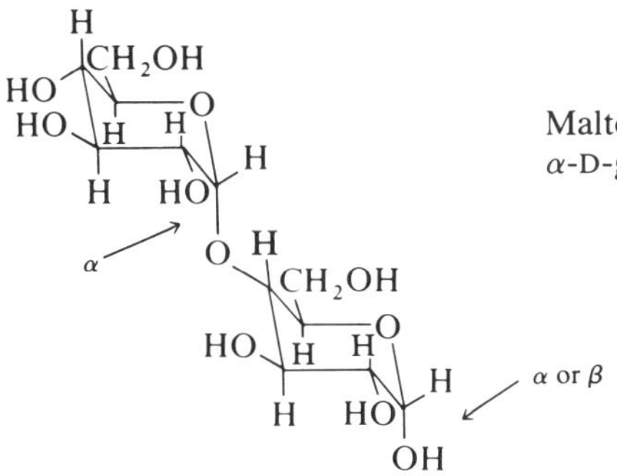

Maltose—glycoside link from $\alpha$-D-glucose

The hydrolysis of these two disaccharides points out an interesting fact concerning enzymes: in most cases, enzymes will only attack *either* an $\alpha$- or $\beta$-link in a di- or polysaccharide. Thus they are quite *specific* in their action.

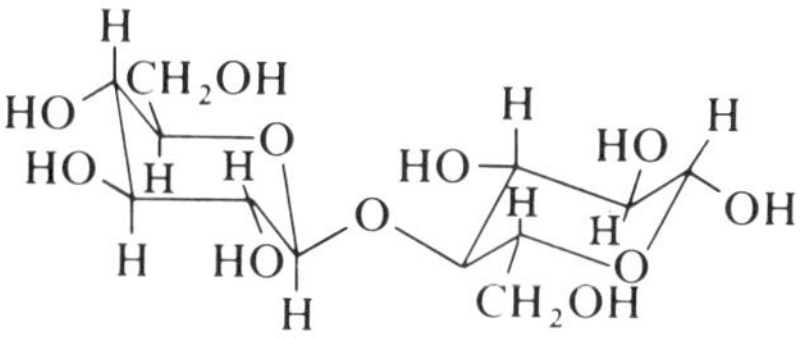

Cellobiose—glycoside link from $\beta$-D-glucose

## THE STRUCTURE AND CHEMISTRY OF POLYSACCHARIDES

Polysaccharides are carbohydrate chains of very high molecular weight. Like the disaccharides, they can be hydrolyzed to individual monosaccharide in acid or enzymatically. Many large polysaccharides, like starch, yield only one monosaccharide upon hydrolysis, indicating that their structure is a repeating unit. Other polysaccharides, however, yield more than one monosaccharide. We will limit our present discussion to the simple polysaccharides composed of single monosaccharide units. Incomplete hydrolysis of polysaccharides may yield oligosaccharides (3–10 units) or disaccharides. For example, the enzyme diastase produces maltose from starch.

The two most important polysaccharides, as far as man is concerned, are *cellulose* and *starch*. Cellulose is the most prevalent organic compound found in nature and is a prime constitutent of plant fiber such as wood and cotton. Starch serves as the major carbohydrate reserve in many plants such as potatoes, corn, rice, and cereal grains. Hydrolysis of both starch and cellulose yields glucose. However, the structure of cellulose may be considered to be repeating cellobiose units in which the individual glucose units have $\beta$-glycosidic links, while starch is a mixture of two polymers, *amylose* (ca. 20%) and *amylopectin* (ca. 80%). Partial hydrolysis of cellulose does yield cellobiose, while partial hydrolysis of starch yields, among other things, maltose. Starch is therefore a mixture with $\alpha$-glycosidic links between glucose units. Amylose is a linear glucose polymer composed of 1,4 links between individual units. Amylopectin also contains 1,6 as well as 1,4 links which makes it a highly branched polymer. The structures of cellulose, amylose, and amylopectin showing these structural relationships are shown below.

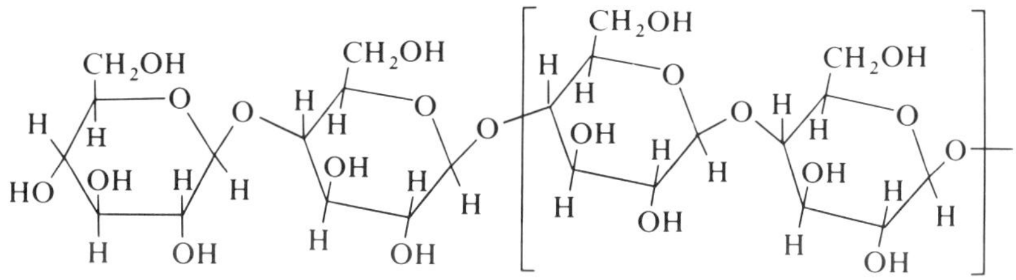

Cellulose
$n = 1800$–$3000$

Cellobiose repeating unit

Maltose repeating unit

Amylose
$n = 60-300$

Amylopectin

[Maltose unit]$_n$

1,6 link

1,4 link

The 1,6 links in amylopectin occur every 24–30 glucose units, while chain lengths in any particular branch contain from 20 to 30 glucose units.

In animal systems, starch is partially broken down by the enzyme *amylase* to maltose and then by the enzyme $\alpha$-glucosidase to glucose. The glucose is transported by the blood stream to the liver and other portions of the body where it is repolymerized to glycogen for storage. Glycogen is similar to amylopectin except that 1,6 links occur with greater frequency, averaging between 8–10 units. When energy is required quickly, glycogen can be converted to glucose quickly and transported by the blood stream in animals to where it is needed.

## USES OF CELLULOSE

Cellulose, in its many forms, is an extremely important commodity in our society. In raw form, such as wood (40–50% cellulose), we depend on its structural strength. Wood pulp produces paper, without which our society could not function. Natural fibers such as cotton and linen clothe much of the world, even in this modern day of synthetic fibers. However, the uses of cellulose do not end with the raw material. Many other useful products can be made from cellulose, as outlined below.

### Cellulose Esters

The alcohol functional groups in cellulose can be esterified. Cellulose nitrate results from treatment with nitric acid. If all three free OH groups are nitrated, the product is known as *guncotton* or *cordite*. Cotton can be nitrated in similar fashion. Both are highly efficient explosives.

Cellulose acetate, resulting from esterification with acetic anhydride, can be extruded through fine holes to produce a fiber called *cellulose acetate rayon*. The acetate can also be extruded in a flat, thin film, which is transparent and useful as wrapping material (*cellophane*). Acetate films were used in the early days of movie photography but had the disadvantage of being both brittle and highly flammable. Many fine films have been lost because of these unfortunate properties.

$X = NO_2$     Cellulose nitrate

$X = OCCH_3$   Cellulose acetate
     $\parallel$
     $O$

## Cellulose Ethers

When cellulose is treated with concentrated base followed by ethyl chloride, or similar halides, cellulose ethers are produced. These ethers are utilized to produce films and coatings. Cellulose xanthate can be prepared by reacting cellulose, sodium hydroxide, and carbon disulfide.

$$\text{Cellulose}-\text{OH} + \text{Na}^{\oplus}\text{OH}^{\ominus} + \text{CS}_2 \longrightarrow \underset{\text{Xanthate}}{\text{Cellulose}-\text{O}-\overset{\displaystyle S}{\overset{\displaystyle \parallel}{\text{C}}}-\text{S}^{\ominus}\text{Na}^{\oplus}}$$

The xanthate yields viscous aqueous solutions known as *viscose*. When these solutions are extruded through fine openings into an acidic solution, the xanthate decomposes to yield fine strands of cellulose called *rayon* (not to be confused with rayon acetate). Rayon in this form is commonly used as a fabric and to a lesser extent as a tire cord. Similarly, *cellophane* is produced by extrusion in sheets.

## CARBOHYDRATE METABOLISM

Carbohydrates are synthesized in plants from carbon dioxide and water. Animals consume carbohydrates from plant sources and either utilize their stored energy directly or store them for future energy needs. The reactions involved in these processes are extremely complex and require the intervention of enzymes and other chemicals such as phosphates. You will recall that alcohols form phosphoric acid esters. The alcohol groups in carbohydrates behave in a similar fashion and the

$$ROH + HO-\overset{\overset{\displaystyle O}{\|}}{\underset{\underset{\displaystyle OH}{|}}{P}}-OH \rightleftharpoons RO-\overset{\overset{\displaystyle O}{\|}}{\underset{\underset{\displaystyle OH}{|}}{P}}-OH + H_2O$$

resulting complex esters are intimately involved in carbohydrate metabolism, particularly adenosine triphosphate (ATP) and acetyl coenzyme-A (acetyl CoA).

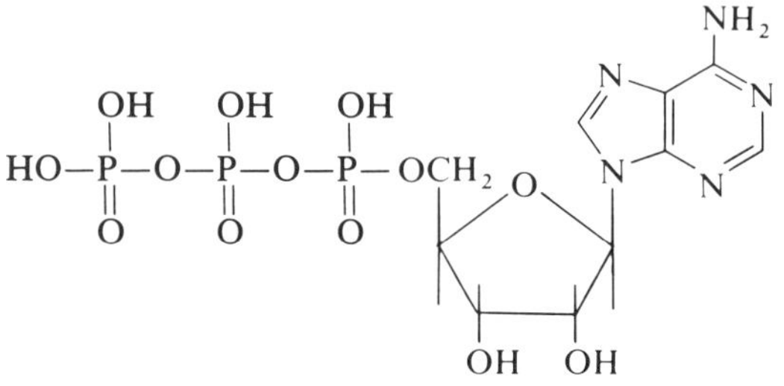

ATP
(a phosphate ester)

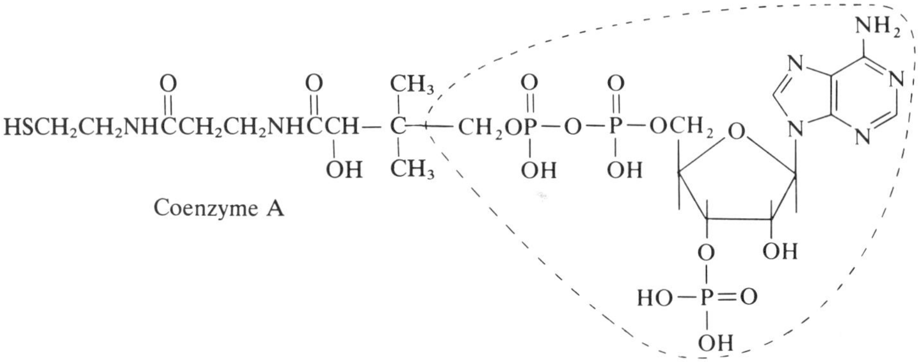

Coenzyme A

Phosphate ester

An interesting aspect of carbohydrate metabolism is the difference in various animal digestive systems that dictates *which* carbohydrates can be metabolized. Most animals, including man, cannot digest carbohydrates having $\beta$-glycosidic links. Thus we cannot digest cellulose. Animals such as cows, horses, termites, and various rodents have protozoa living in their digestive tracts that do have enzymes to break down these $\beta$-links and thus can survive on a diet extremely high in cellulose. Starch, however, has $\alpha$-glycosidic links and can be used as a source of glucose for those animals whose enzyme hydrolyze $\alpha$- rather than $\beta$-linked carbohydrates.

Once glucose is obtained from starch in man, it can be transported by the bloodstream to various portions of the body and stored as glycogen, as discussed previously. The human body has a finite capacity for storing glycogen, and when this is exceeded, further carbohydrate storage occurs as fat deposits. When energy is needed, glycogen is reconverted to glucose enzymatically, and the glucose thus released feeds the various biochemical cycles in the body by complex enzymatic processes involving fructose diphosphate, pyruvic acid ($CH_3COCOOH$), and acetyl coenzyme A. These processes are outlined below in Fig. 12-2, but the details are beyond the scope of this text and are more properly studied in a course in biochemistry.

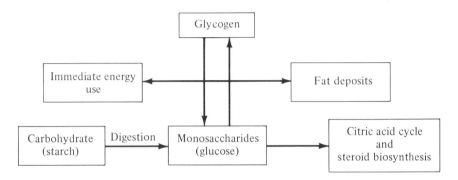

**Figure 12-2.** Simplified scheme for carbohydrate metabolism.

## WHY SUGARS TASTE SWEET

The phenomenon of sweetness has always been intriguing to humans. We have a fatal fondness for sweet things, much to the detriment of our teeth and waistlines. The search for sugar substitutes during the past two decades has produced saccharine, cyclamates, and many other sweet tasting chemicals, and an analysis of their structural similarities has led to a better understanding of the nature of sweetness.

Sugars are hydrogen-bonded in solution because of the large number of polar groups. It has been suggested that variations in sugar sweetness are due to differences in intramolecular hydrogen bonding. In fact there is strong evidence that intramolecular hydrogen bonding influences the intensity of sweetness, while intermolecular H bonding affects the rate at which sweetness is detected. Our tongues have various receptor sites (taste buds) which send signals to the brain when activated by external chemicals. It has only been within the past ten years, however, that we have begun to understand the nature of this chemical interaction with the taste bud receptors.

It appears that the receptor responds to both acidic (AH) and basic (B) components in a molecule. Both sites must be stimulated in order for sweetness to be perceived. The sweet molecule must have a slightly acidic proton and an electronegative site within a certain distance and in a correct conformation. Several examples of compounds meeting these criteria are shown below. It has also been

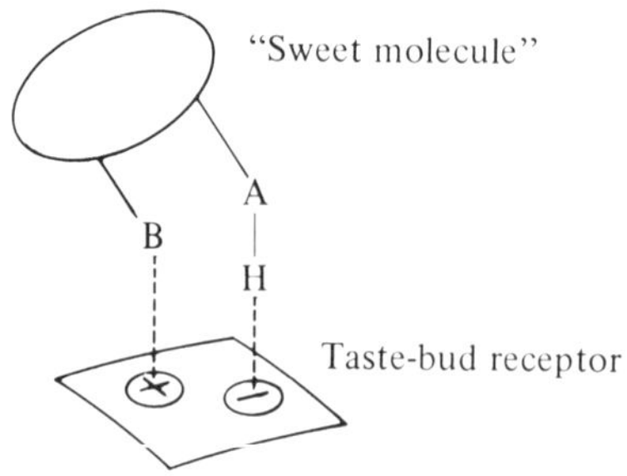

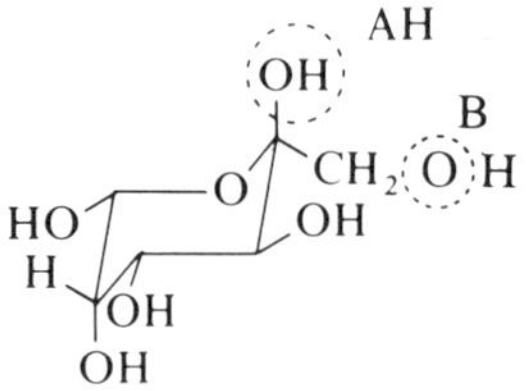

β-D-Fructose

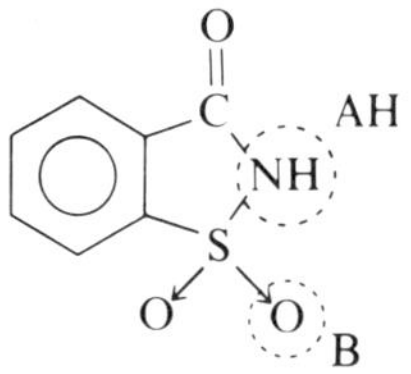

Saccharine

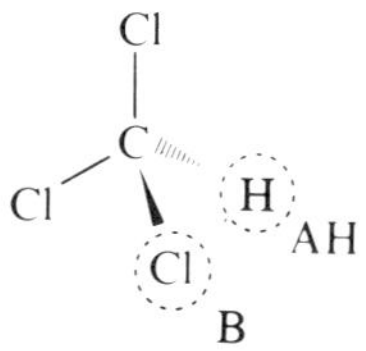

Chloroform

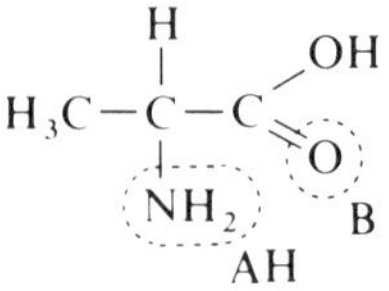

Alanine

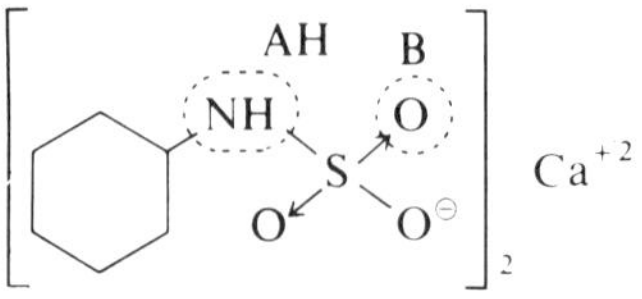

Calcium cyclamate

suggested that the taste bud receptor site is also an AH/B system as well, so that the interaction between the sweet molecule and the receptor can actually be represented as follows:

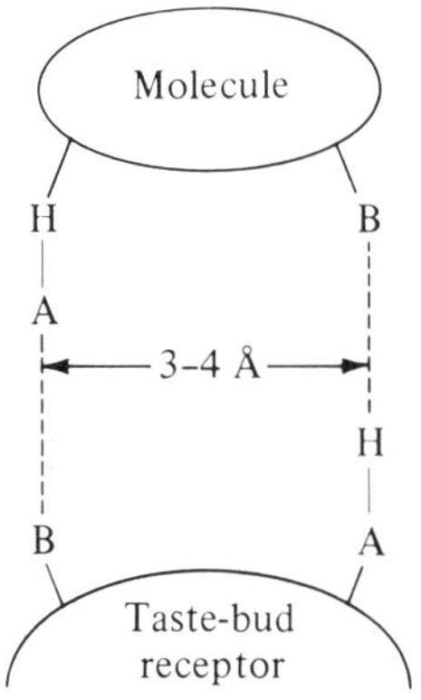

256

The model system shown above is two-dimensional, but there is evidence that the interaction is actually three-dimensional. For example, enantiomers of many chiral sweet molecules actually taste bitter. Much research remains to be done in this area, but the current theories hold much promise.

## REVIEW OF NEW TERMS AND CONCEPTS

Define each term and, if possible, give an example of each:

| | |
|---|---|
| Carbohydrate | Aldose |
| Ketose | Monosaccharide |
| Oligosaccharide | Polysaccharide |
| Absolute carbohydrate configuration -D and L (R and S) | Intramolecular hemiacetals and hemiketals |
| Haworth projection | Pyranose |
| Furanose | α and β forms of carbohydrates |
| Mutarotation | Anomeric carbon |
| Osazone formation | Glycoside |
| Nucleoside | Reducing and nonreducing sugars |
| Disaccharide | Starch |
| Cellulose | Glycogen |
| Rayon | Cellophane |

## PROBLEMS: CHAPTER I-12

**1.** Draw a typical structure for each of the following:

(a) an aldopentose        (b) a ketohexose
(c) an aldohexose         (d) a β-furanoside
(e) an α-glycoside        (f) an α-pyranoside

**2.** Draw formulas (Haworth and chair) for:

(a) D-glucose        (b) D-galactose        (c) D-mannose

**3.** Write equations for each of the following:

(a) D-galactose + $HNO_3$        (b) D-glucose + $Ac_2O/HOAc$
(c) D-threose + HCN              (d) lactose + $H_3O^+/\Delta$
(e) lactose + Fehling's solution (f) sucrose + Tollens' reagent

**4.** We have used D and L designations in this chapter because much of the literature of biochemistry still uses this system. Answer the following questions concerning absolute configuration.

(a) Are all D sugars also $R$, or are some $S$?

(b) Determine the $R$ and $S$ configurations for each chiral carbon in D and L threose and D and L erythrose.

(c) Do $\alpha$ and $\beta$-D-galactose have different $R$ and $S$ configurations at the anomeric carbon?

**5.** Draw structures for two aldopentoses that would yield the same osazone. Give another aldopentose structure that would yield a different osazone.

**6.** Explain why a reducing sugar can undergo mutarotation but a nonreducing sugar cannot.

**7.** Aldohexoses can be prepared from aldopentoses by HCN addition, followed by a complex reaction series, which converts the —CN functional group to —CHO. Suggest an aldopentose (see Figure 12-1) that could be used for the synthesis of D-galactose.

**8.** We have previously described the structure of sucrose as $\alpha$-D-glucose-(1,2)-$\beta$-D-fructose. A more accurate name would be $\alpha$-D-glucopyranosyl-$\beta$-D-fructo-furanoside. Explain why this name is more descriptive. Using this nomenclature, name both lactose and maltose.

**9.** How many disaccharides could you imagine from the combination of glucose and galactose? Draw their structures and name them.

**10.** How would you explain the taste sensations produced by the following compounds?

Bitter                    Sweet

# 13

# Amines and Amine Derivatives

We have seen in previous chapters that compounds such as phenol and carboxylic acids function as organic acids. Amines ($RNH_2$, $R_2NH$, $R_3N$) are organic bases. They are important synthetic intermediates and are widely distributed in nature. Amino acids and proteins are essential to life, and the amino group is also found in many other naturally occurring and synthetic pharmaceuticals. In this chapter we will investigate amine synthesis and the reaction chemistry of the amino group. We will also learn that other nitrogen-containing functional groups may be derived from amines, and we will survey some of the more important aspects of their chemistry.

## THE NATURE OF THE AMINO GROUP AND THE CLASSIFICATION OF AMINES

Amines may be considered to be derivatives of ammonia. Both alkyl and aryl groups may be substituted for hydrogen, and the number of hydrogens replaced determines the classification of amines as primary (1°—one hydrogen replaced), secondary (2°—two hydrogens replaced), and tertiary.

$NH_3$—Ammonia (parent)
$RNH_2$—1° Amine
$R_2NH$—2° Amine
$R_3N$—3° Amine

Example:

$$CH_3CH_2NH_2 \quad 1° \text{ Amine}$$

$$\langle\bigcirc\rangle-NHCH_2CH_2CH_3 \quad 2° \text{ Amine}$$

3° Amine

N
|
$CH_3$

Amines, like the parent compound, ammonia, are good nucleophiles and Lewis bases. As we shall see in the section on amine reactions, these properties account for much of the chemistry of the amino group.

$\ddot{N}$ ⟵ ——— Lone pair available for reaction

$H_3N\colon$ and $R_3N\colon$ are Lewis bases and good nucleophiles

## NOMENCLATURE OF AMINES

Simple amines are named by identifying the alkyl groups attached to the nitrogen and adding the word *amine*. If the alkyl chain is complex, the amino group is treated as a substituent such as chloro or methyl.

Examples:

| | |
|---|---|
| $CH_3CH_2NH_2$ | Ethylamine |
| $CH_3CH_2NHCH_3$ | Methylethylamine |
| $\overset{6}{C}H_3\overset{5}{C}H\overset{4}{C}H_2\overset{3}{C}H\overset{2}{C}H_2\overset{1}{C}H_3$ <br> $\quad\;\; | \qquad\;\; |$ <br> $\quad\;\; CH_3 \quad\; NH_2$ | 3-Amino-5-methylhexane |
| $H_2NCH_2CH_2CH_2CH_2CH_2CH_2NH_2$ | 1,6-Diaminohexane |
| $CH_3CHCH_2CH_2CH_3$ <br> $\quad\;\; |$ <br> $\quad\;\; N(CH_3)_2$ | 2-*N,N*-Dimethylaminopentane |

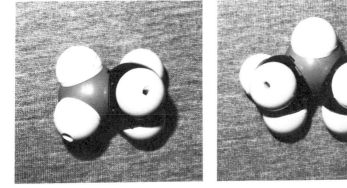

| | | |
|:---:|:---:|:---:|
| $CH_3NH_2$ | $(CH_3)_2NH$ | $(CH_3)_3N$ |

**Figure 13-1.** Typical 1°, 2°, and 3° amines.

Aromatic amines are named as derivatives of the parent, aniline.

Aniline        2,4-Dibromoaniline        4-Bromo-*N*-methylaniline

## PHYSICAL PROPERTIES AND STRUCTURE OF AMINES

The lower-molecular-weight amines, such as methylamine and dimethylamine are gases, like ammonia. One can conclude from this that amines are not as highly associated as the corresponding oxygen compounds, nor as polar.

| Compound | Mol. wt. | B.p. (°C) | Comment |
|---|---|---|---|
| $NH_3$ | 17 | −33 | $H_2O$ more associated |
| $H_2O$ | 18 | 100 | |
| $CH_3NH_2$ | 31 | −7.5 | $CH_3OH$ more associated |
| $CH_3OH$ | 32 | 64.5 | |
| $(CH_3)_2NH$ | 45 | 7.7 | $(CH_3)_2NH$ more associated |
| $(CH_3)_2O$ | 46 | −23.7 | |

Primary and secondary amines do contain polar N—H bonds, and may hydrogen-bond with hydroxylic solvents; thus amines containing up to six carbons have some solubility in water. Amines have characteristically unpleasant odors, ranging from ammonia-like for the lower-molecular-weight examples, to fishy for the higher-molecular-weight compounds. The "fishy" odor we associate with decaying fish or other animal flesh is really a combination of the odors of several amines and diamines such as putrescine ($NH_2CH_2CH_2CH_2CH_2NH_2$) and cadaverine ($NH_2CH_2CH_2CH_2CH_2CH_2NH_2$) resulting from the decomposition of various amino acids in protein tissue.

The nitrogen atom in aliphatic amines appears to be $sp^3$-hybridized with the H—N—R bond angles of approximately 108°, while the nitrogen lone pair of

$sp^3$-hybridized
nitrogen

electrons occupies the vacant $sp^3$ lobe. Thus amines appear to be pyramidal.

Lone pair occupies
*vacant $sp^3$ lobe*

## THE BASICITY OF AMINES AND SALT FORMATION

Amines are Lewis bases because of the presence of the unshared electron pair. They are also proton acceptors, forming ammonium salts in the process.

$$R\ddot{N}H_2 + H^{\oplus}X^{\ominus} \longrightarrow R\overset{\oplus}{N}H_3, X^{\ominus}$$

1° Amine                    Ammonium salt

$$R\ddot{N}H_2 + H_2O \underset{}{\overset{K_b}{\rightleftharpoons}} R\overset{\oplus}{N}H_3, OH^{\ominus}$$

Equilibrium in water

Methyl and dimethyl amines are stronger bases than the ammonia itself because of the electron-releasing character of alkyl groups, which tends to increase electron density on the nitrogen. However, when three bulky R groups are attached to the

Alkyl groups donate
electron density to N

nitrogen (3° amine), the accessibility of the electron-pair is severely reduced, and such amines are said to be sterically hindered; although they are stronger bases than ammonia, they are weaker bases than either 1° or 2° amines.

Three large R groups hinder the approach of reagents to the lone-pair

| Amine | $K_b$ |
|---|---|
| $NH_3$ | $1.8 \times 10^{-5}$ |
| $CH_3NH_2$ | $44.0 \times 10^{-5}$ |
| $(CH_3)_2NH$ | $51.0 \times 10^{-5}$ |
| $(CH_3)_3N$ | $6.3 \times 10^{-5}$ |

Order of base strengths:
$(CH_3)_2NH > CH_3NH_2 > (CH_3)_3N > NH_3$

When amines react with hydrogen halide, quaternary ammonium salts are formed which are stable crystalline solids. The free amine can be regenerated by treating the salt with base. Salts can be named by identifying the R groups and adding "ammonium halide."

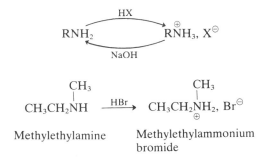

Methylethylamine            Methylethylammonium bromide

**Exercise 13-1:**   Would you expect aniline $\left(\bigcirc\!\!-\!NH_2\right)$ to be a stronger or weaker base than methylamine? Explain your answer on the basis of relative nitrogen electron densities.

# THE STEREOCHEMISTRY OF AMINES

Pyramidal nitrogen should be chiral if three different R groups are attached to the nitrogen, the mirror images being nonsuperimposable. However, simple tertiary amines cannot be separated into $R$ and $S$ forms. It has been experimentally verified that nitrogen continuously undergoes an "umbrella" inversion reaction which in effect continuously racemizes a chiral nitrogen center. This process occurs at an extremely rapid rate, many thousands of times per second.

Mirror images are nonsuperimposable

Pyramidal          Planar                Pyramidal
                   transition state

When pyramidal nitrogen reacts with HX, tetrahedral nitrogen results. This effectively stops the nitrogen inversion process, and quaternary ammonium salts *can* be separated into $R$ and $S$ forms.

Pyramidal N—rapidly          Tetrahedral N—frozen
interconverting

---

**Exercise 13-2:** Suppose we convert methylethyl-*n*-propyl amine to a quaternary salt by reaction with HBr, and separate the salt into $R$ and $S$ forms. Why can't we get the pure $R$ and $S$ amines back by treating each salt with aqueous base?

---

# PREPARATIONS OF AMINES

## Alkylation of Ammonia and Amines

We have previously seen that ammonia or amines will form quaternary salts with hydrogen halide. A similar displacement reaction ($S_N2$) also occurs with alkyl halide to produce an ammonium salt. If we then treat this salt with aqueous base, a primary amine is liberated. In essence, we have *alkylated* the amine and the overall process is known as *alkylation*. This process can be modified to produce 2° and 3° amines as well as quaternary ammonium salts.

(1)  $H_3N\overset{..}{} + R-X \longrightarrow H_3\overset{\oplus}{N}-R, X^\ominus$

(2)  $H_3\overset{\oplus}{N}-R, X^\ominus + NaOH \longrightarrow H_2NR + Na^\oplus X^\ominus + H_2O$
      1° Amine

$\quad$ X = Cl, Br, I
$\quad$ R = alkyl

$R\overset{..}{N}H_2 + R'X \longrightarrow \underset{\underset{R'}{|}}{R\overset{\oplus}{N}H_2}, X^\ominus \xrightarrow{NaOH} R\overset{..}{N}HR'$
$\qquad\qquad\qquad\qquad\qquad\qquad\qquad$ 2° Amine

$RNHR' + R''X \longrightarrow \underset{\underset{R'}{|}}{R\overset{\oplus}{N}HR''}, X^\ominus \xrightarrow{NaOH} \underset{\underset{R'}{|}}{R-\overset{..}{N}-R''}$
$\qquad\qquad\qquad\qquad\qquad\qquad\qquad$ 3° Amine

$\underset{\underset{R'}{|}}{R\overset{..}{N}R''} + R'''X \longrightarrow \underset{\underset{R'}{|}}{\overset{\overset{R'''}{|}}{R-N-R''^\oplus}}, X^\ominus \xrightarrow{NaOH}$ No further reaction
$\qquad\qquad\qquad\qquad$ Quaternary
$\qquad\qquad\qquad\qquad$ ammonium
$\qquad\qquad\qquad\qquad$ salt

Examples:

$CH_3CH_2NH_2 \xrightarrow{CH_3Br} \underset{\oplus}{CH_3CH_2\overset{\overset{CH_3}{|}}{N}H_2}, Br^\ominus \xrightarrow{NaOH} CH_3CH_2NHCH_3$
Ethylamine $\qquad\qquad\qquad\qquad\qquad\qquad\qquad\qquad\qquad$ Methylethylamine

$(CH_3)_2NH + CH_2CHCH_2CH_2CH_2CH_3 \longrightarrow CH_3\overset{\overset{}{|}}{C}HCH_2CH_2CH_2CH_3$
$\qquad\qquad\qquad\quad \overset{\overset{}{|}}{Br} \qquad\qquad\qquad\qquad\qquad \underset{\underset{Br^\ominus}{\overset{\oplus}{}}}{\overset{\overset{}{|}}{NH(CH_3)_2}} \Big| NaOH$

$\qquad\qquad\qquad\qquad\qquad\qquad\qquad\qquad CH_3\overset{\overset{}{|}}{C}HCH_2CH_2CH_2CH_3$
$\qquad\qquad\qquad\qquad\qquad\qquad\qquad\qquad\quad \overset{\overset{}{|}}{NMe_2}$
$\qquad\qquad\qquad\qquad\qquad\qquad\qquad 2\text{-(N,N-Dimethylamino)hexane}$

---

**Exercise 13-3:**  Would you expect the following reaction sequences to produce the products as written?  If not, why not?

(a) $CH_3CH=CHBr \xrightarrow{NH_3} CH_3CH=CH-\overset{\oplus}{N}H_3, Br^\ominus \xrightarrow{NaOH} CH_3CH=CHNH_2$

(b) ⟨O⟩—Cl $\xrightarrow{CH_3NH_2}$ ⟨O⟩—$\overset{\oplus}{N}H_2CH_3$, Cl$^{\ominus}$ $\xrightarrow{NaOH}$ ⟨O⟩—NHCH$_3$

## Reduction of Nitro Compounds

Aniline is produced commercially by chemical reduction of nitrobenzene. Iron and hot steam are used by industry as the reducing agent, but Sn/HCl or Zn/HCl is more convenient on a laboratory scale. Lithium aluminum hydride also works well. Aliphatic nitro compounds also reduce under these conditions to yield 1° aliphatic amines.

⟨O⟩—NO$_2$ $\xrightarrow{[R]}$ ⟨O⟩—NH$_2$

$(CH_3)_2CHCH_2NO_2 \xrightarrow{[R]} (CH_3)_2CHCH_2NH_2$

[R] = Fe/steam, Sn/HCl, Zn/HCl, LiAlH$_4$, or Pd/H$_2$

---

**Exercise 13-4:**   Which of the preceding reducing agents would you use to carry out the following reduction?  Are there any of these methods you would *not* use?

$$\begin{array}{c} CH_3 \\ \diagdown \\ \phantom{xx}C=CHCH_2NO_2 \\ \diagup \\ CH_3 \end{array} \longrightarrow \begin{array}{c} CH_3 \\ \diagdown \\ \phantom{xx}C=CHCH_2NH_2 \\ \diagup \\ CH_3 \end{array}$$

---

## Reduction of Nitriles

Nitriles, usually formed from alkyl halides by $S_N2$ attack of cyanide ion, may also be reduced to 1° amines.  Normally this reduction is accomplished either by catalytic hydrogenation or with LiAlH$_4$.

$$RC\equiv N \xrightarrow{[R]} RCH_2NH_2 \quad [R] = Pd/H_2 \text{ or LAH}$$
$$\text{Nitrile} \qquad\qquad 1° \text{ Amine}$$

Example:

$$(CH_3)_2CHCH_2C\equiv N \xrightarrow[Pd]{H_2} (CH_3)_2CHCH_2CH_2NH_2$$

⟨O⟩—CH$_2$C≡N $\xrightarrow[\text{2. H}_2\text{O}]{\text{1. LAH}}$ ⟨O⟩—CH$_2$CH$_2$NH$_2$

## Reduction of Amides

Amides may be reduced by LAH to yield either 1°, 2°, or 3° amines. Since amides are readily prepared from carboxylic acids (see Chapter I-12), this is perhaps the most general and best method for laboratory preparation of amines.

$$R-\overset{\overset{\displaystyle O}{\|}}{C}NH_2 \xrightarrow[\text{2. H}_2\text{O}]{\text{1. LAH}} RCH_2NH_2 \quad 1° \text{ Amine}$$

$$R-\overset{\overset{\displaystyle O}{\|}}{C}NHR' \xrightarrow[\text{2. H}_2\text{O}]{\text{1. LAH}} RCH_2NHR' \quad 2° \text{ Amine}$$

$$R-\overset{\overset{\displaystyle O}{\|}}{C}-NR_2' \xrightarrow[\text{2. H}_2\text{O}]{\text{1. LAH}} RCH_2NR_2' \quad 3° \text{ Amine}$$

Example:   Prepare $(CH_3)_2NCH_2CH_2CH_2CH(CH_3)_2$

$$(CH_3)_2CHCH_2CH_2\overset{\overset{\displaystyle O}{\|}}{C}OH \xrightarrow{SOCl_2} (CH_3)_2CHCH_2CH_2\overset{\overset{\displaystyle O}{\|}}{C}Cl$$

4-Methylpentanoic acid

$\downarrow (CH_3)_2NH$

$$(CH_3)_2CHCH_2CH_2CH_2NMe_2 \xleftarrow[\text{2. H}_2\text{O}]{\text{1. LAH}} (CH_3)_2CHCH_2CH_2\overset{\overset{\displaystyle O}{\|}}{C}NMe_2$$

1-(N,N-Dimethylamino)-
4-methylpentane

N,N-Dimethyl-
4-methylpentanamide

## Reductive Amination via Imines

In Chapter I-10, we learned that ammonia and amines react reversibly with carbonyl groups to produce *amino alcohols* and *imines*. This is a dynamic equili-

$$R-\overset{\overset{\displaystyle O}{\|}}{C}-R' + NH_3 \rightleftharpoons R-\overset{\overset{\displaystyle OH}{|}}{\underset{\underset{\displaystyle NH_2}{|}}{C}}-R' \xrightarrow[]{-H_2O} R-\overset{\overset{\displaystyle}{}}{\underset{\underset{\displaystyle NH}{\|}}{C}}-R' + H_2O$$

Amino alcohol            Imine

brium which may be shifted to favor the imine by water removal, usually by azeotropic distillation. Imines may be catalytically reduced like most unsaturated links producing amines. Primary amines are formed when ammonia is utilized, while 2° amines result when $RNH_2$ is used to form the initial imine.

267

$$\underset{\substack{R \\ \diagdown \\ R \diagup \\ \text{Ketone}}}{C=O} + NH_3 \rightleftharpoons \left[\underset{\substack{R \\ \diagdown \\ R \diagup \\ \text{Imine}}}{C=NH}\right] \xrightarrow[\text{Cat.}]{H_2} \underset{\substack{R \\ \diagdown \\ R \diagup \\ 1° \text{ Amine}}}{CHNH_2}$$

$$\underset{\substack{R \\ \diagdown \\ H \diagup \\ \text{Aldehyde}}}{C=O} + R'NH_2 \rightleftharpoons \left[\underset{\substack{R \\ \diagdown \\ H \diagup \\ \text{Imine}}}{C=NR'}\right] \xrightarrow[\text{Cat.}]{H_2} \underset{\substack{R \\ \diagdown \\ H \diagup \\ 2° \text{ Amine}}}{CHNHR'}$$

Examples:

$$CH_3CH_2CHO \underset{\text{Propanal}}{\overset{NH_3}{\rightleftharpoons}} [CH_3CH_2CH=NH] \xrightarrow[Ni]{H_2} \underset{\text{n-Propylamine}}{CH_3CH_2CH_2NH_2}$$

$$\underset{CH_3CH_2\overset{O}{\overset{\|}{C}}CH_3}{} + \overset{\bigcirc}{}\text{—}CH_2NH_2 \rightleftharpoons \left[\underset{\substack{CH_3CH_2 \\ \diagdown \\ CH_3 \diagup}}{C=NCH_2\phi}\right] \xrightarrow[Ni]{H_2} \underset{\substack{CH_3CH_2CHCH_3 \\ | \\ NHCH_2\phi \\ \textit{sec}\text{-Butylbenzylamine}}}{}$$

### Hofmann Degradation of Amides

Amides may be converted to amines by an oxidative procedure known as Hofmann degradation. In this process, an amide is treated with alkaline bromine in aqueous solution. A series of unstable intermediates eventually produces an iso-cyanate which is converted to a thermally unstable carbamic acid. Decarboxylation of the carbamic acid yields the amine.

$$R\text{—}\overset{O}{\overset{\|}{C}}\text{—}NH_2 \xrightarrow[NaOH]{Br_2} \left[R\text{—}\overset{O}{\overset{\|}{C}}NHBr\right] \longrightarrow [R\text{—}N=C=O]$$

Isocyanate

$$\Big\downarrow H_2O$$

$$R\text{—}NH_2 + CO_2\uparrow \xleftarrow[(-CO_2)]{\Delta} [R\text{—}NHCOOH]$$

Carbamic acid

$$\Big\downarrow OH^-$$

$$CO_3^{-2}$$

Example:

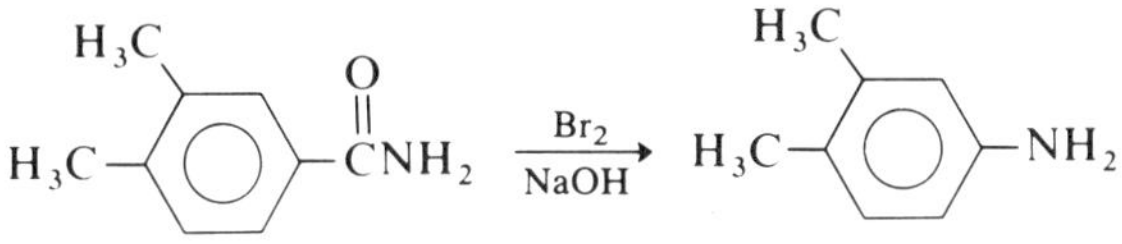


## REACTIONS OF AMINES

### Salt Formation

Primary, secondary, and tertiary amines all form salts with HX or RX. These salts are ionic and are soluble in water. Free amines can be released from this aqueous solution by neutralization with sodium hydroxide, followed by either extraction or steam distillation. This process allows amines to be readily separated from other organic material which is not soluble in acid.

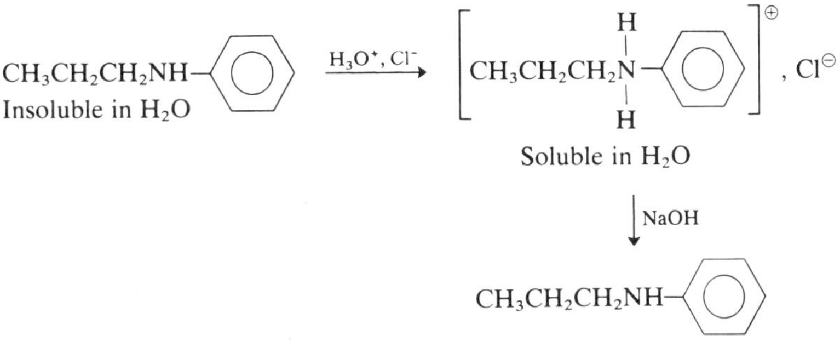


$$CH_3CH_2CH_2\overset{CH_3}{\underset{CH_3}{N}}\!: \; + \; CH_3CH_2Br \longrightarrow \left[CH_3CH_2CH_2\overset{CH_3}{\underset{CH_3}{N}}CH_2CH_3\right]^{\oplus}, \; Br^{\ominus}$$

Soluble in $H_2O$

Quaternary ammonium hydroxides can be prepared from solutions of quaternary salts by precipitating the halide with $Ag_2O$. The resulting basic solutions are 100% ionic and are often used when a strong organic base is required.

$$CH_3CH_2CH_2\overset{CH_3}{\underset{CH_3}{N}}CH_3^{\oplus}, \; Br^{\ominus} \xrightarrow[H_2O]{Ag_2O} AgBr\downarrow + CH_3CH_2CH_2\overset{CH_3}{\underset{CH_3}{N}}CH_3^{\oplus}, \; OH^{\ominus}$$

Tetraalkylammonium halide

100% ionic in $H_2O$
Tetraalkylammonium hydroxide

**Exercise 13-5:** $NH_4OH$ is generally regarded as a weak base ($K_b \sim 1.8 \times 10^{-5}$), yet $(CH_3)_4NOH$ is a very strong base. Explain this seeming anomaly.

## Hofmann Elimination

When quaternary ammonium hydroxides are heated strongly, they decompose forming an olefin and a tertiary amine. When this reaction is used synthetically it is referred to as *Hofmann elimination.* In order for this elimination to occur, at least

$$CH_3CH_2CH_2\overset{\overset{\displaystyle CH_3}{|}}{\underset{\underset{\displaystyle CH_3}{|}}{N}}CH_3^{\oplus}, OH^{\ominus} \xrightarrow{\Delta} CH_3CH=CH_2 + NMe_3 + H_2O$$

one of the alkyl groups must be an ethyl or larger hydrocarbon group. This reaction has been utilized to prepare compounds that are difficult to prepare by other means, such as the $Z$- and $E$-1,3,5-hexatriene.

**Exercise 13-6:** Write structures for the products of Hofmann elimination of the following compounds. Write complete reaction sequences showing all intermediates.

(a)  $CH_3CH_2CH_2CH_2CH_2N(CH_3)_2$

(b)

## Amide and Sulfonamide Formation

In the previous chapter (I-12) we learned that amides could be prepared from 1° or 2° amines reacting with acyl derivatives. Tertiary amines, however, do not react.

$$R-\overset{\overset{\text{O}}{\|}}{C}-Cl + H_2NR' \longrightarrow R-\overset{\overset{\text{O}}{\|}}{C}NHR'$$

Acyl halide                                N-Substituted amide

$$R-\overset{\overset{\text{O}}{\|}}{C}-O-\overset{\overset{\text{O}}{\|}}{C}-R + HNR'_2 \longrightarrow R-\overset{\overset{\text{O}}{\|}}{C}-NR'_2$$

Anhydride                               N,N-Disubstituted amide

R, R' = alkyl, aryl, or H

Examples:

$$CH_3CH_2CH_2CH_2\overset{\overset{\text{O}}{\|}}{C}Cl \xrightarrow{\text{EtNH}_2} CH_3CH_2CH_2CH_2\overset{\overset{\text{O}}{\|}}{C}NHEt$$

N-Ethylpentanamide

$$\langle\bigcirc\rangle-NH_2 \xrightarrow{CH_3\overset{\text{O}}{C}O\overset{\text{O}}{C}CH_3} \langle\bigcirc\rangle-NH\overset{\overset{\text{O}}{\|}}{C}CH_3$$

Acetanilide

$$CH_3CH_2CH_2\overset{\overset{\text{O}}{\|}}{C}Cl + Me_3N \longrightarrow \text{No reaction}$$

Similar reactions are observed between amines and sulfonyl halides. Sulfonamide formation is the basis of an amine classification scheme known as the *Hinsberg test* for amines, outlined below. If an amine gives a precipitate with

$$RSO_2Cl + H_3N \longrightarrow RSO_2NH_2 \quad \text{Sulfonamide}$$

$$RSO_2Cl + H_2NR' \longrightarrow RSO_2NHR' \quad \text{N-substituted sulfonamide}$$

$$RSO_2Cl + HNR'_2 \longrightarrow RSO_2NR' \quad \text{N,N-Disubstituted sulfonamide}$$

$$RSO_2Cl + NR_3 \longrightarrow \text{No reaction}$$

benzene sulfonyl chloride, then it is either a 1° or 2° amine—3° amines do not react. The sulfonamides derived from 1° amines are soluble in base, but those from 2°

1° amine:

$$RNH_2 + \langle\bigcirc\rangle-SO_2Cl \longrightarrow \overset{\text{Ppt.}}{\langle\bigcirc\rangle-SO_2NHR} \xrightarrow{\text{NaOH}} \langle\bigcirc\rangle-SO_2\overset{\overset{Na^{\oplus}}{..}}{\underset{..}{N}}R$$

Soluble in base

2° amine:

$$R_2NH + \left\langle \bigcirc \right\rangle - SO_2Cl \longrightarrow \left\langle \bigcirc \right\rangle - SO_2NR_2 \xrightarrow{\text{Ppt.}} \xrightarrow{\text{NaOH}} \text{N.R.}$$

Insoluble in base

3° amine:

$$R_3N + \left\langle \bigcirc \right\rangle - SO_2Cl \longrightarrow \text{N.R.}$$

amines are not. As a final confirmation, the 1° sulfonamide can be regenerated from basic solution by acidification:

$$\left\langle \bigcirc \right\rangle - SO_2\overset{\ominus}{N}R \overset{Na^{\oplus}}{} \xrightarrow{\text{HCl}} \left\langle \bigcirc \right\rangle - SO_2NHR\downarrow$$

---

**Exercise 13-7:** Show how the following isomeric compounds could be distinguished from one another by applying the Hinsberg test:

$$H_3C - \left\langle \bigcirc \right\rangle - NH_2 \qquad H_3C - \left\langle \bigcirc \right\rangle - NHCH_3 \qquad \left\langle \bigcirc \right\rangle - N(CH_3)_2$$
$$\qquad\qquad\quad CH_3$$

---

## Reactions with Nitrous Acid (HONO)

Nitrous acid reacts with 1°, 2°, and 3° amines, but quite differently in each case. This is the basis for an additional method of distinguishing between various types of amines. Nitrous acid is unstable at room temperature, and is usually generated *in situ* by reacting sodium nitrite with a mineral acid. Primary amines react to form *alkyl diazonium ions* which are thermally unstable. When they decompose, nitrogen gas is

$$Na^{\oplus}\overset{\ominus}{O}NO \xrightarrow{H^+X^-} H-O-N=O + Na^+X^-$$

Nitrous acid

eliminated, appearing as small bubbles in the test solution. The carbonium ion formed by the decomposition of the diazonium intermediate then reacts by a variety of mechanisms, yielding a mixture of olefins, alcohols, substitution products, and other, more complex products.

$$RNH_2 + HONO \longrightarrow \text{[Several intermediates]} \longrightarrow [R-N\equiv N^{\oplus}], X^{\ominus}$$

1° Amine                                                                    Alkyl diazonium ion

$$\text{Mixture of products} \xleftarrow{H_2O} [R^{\oplus}] + N_2\uparrow$$

Secondary amines form *N-nitrosoamines* rather than diazonium salts. These products are yellow oils which slowly separate from solution; no nitrogen gas is given off. N-nitrosamines have recently been recognized as being carcinogenic, and for this reason the use of sodium nitrite as a meat preservative has come under attack.

$$\begin{array}{ccc} R-NH + HONO & \longrightarrow & R-N-NO + H_2O \\ \quad | & & \quad | \\ \quad R' & & \quad R' \end{array}$$

2° Amine                          N-Nitrosoamine

Tertiary amines do not react directly with the nitroso group of HONO, but they do dissolve in the acidic solution forming a salt.

$$R_3N + HONO \longrightarrow [R_3\overset{\oplus}{N}H][ONO^{\ominus}]$$

In aqueous solution

Examples:

$$CH_3CH_2CH_2CH_2NH_2 \xrightarrow[\text{(HCl)}]{HONO} [CH_3CH_2CH_2CH_2N\equiv N^{\oplus}]Cl^{\ominus}$$

$$\begin{array}{l} CH_3CH_2CH_2CH_2OH \\ \quad + \\ CH_3CH_2CH=CH_2 \\ \quad + \\ \text{Other products} \end{array} \xleftarrow{H_2O} [CH_3CH_2CH_2CH_2^{\oplus}] + N_2\uparrow$$

$$\begin{array}{c} \qquad\qquad CH_3 \\ \qquad\qquad | \\ CH_3CH_2N: \xrightarrow{HONO} \left[ CH_3CH_2\overset{\quad CH_3 \atop |}{N}H \right]^{\oplus}, ONO^{\ominus} \\ \qquad | \\ \qquad\qquad CH_3 \qquad\qquad\qquad\quad CH_3 \end{array}$$

## Aromatic Diazonium Salts and Their Reactions

Aromatic amines behave similarly to aliphatic amines when treated with nitrous acid, with one important difference. Benzenediazonium halides are considerably more stable than their alkyl counterparts and may be employed as chemical reagents.

Benzenediazonium chloride
stable in solution at 0°

**Exercise 13-8:**  Can you explain why benzenediazonium halides are more stable than the corresponding alkyl diazonium salts? *Hint:* Consider how molecules stabilize charge.

Nitrogen can be displaced from aromatic diazonium salts by nucleophiles; this leads to a number of synthetic possibilities as outlined below. In this process, the nitrogen is eliminated as a gas. The use of cuprous halides as nucleophiles in these substitutions is known as the *Sandmeyer reaction*.

Reactions of Benzenediazonium Salts

Fluorobenzene is obtained by a careful heating of the fluoroborate salt. This process, in which both nitrogen and boron trifluoride are expelled as gases, is known as the *Schiemann reaction.* Another useful synthetic reaction of diazonium salts is reduction by hypophosphorous acid, $H_3PO_2$. This reaction replaces the $-N_2^+$ group with hydrogen. In essence, then, the amino group may be used to direct ring substitution, then be removed at a later stage, as illustrated below in the synthesis of 1,3-dibromobenzene.

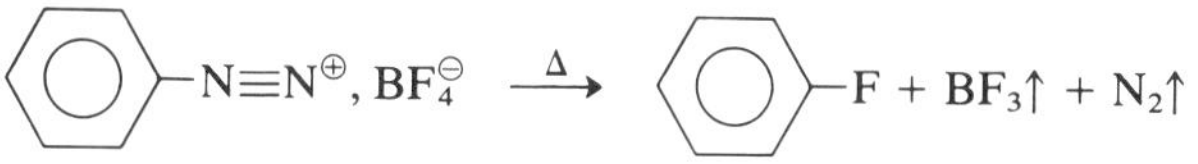

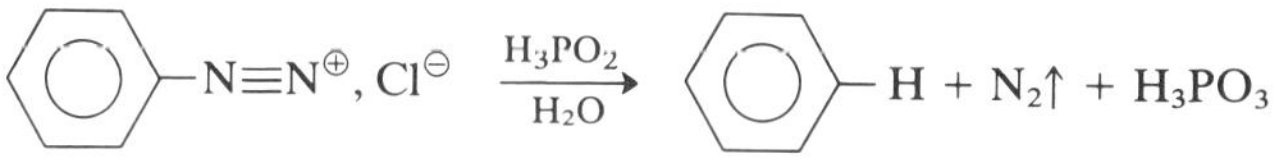

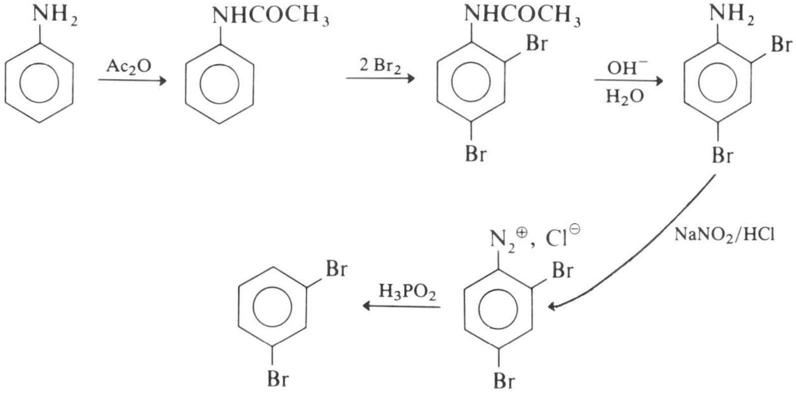

---

**Exercise 13-9:** Prepare each of the following compounds from either benzene or toluene using a route involving a diazonium salt:

(a)   Cl

(b)   OH

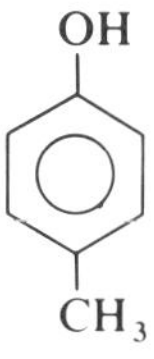

(c)   C≡N

---

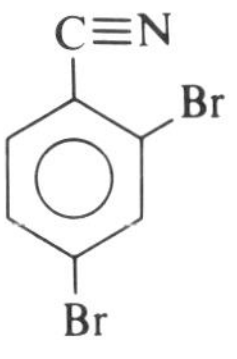

Although the substitution reactions of diazonium salts are important to the laboratory organic chemist, one of the most important industrial uses of these salts is the manufacture of *azo dyes.* Diazonium salts couple with aromatic rings containing activating groups (OH, OR, $NR_2$, etc.) usually in the para position. Azo compounds

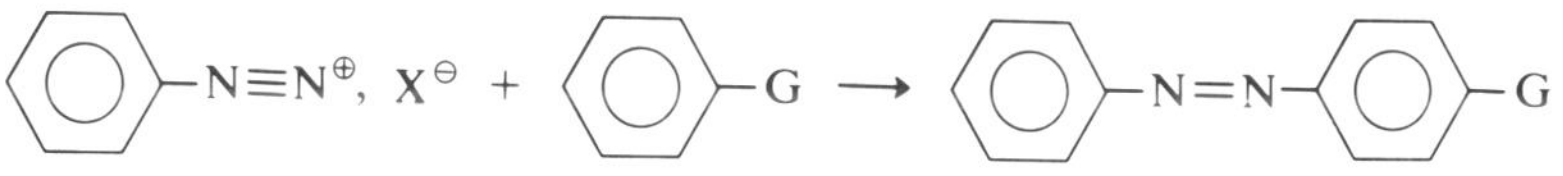

are highly colored and are used both as acid–base indicators and as dyes. Some of the
more familiar ones are shown below:

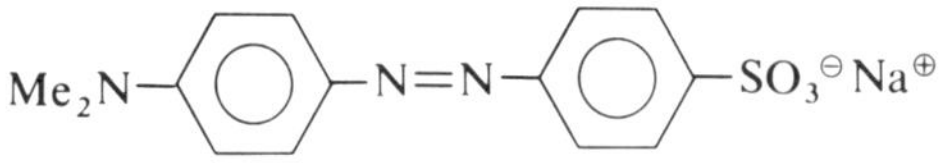

$$Me_2N-\phantom{}\!\!\bigcirc\!\!-N=N-\!\!\bigcirc\!\!-SO_3^{\ominus}Na^{\oplus}$$

Methyl orange—red in acid solution, yellow in base

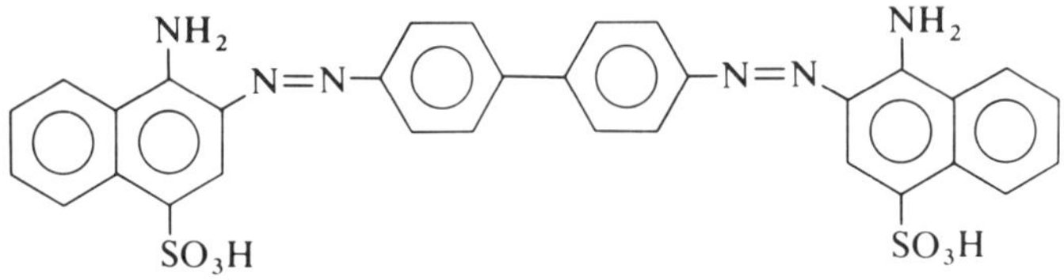

Congo red—blue in acid solution, red in base

---

**Exercise 13-10:**   Write structures for congo red as it exists in (a) acid solution and (b)
basic solution.

---

## Aromatic Substitution of Aniline

The amino group is highly activating toward electrophilic substitution, and is an
*o,p*-director.  Bromination, in fact, yields 2,4,6-tribromoaniline.

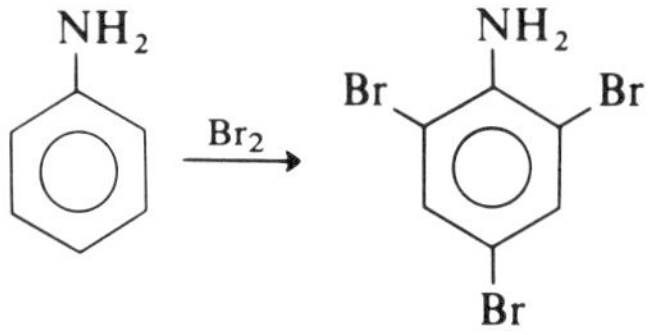

Reaction does not stop at
mono- or disubstitution

If the amino group is acylated, however, the activity decreases, and it is possible to
monobrominate the ring in the *para* position.  One can convert the *o,p*-directing

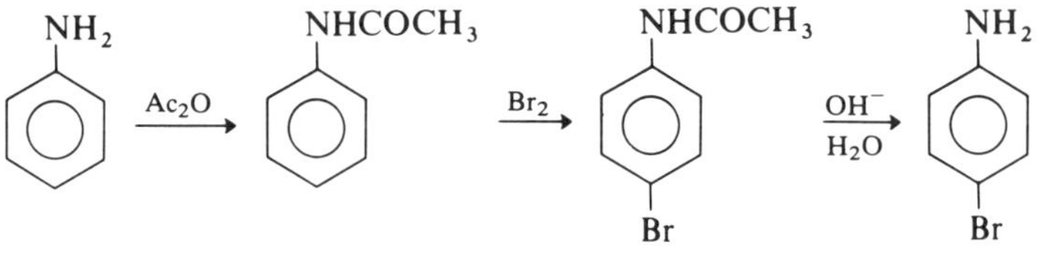

ability of an amino group to that of an *m*-director by converting the amino nitrogen
into a positively charged quaternary salt.

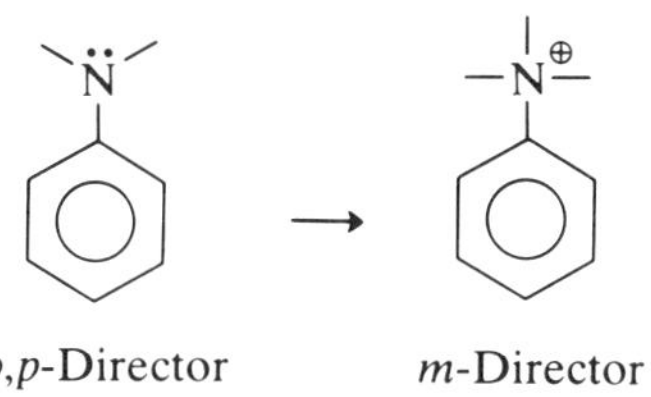
N
N
o,p-Director
m-Director

Sulfonation of aniline with sulfuric acid produces a salt, anilinium hydrogen sulfate. Strong heating of the salt (200°) produces sulfanilic acid. The last reaction may be thought of as an "internal" electrophilic substitution.

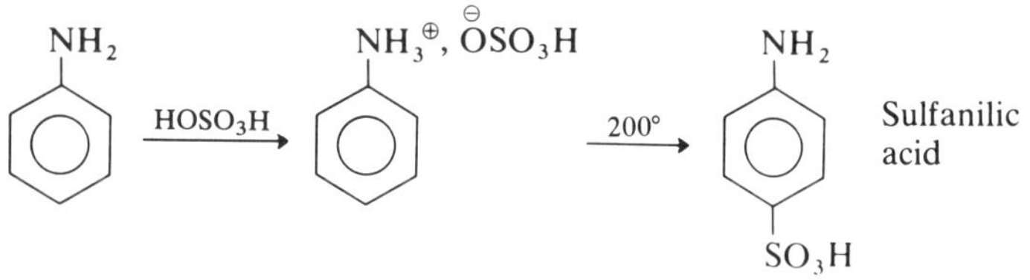
NH2
HOSO3H
NH3⊕, ⊖OSO3H
200°
NH2
SO3H
Sulfanilic
acid

**Exercise 13-11:** Comment on the relative solubilities of sulfanilic acid in (a) acid, (b) base, and (c) water.

## CYCLIC AMINES—HETEROCYCLIC RINGS CONTAINING NITROGEN

Rings containing atoms other than carbon are referred to as *heterocyclic* rings. The most common hetero atoms are N, O, and S. Nitrogen heterocycles are quite common in natural products, and the more important parent rings are shown below:

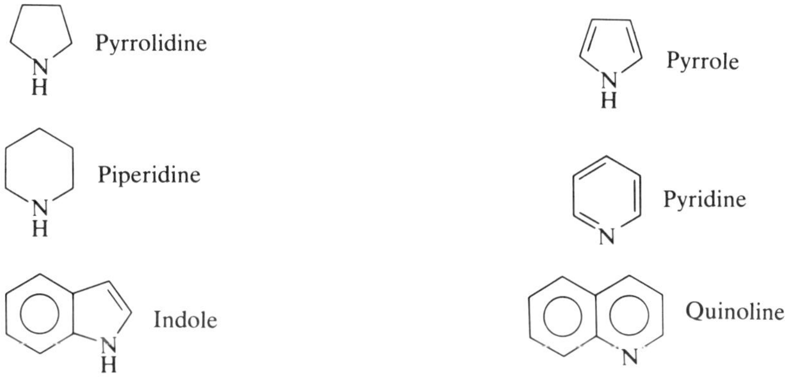
N
H
Pyrrolidine
N
H
Pyrrole
N
H
Piperidine
N
Pyridine
N
H
Indole
N
Quinoline

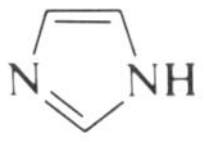

Imidazole

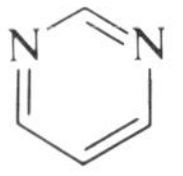

Pyrimidine

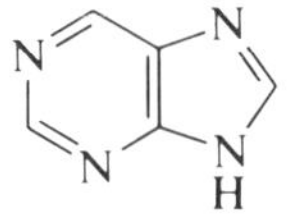

Purine

The importance of these ring systems in both plants and animals will be discussed in more detail in Chapter II-8.


## REVIEW OF NEW TERMS AND CONCEPTS

Define each term and, if possible, give an example of each:

Amine( 1°, 2°, and 3°)                     Ammonium salt
Amine "umbrella" inversion                 Amine alkylation
Imines                                     Reductive amination
Hofmann degradation                        Quaternary ammonium salt
Quaternary ammonium hydroxide              Hofmann elimination
Sulfonamide                                Hinsberg test
Diazonium salt                             Nitrosoamine
Sandmeyer reaction                         Schiemann reaction
Azo dyes                                   Nitrogen heterocycle


## PROBLEMS: CHAPTER I-13

**1.** Name the following compounds by the IUPAC system:

(a) $CH_3CH_2CH_2N(CH_3)_2$

(b) 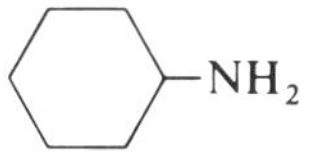 $-NH_2$

(c) $(CH_3)_4N^{\oplus}, Br^{\ominus}$

(d) $CH_3CHCH_2CH=CH_2$
       |
      $NMe_2$

(e)

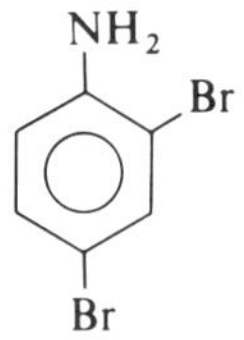

(f) 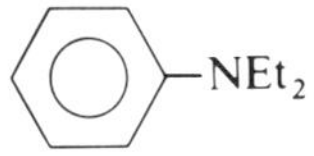 $-NEt_2$

**2.** For the following pairs of compounds choose the stronger base and explain your choice:

(a) $(CH_3CH_2)_2NH$   and   $CH_3NH_2$

(b)

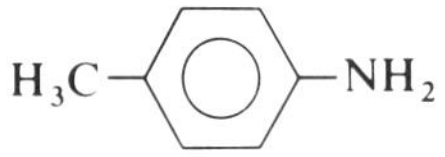

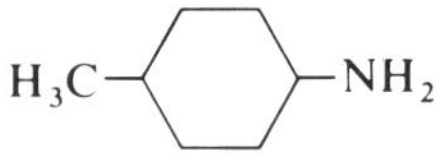

(c)

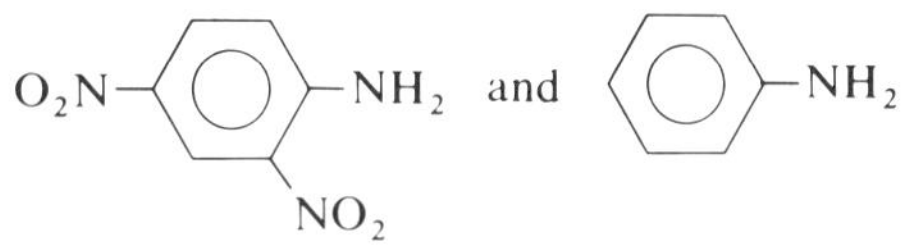

$$O_2N-\bigcirc-NH_2 \quad \text{and} \quad \bigcirc-NH_2$$
$$\qquad\quad \underset{NO_2}{|}$$

(d)

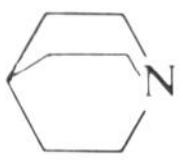

$$\bigcirc N \quad \text{and} \quad (CH_3CH_2)_3N$$

**3.** Draw Fisher projections for a pair of enantiomeric quaternary ammonium salts. Could we use an $R$, $S$ system to define absolute configuration for these structures?

**4.** Complete the following reactions showing only organic products:

(a) $(CH_3CH_2CH_2)_2NH \xrightarrow{H_2SO_4}$

(b)

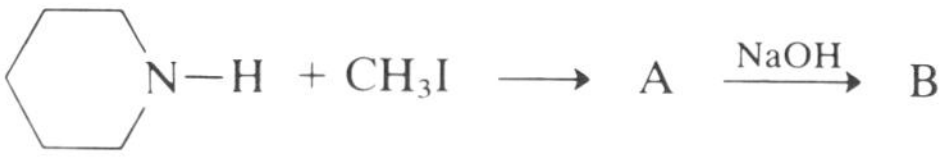

$$\bigcirc N-H + CH_3I \longrightarrow A \xrightarrow{NaOH} B$$

(c)

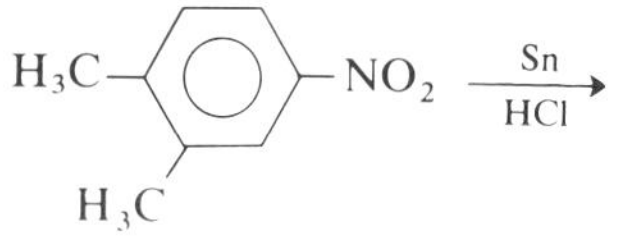

$$H_3C-\bigcirc-NO_2 \xrightarrow[HCl]{Sn}$$
$$\quad\ \underset{H_3C}{|}$$

(d)

$$\overset{O}{\overset{||}{CH_3CH_2CH_2CNHCH_3}} \xrightarrow[2.\ H_2O]{1.\ LAH}$$

(e)

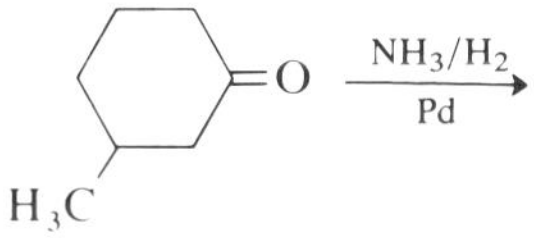

$$\bigcirc=O \xrightarrow[Pd]{NH_3/H_2}$$
$$\underset{H_3C}{|}$$

(f)

$$\bigcirc-CH_2CH_2CH_2\overset{O}{\overset{||}{C}}NH_2 \xrightarrow[NaOH]{Br_2}$$

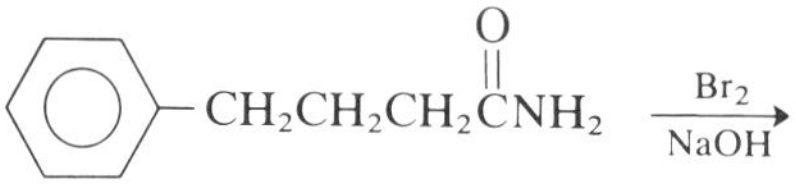

(g)

$$(CH_3)_2CHCH_2\overset{O}{\overset{||}{C}}Cl + (CH_3)_2NH \longrightarrow$$

(h)

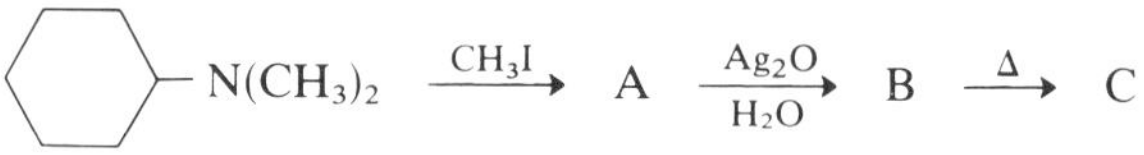

$$\bigcirc-N(CH_3)_2 \xrightarrow{CH_3I} A \xrightarrow[H_2O]{Ag_2O} B \xrightarrow{\Delta} C$$

(i) $CH_3CH_2CH_2CH_2NH_2 \xrightarrow{\phi SO_2Cl} A \xrightarrow{Aq.NaOH}$

(j)

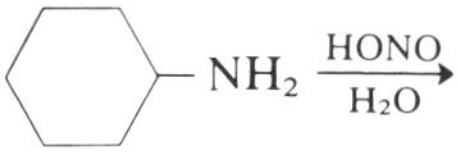

$$\bigcirc-NH_2 \xrightarrow[H_2O]{HONO}$$

(k)

$$H_3C-\bigcirc-NH_2 \xrightarrow{HONO} \xrightarrow{KCN}$$

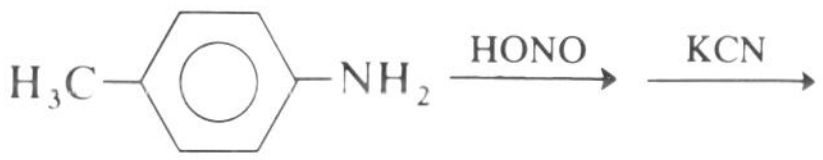

(l)

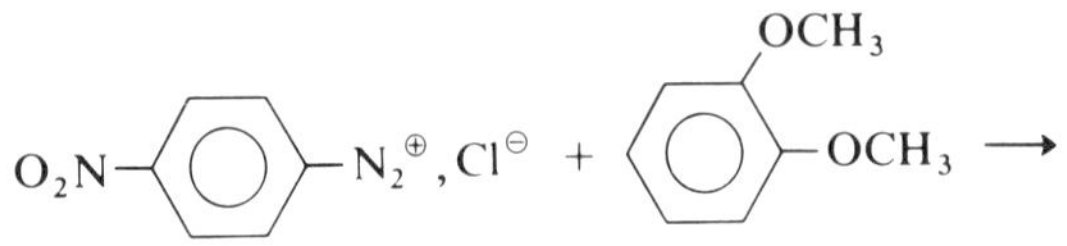

(m)

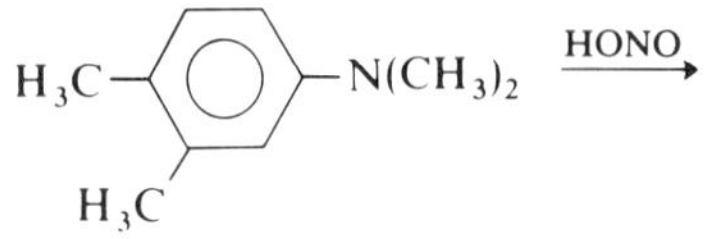

**5.** Although $—NMe_2$ is an *ortho,para* director, many common electrophilic substitution reactions in acid solution produce large amounts of *meta*-substituted product. How would you explain this seeming contradiction?

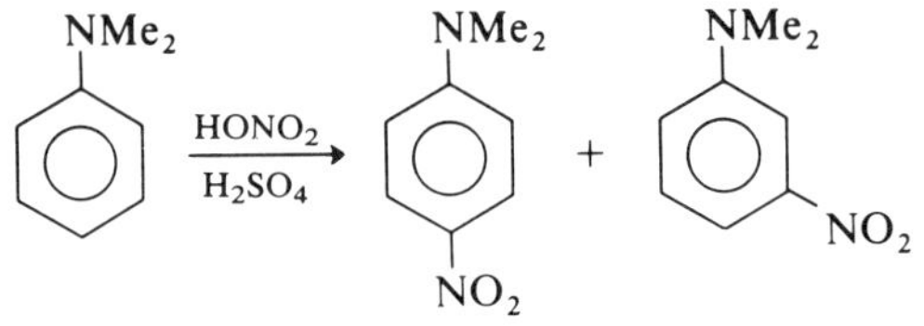

**6.** Most free amines found in nature are the end product of the decarboxylation of an $\alpha$-amino acid.

$$\underset{\displaystyle NH_2}{RCHCOOH} \xrightarrow{-CO_2} RCH_2NH_2 + CO_2$$

(a) What amines would you expect to be formed from the following amino acids?

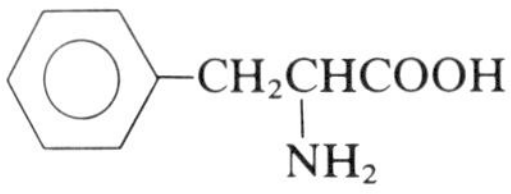

Phenylalanine

$HOCH_2CH(NH_2)COOH$

Serine

$H_2N(CH_2)_4CH(NH_2)COOH$

Lysine

(b) Isopentyl amine is the most widespread free amine found in nature, and histamine is found in stinging nettles. Suggest structures for their amino-acid precursors.

$(CH_3)_2CHCH_2CH_2NH_2$

*i*-Pentylamine

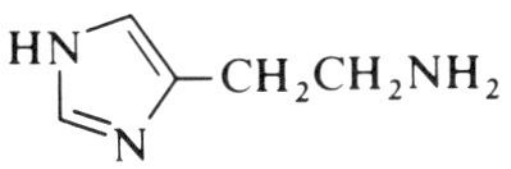

Histamine

**7.** Carry out the following conversions as shown:

(a) $(CH_3)_2CHCH_2COOH \longrightarrow (CH_3)_2CHCH_2NH_2$

(b)

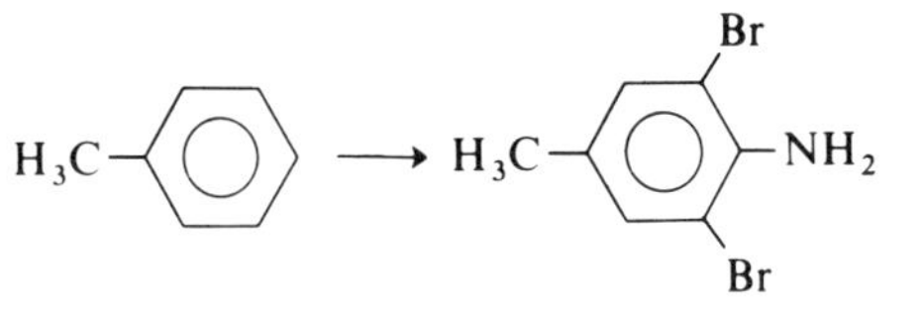

(c)

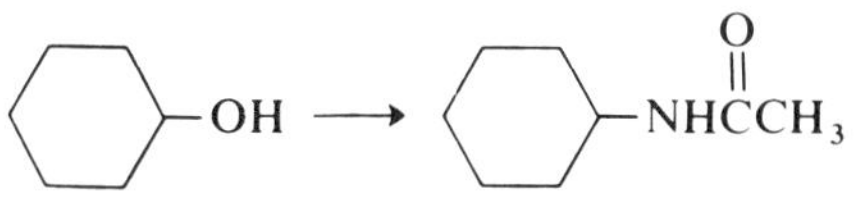

(d)

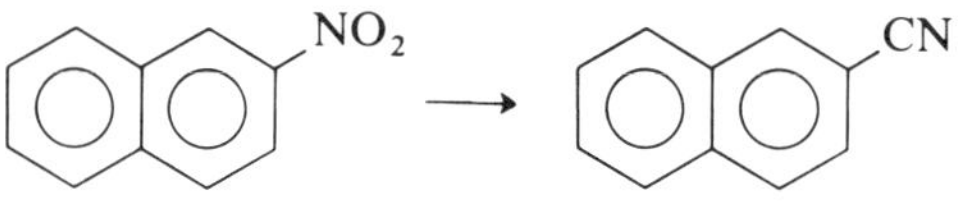

**8.** Suggest a method of making congo red from benzidine and $\alpha$-naphthyalamine:

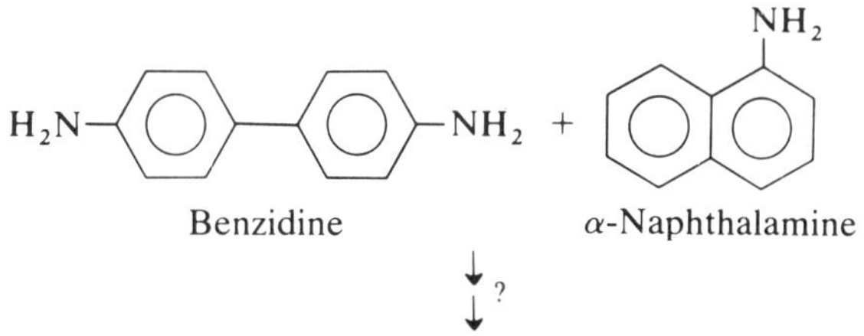

Benzidine            $\alpha$-Naphthalamine

↓ ?

Congo Red

**9.** How would you distinguish between the following pairs of compounds?

(a) $(CH_3CH_2CH_2CH_2)_2NH$   and   $(CH_3CH_2CH_2CH_2)_3N$

(b)

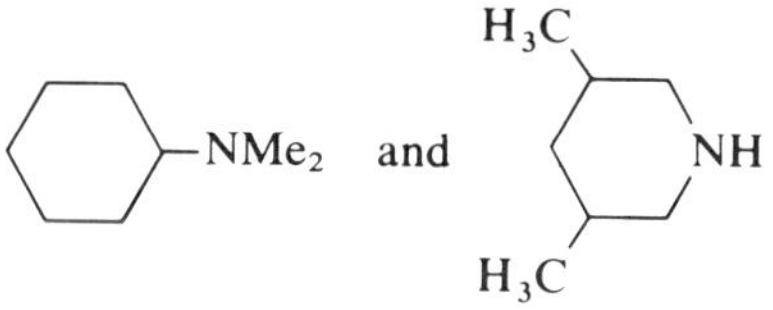

**10.** (a)  An unknown compound, A ($C_{10}H_{19}NO_2$), is soluble in dilute HCl, but does not react with benzenesulfonyl chloride. Refluxing in sodium hydroxide yields B ($C_3H_8O$), and C ($C_7H_{13}NO_2$). B is soluble in water, but reacts very slowly with Lucas reagent. C is soluble in both acid and base, and has a neutralization equivalent of 143. Compound A reacts with methyl iodide to give a salt, D ($C_{11}H_{22}NO_2I$). D gives a yellow precipitate with silver nitrate in aqueous solution. Heating the aqueous solution yields E ($C_{11}H_{21}NO_2$). Give structures for A, B, C, D, and E, and write out all the reaction sequences.

(b)  Give a synthesis for A starting from the compound given below and any carbon compounds containing 3 carbons or fewer.

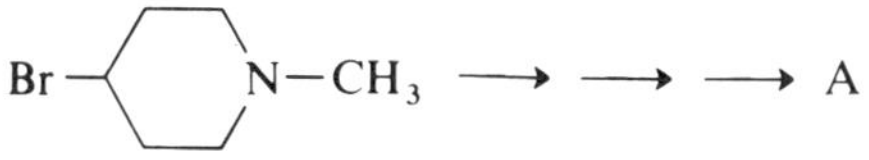

# 14

# Amino Acids
# and Proteins

Proteins and amino acids are essential to life. They are structural components of cell membranes; of cytoplasm; and of hair, nails, scales, feathers, horns, and hooves; as well as of muscle, cartilage, and tendon tissue. Functionally, they are found as enzymes, which facilitate almost all chemical reactions in living systems, including those involved in replication, biodegradation, energy production, and structural synthesis. Antibodies and some regulatory hormones are proteins as well. In this chapter we will examine how protein structure is dependent on chiral $\alpha$-amino acids. We will learn how proteins function in living systems and how protein structure is determined by the nature of the amide bond.

## THE STRUCTURE AND CONFORMATION
## OF AMINO ACIDS

Amino acids are simply carboxylic acids having at least one amino group in addition to the carboxylic acid functional group. However, $\alpha$-amino acids, in which the amino group is bonded to the $\alpha$-carbon, are by far the most important representatives of this class of compounds. If the residue is anything other than hydrogen, the carbon containing the amino group ($\alpha$-C) is chiral.

$$
\begin{array}{c}
\overset{\displaystyle H}{\underset{\displaystyle NH_2}{R-C-COOH}}
\end{array}
\quad \text{An } \alpha\text{-amino acid}
$$

H — α-Carbon
Residue — α-Amino group

The residues may be simple alkyl groups, or they may be quite complex side chains containing other functional groups (OH, SH, NH$_2$) or heterocyclic rings.

Most free amino acids, and amino acids bound in protein chains, belong to the *S* series. In the older literature, naturally occurring amino acids were defined as being L configuration, compared to L-glyceraldehyde.

$$
\begin{array}{c}
CHO \\
H_2N\!-\!\!\!-\!\!\!-\!H \\
R
\end{array}
\quad \text{equivalent to} \quad
\begin{array}{c}
CHO \\
HO\!-\!\!\!-\!\!\!-\!H \\
CH_2OH
\end{array}
$$

L-Amino acid          L-Glyceraldehyde

$$
\begin{array}{c}
COOH \\
H_2N\!\blacktriangleright\!C\!\blacktriangleleft\!H \\
R
\end{array}
\equiv
\begin{array}{c}
COOH \\
R \quad NH_2
\end{array}
\quad S \text{ Configuration}
$$

Absolute configuration of naturally occurring amino acids

Amino acids have both acidic and basic components. One would expect, then, that proton transfer, either intermolecular or intramolecular, would occur. This

$$
\underset{\text{Neutral form}}{\overset{\displaystyle R-CHCOOH}{\underset{\displaystyle NH_2}{\phantom{x}}}}
\xrightarrow[\substack{\text{molecular} \\ \text{H-transfer}}]{\text{inter- or intra-}}
\underset{\substack{\text{Dipolar or zwitterion} \\ \text{form}}}{\overset{\displaystyle R-CH-COO^{\ominus}}{\underset{\displaystyle _{\oplus}NH_3}{\phantom{x}}}}
$$

transfer converts the amino acid into an ionic form, containing both positive and negative charges, which is referred to as a *dipolar ion* or *zwitterion* (G: *zwitter* = double). The zwitterionic form of amino acids is thought to be more prevalent than the neutral form. This explains, for example, why amino acids are highly crystalline solids having high melting points, and also why they are more soluble in water than in organic solvents. One would expect amino acids to be soluble in both acid and base, since there will be an ionized functional group in either case. The actual form of an

$$
\underset{\substack{\text{Amino acid in} \\ \text{basic solution}}}{\overset{\displaystyle R-CH-COO^{\ominus}}{\underset{\displaystyle NH_2}{\phantom{x}}}}
\xleftarrow{OH^{\ominus}}
\underset{\text{Zwitterion}}{\overset{\displaystyle R-CHCOO^{\ominus}}{\underset{\displaystyle _{\oplus}NH_3}{\phantom{x}}}}
\xrightarrow{H^{\oplus}}
\underset{\substack{\text{Amino acid in} \\ \text{acidic solution}}}{\overset{\displaystyle R-CH-COOH}{\underset{\displaystyle _{\oplus}NH_3}{\phantom{x}}}}
$$

amino acid solution, then, will be dependent on the hydrogen ion concentration, or pH, and will be different for each acid, primarily because of the effect of the R residue. If we place an amino-acid solution in an electric field with both positive and negative electrodes, the acid will migrate to one or the other electrode, depending on its ionic charge and on the pH of the solution. If we vary the pH we will eventually find a proton concentration for which no migration occurs; at this pH the amino acid is in the zwitterion form, and that pH is referred to as the *isoelectric point*, for which each amino acid has its own characteristic value.

## COMMON NATURALLY OCCURRING AMINO ACIDS

Amino acids found in nature fall into several distinct categories. One can divide the family into several subclasses by examining the properties of the residues. Some of these residues contain OH or SH functional groups, while others contain heterocyclic rings, a second COOH group, etc. In Table 14-1, the acids have been listed according to whether the residues are neutral, acidic, or basic. Structural features of importance have also been noted.

## ESSENTIAL AMINO ACIDS

Many of the amino acids in Table 14-1 are referred to as *essential amino acids*; this means that these particular acids are essential for human nutrition and cannot be synthesized by the body. Therefore they must be contained in the food intake if good health is to be maintained. These essential acids are listed below.

| | |
|---|---|
| Valine | Methionine |
| Leucine | Threonine |
| Isoleucine | Lysine |
| Phenylalanine | Histidine |
| Tryptophan | Arginine |

About 1 to 2 grams of each of these acids is required daily and is usually obtained from protein sources.

## SYNTHESIS OF AMINO ACIDS

Although all the amino acids listed in Table 14-1 are available commercially, there are two general syntheses we should review at this time. Both utilize readily available starting materials and reactions we have already studied. The first involves ammonolysis of an $\alpha$-bromo carboxylic acid. The $\alpha$-bromoacids are usually obtained

$$\underset{\underset{Br}{\vert}}{RCHCOOH} \xrightarrow[NH_3]{Excess} \underset{\underset{NH_2}{\vert}}{RCHCOOH}$$

by bromination of carboxylic acids in the presence of phosphorus, a process known as the *Hell–Volhard–Zelinsky reaction.*

$$RCH_2COOH \xrightarrow[P]{Br_2} \underset{\underset{Br}{\vert}}{RCHCOOH}$$

The second amino acid synthesis is a variation of HCN addition to aldehydes. If the addition is carried out in the presence of ammonia, we obtain an aminonitrile instead of a hydroxynitrile. Acid-catalyzed hydrolysis of the intermediate yields the

$$\underset{}{\overset{\overset{\displaystyle O}{\parallel}}{R-C-H}} \xrightarrow[NH_3]{HCN} \underset{\underset{NH_2}{\vert}}{R-CHC\equiv N} \xrightarrow{H_3O^+} \underset{\underset{NH_2}{\vert}}{RCHCOOH}$$

desired $\alpha$-amino acid. The overall process is known as the *Strecker synthesis.* In both these amino acid syntheses, we are creating a potential chiral center, and the acids are obtained as racemic mixtures. In order to obtain the pure $R$ or $S$ acid, a resolution must be performed as described in Chapter I-4.

---

**Exercise 14-1:**  Carry out the following conversions:

(a)

(b) $(CH_3)_2CHCH_2OH \longrightarrow (CH_3)_2CHCH_2\underset{\underset{NH_2}{\vert}}{CH}COOH$

---

## ANALYSIS OF AMINO ACIDS

Amino acids have both amino ($NH_2$) and carbonyl groups (COOH), and much of their reaction chemistry is related to the individual reactions of these two functional groups. For example, in the previous chapter we learned that nitrous acid reacts with

Table 14-1. Naturally Occurring Amino Acids

| Amino acid | Abbreviation | Structure | Special features |
|---|---|---|---|
| Neutral Amino Acids | | | |
| Glycine | Gly | H—CH—COOH<br>⎮<br>$NH_2$ | Only nonchiral amino acid |
| Alanine | Ala | $CH_3$—CH—COOH<br>⎮<br>$NH_2$ | R = alkyl |
| Valine | Val | $(CH_3)_2CH$—CHCOOH<br>⎮<br>$NH_2$ | |
| Leucine | Leu | $(CH_3)_2CHCH_2$—CHCOOH<br>⎮<br>$NH_2$ | |
| Isoleucine | ILeu | $CH_3CH_2CH$—CHCOOH<br>⎮ ⎮<br>$CH_3$ $NH_2$ | |
| Serine | Ser | $HOCH_2$—CHCOOH<br>⎮<br>$NH_2$ | R contains OH group |
| Threonine | Thr | $CH_3CH$—CHCOOH<br>⎮ ⎮<br>OH $NH_2$ | |
| Cysteine | CySH | $HSCH_2$—CHCOOH<br>⎮<br>$NH_2$ | R contains $S$ functional group |
| Cystine | $(CyS)_2$ | $(SCH_2$—CHCOOH$)_2$<br>⎮<br>$NH_2$ | |
| Methionine | Met | $CH_3S(CH_2)_2$—CHCOOH<br>⎮<br>$NH_2$ | |
| Phenylalanine | Phe | $C_6H_5$—$CH_2CHCOOH$<br>⎮<br>$NH_2$ | R contains aromatic ring |
| Tyrosine | Tyr | HO—$C_6H_4$—$CH_2CHCOOH$<br>⎮<br>$NH_2$ | |
| Tryptophane | Try | (indole)—$CH_2CHCOOH$<br>⎮<br>$NH_2$ | |

Table 14-1 (*continued*)

| Amino acid | Abbreviation | Structure | Special features |
|---|---|---|---|
| **Neutral Amino Acids** | | | |
| Proline | Pro | | Cyclic Amino Acid; amino group part of ring |
| Hydroxyproline | HPro | | |
| Asparagine | Asn | $O$ $\parallel$ $H_2NCCH_2-CHCOOH$ $\mid$ $NH_2$ | R contains amide group |
| Glutamine | Gln | $O$ $\parallel$ $H_2NCCH_2CH_2-CHCOOH$ $\mid$ $NH_2$ | |
| **Acidic Amino Acids** | | | |
| Aspartic acid | Asp | $HOOCCH_2-CHCOOH$ $\mid$ $NH_2$ | R contains COOH group |
| Glutamic acid | Glu | $HOOCCH_2CH_2-CHCOOH$ $\mid$ $NH_2$ | |
| **Basic Amino Acids** | | | |
| Lysine | Lys | $H_2N(CH_2)_4-CHCOOH$ $\mid$ $NH_2$ | R contains amino group or ring |
| Arginine | Arg | $H_2N$ $\diagdown$ $C-NH(CH_2)_3CHCOOH$ $\diagup$ $\mid$ $HN$ $NH_2$ | |
| Histidine | His | $CH_2CHCOOH$ $\mid$ $NH_2$ | |

primary amines to yield nitrogen. This reaction can be carried out quantitatively for amino acids and proteins and is known as the *Van Slyke determination.*

$$\underset{\underset{NH_2}{|}}{RCHCOOH} \xrightarrow{HONO} \underset{\underset{OH}{|}}{RCHCOOH} + N_2\uparrow$$

Collect and measure

---

**Exercise 14-2:**   Are there any amino acids in Table 14-1 that would not release nitrogen when treated with HONO? If so, explain why they don't.

---

Another valuable reagent for the analysis of amino acids is ninhydrin. This reagent reacts with ammonia or any amine having a $-CH-NH_2$ structure, forming a purple complex. When proteins are hydrolyzed into their constituent amino acids,

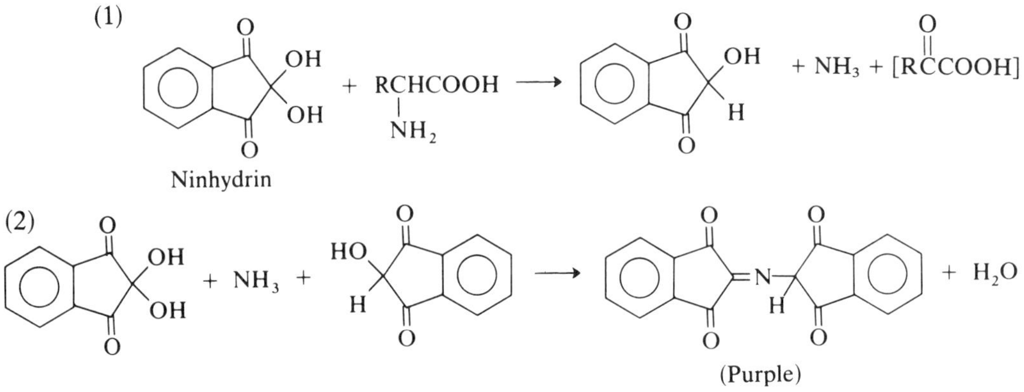

(1)

Ninhydrin

(2)

(Purple)

they may be separated from one another by various chromatography techniques (column, thin-layer, etc). Since the ninhydrin reaction is extremely sensitive (about $10^{-6}$ moles can be detected), it is often used to tell when an amino acid elutes from a column or where it might be on a thin-layer plate. Automatic amino-acid analyzers (available commercially) are capable of separating mixtures chromatographically and identifying each individual component.

## OTHER REACTIONS OF AMINO ACIDS

Most of the reactions commonly associated with carboxyl or amino group chemistry also apply to amino acids. For example, amino acids form esters, amides, and acyl halides.

$$RCHCOOH + R'OH \xrightarrow{\text{H}_+} RCHCOOR' + H_2O$$

with $NH_2$ groups on both left and right species.

$$RCHCOOH + SOCl_2 \longrightarrow \left[ RCHC\overset{O}{\overset{\|}{C}}-Cl \right] \xrightarrow[\text{reaction}]{\text{Further}}$$

with $NH_2$ groups.

$$RCHCOOH + NH_3 \xrightarrow{\Delta} RCHCONH_2$$

with $NH_2$ groups.

By far the most important reaction of amino acids, however, is the formation of peptides. A peptide bond is formed when the amino group of one amino acid forms an amide link to the carboxyl group of a second amino acid. If the two amino acids

$$RCHCOOH + H_2NCHCOOH \xrightarrow{\Delta}$$

NH$_2$                    R'
1st Amino acid    2nd Amino acid

$$R-CH-\begin{bmatrix} O \\ \| \\ C-NH \end{bmatrix}-CHCOOH$$
NH$_2$                              R'

Peptide link

$$R'CH-\begin{bmatrix} O \\ \| \\ C-NH \end{bmatrix}-CHCOOH$$
NH$_2$                              R

have different R residues, then two products are possible. In the next section, we will discuss this class of compounds in detail.

## PEPTIDES

Peptides are characterized by the number of amino acid residues in their structure. If two amino acids are joined by a peptide bond, the molecule is a *dipeptide*; if three amino acids are joined together, we have a *tripeptide*. If the peptide contains more than 10 amino acid residues, the molecule is referred to as a *polypeptide*. Proteins are large polypeptides which frequently contain more than a hundred acid units in various combinations of the common amino acids in Table 14-1. Each polypeptide, regardless of its size, has an amino group at one end (N-terminus) and a carboxyl group at the other (C-terminus). The structure may be represented by drawing complete structures, or by indicating each amino acid component by its abbreviation.

In all cases, the N-terminus is placed to the left, while thc C-terminus is on the right and the structure is read from left to right.

$$H_2NCH-\overset{\overset{\displaystyle O}{\|}}{C}-NHCH-\overset{\overset{\displaystyle O}{\|}}{C}-NH-CH-\overset{\overset{\displaystyle O}{\|}}{C}-OH$$

$$\quad\; R_1 \qquad\qquad R_2 \qquad\qquad R_3$$

N-terminus                                                        C-terminus

Amino acid     Amino acid                 Amino acid
   # 1              # 2                         # 3

Tripeptide containing three different amino acids

$$H_2NCH-\overset{\overset{\displaystyle O}{\|}}{C}-NH-CH_2\overset{\overset{\displaystyle O}{\|}}{C}-NH-CHCOOH$$

$$\quad\; CH_3 \qquad\qquad\qquad\qquad\qquad CH-CH_2CH_3$$

$$\qquad\qquad\qquad\qquad\qquad\qquad\qquad CH_3$$

Alanine       Glycine       Isoleucine

Alanylglycylisoleucine
Ala-Gly-Ileu

$$H_2NCH-\overset{\overset{\displaystyle O}{\|}}{C}-NHCH-\overset{\overset{\displaystyle O}{\|}}{C}-NH-CH-\overset{\overset{\displaystyle O}{\|}}{C}-NH-CHCOOH$$

$$\quad\; CH_3 \qquad\quad CH_3 \qquad\quad CH_2CH(CH_3)_2 \qquad CH_2OH$$

Alanylalanylleucylserine
Ala-Ala-Leu-Ser

The "backbone" of the polypeptide chain is formed by the $N-C-\overset{\overset{\displaystyle O}{\|}}{C}-$ repeating unit, while the chemical diversity may be attributed to the R group side chains.

Although most proteins are quite large, many simple polypeptides containing 3–10 amino-acid units play important biochemical roles in both plants and animals. *Glutathione*, L-glutamyl-L-cysteinylglycine, is found in yeast and is widely distributed in higher plants and animals as a reducing agent, in the biosynthesis of prostaglandins, for example. *Carnosine* and *anserine* are components of vertebrate

$$H_2NCH_2CH_2\overset{\overset{\displaystyle O}{\|}}{C}NHCHCOOH$$
$$\qquad\qquad\qquad CH_2$$

β-Alanyl-L-histidine
(carnosine)

$$H_2NCH_2CH_2\overset{\overset{\displaystyle O}{\|}}{C}NHCHCOOH$$
$$\qquad\qquad\qquad CH_2$$

β-Alanyl-1-methyl-L-histidine
(anserine)

muscle tissue. At this time, you should recall that most of the amino acids are chiral and that each center can be either $R$ or $S$ (D or L). In larger polypeptides, D acids are sometimes incorporated, even though the majority of naturally occurring amino acids are L, as in the antibacterial agents Gramicidin S and Bacitracin. It is interesting to note that both these polypeptides contain ornithine, an amino acid not found in animals or plants but only in bacteria.

```
        L-Orn                        CH₃
       /      \                       |
   L-Val       L-Leu                 CH₂
     |           |                    |
   L-Pro       D-Phe      CH₃—CH    S—CH₂   O
     |           |          |      /      \  ‖
   D-Phe       L-Pro    H₂N—CH— C      CH—C— L-Leu-D-Glu·L-Ile
     |           |               \    /                L-Leu
   L-Leu       L-Val              N        L-Hist·L-Asp·L-Cys·D-Asp
       \      /                                        D-Orn
        L-Orn                                          L-Ile
                                                       D-Phe
     Gramicidin-S
    (a cyclic polypeptide)                        Bacitracin
```

$$H_2NCH_2CH_2CH_2CHCOOH$$
$$|$$
$$NH_2$$

Ornithine (Orn)

---

**Exercise 14-3:**   (a) How many different polypeptides can you make from the following four amino acids: L-Ala, L-Hist, L-Pro, and L-Gly (each acid can appear only once). (b) How many polypeptides are possible if *one* of these acids may be D or L?

---

# PROTEIN STRUCTURE

## Primary Structure

There are many aspects of protein structure. Perhaps the most fundamental of the structural features is the amino acid sequence of the protein chain. This sequence is referred to as the *primary structure*. *Total hydrolysis* of a protein will reveal the number and relative amounts of amino acids, but it will not provide any information concerning the order in which they were bonded together. *Partial hydrolysis* yields fragments (oligopeptides) which still retain sequence information. When these fragments are separated chromatographically and identified, they provide pieces to the puzzle of the overall structure. For example, if the tetrapeptide Ala-Phe-Leu-Gly is partially hydrolyzed (Fig. 14-1), the dipeptide fraction will contain Ala-Phe, Phe-Leu, and Leu-Gly, but not Ala-Leu, Phe-Gly, or Ala-Gly. For simple poly-

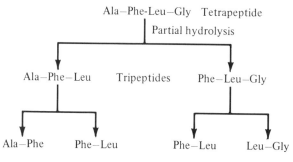

**Figure 14-1.** Partial hydrolysis of polypeptides yields identifiable fragments from which the original structure may be inferred.

peptides this information alone may enable the scientist to ascertain the peptide structure.

Partial hydrolysis is usually carried out enzymatically by *proteases*. These enzymes are specific in that they will only cleave peptide bonds between certain amino acids. Three common proteases and their specificities are outlined below:

*Pepsin:* Cleaves the peptide bond at the amino position of Phe, Try, and Tyr

*Trypsin:* Cleaves the peptide bond at the carbonyl position of Lys, Arg, and His

*Chymotrypsin:* Cleaves the peptide bond at the carbonyl position of Phe, Tyr and Try

Examples:

$$\text{Gly-Ala-Lys-Ala-Gly} \xrightarrow{\text{Trypsin}} \text{Gly-Ala-Lys} + \text{Ala-Gly}$$
$$\uparrow$$

$$\text{Gly-Ala-Phe-Ala-Gly} \xrightarrow{\text{Pepsin}} \text{Gly-Ala} + \text{Phe-Ala-Gly}$$
$$\uparrow$$

---

**Exercise 14-4:** In the above examples, why don't we also obtain the following splitting?

$$\text{Gly-Ala-Lys-Ala-Gly} \xrightarrow{\text{Trypsin}} \text{Gly-Ala} + \text{Lys-Ala-Gly}$$
$$\uparrow$$

$$\text{Gly-Ala-Phe-Ala-Gly} \xrightarrow{\text{Pepsin}} \text{Gly-Ala-Phe} + \text{Ala-Gly}$$
$$\uparrow$$

---

The C-terminus and N-terminus of individual polypeptides may be determined by techniques known as *end-group analysis.* The identity of the N-terminus acid may be identified by reaction of the peptide with 2,4-dinitrofluorobenzene (DNFB— Sanger's reagent).

$$O_2N-\langle \bigcirc \rangle-F + H_2NCH-\overset{\overset{\displaystyle O}{\|}}{C}-\boxed{\text{peptide}} \longrightarrow O_2N-\langle \bigcirc \rangle-NH-CH-\overset{\overset{\displaystyle O}{\|}}{C}-\boxed{\phantom{x}}$$

with NO$_2$ groups, R substituents, and "N-terminus acid" label

$$\downarrow \text{HCl}$$

$$O_2N-\langle \bigcirc \rangle-\underset{\underset{\displaystyle R}{|}}{NHCHCOOH} + \text{amino acids}$$

with NO$_2$

Yellow derivative

After reaction with Sanger's reagent, the peptide derivative is hydrolyzed in mineral acid and the yellow derivatives isolated and identified by comparison to known samples.

**Exercise 14-5:** When lysine is part of the polypeptide chain, N-terminus analysis is difficult to interpret. Why?

C-terminus amino acids may be identified by enzymatic cleavage with *carboxypeptidase.* This enzyme cleaves the peptide bond at the C-terminus, forming a new C-terminus and freeing the terminal amino acid, which can then be separated and identified.

$$\boxed{\text{peptide}}-\overset{\overset{\displaystyle O}{\|}}{C}NHCH\overset{\overset{\displaystyle O}{\|}}{C}OH \xrightarrow{\underset{\text{peptidase}}{\text{Carboxy-}}} \boxed{\phantom{xxx}}-\overset{\overset{\displaystyle O}{\|}}{C}-OH + H_2NCHCOOH$$

with R substituents

A similar enzymatic process may also be utilized for N-terminus analysis. *Aminopeptidase* cleaves the peptide bond, freeing the N-terminus acid and forming a new N-terminus. In both carboxypeptidase and aminopeptidase reactions, one must

$$H_2NCH\overset{\overset{\displaystyle O}{\|}}{C}-\boxed{\text{peptide}} \xrightarrow{\text{Aminopeptidase}} H_2NCH\overset{\overset{\displaystyle O}{\|}}{C}OH + H_2N-\boxed{\text{peptide}}$$

with R substituents

follow the release of amino acids as a function of time, as the reaction does not stop after the first cleavage. This is usually done by removing aliquots from the reaction at various times and analyzing the released amino acids by chromatographic techniques. Many large proteins such as ribonuclease (MW = 13,700), cytochrome C (12,400), lysozyme (14,400), papain (21,000), trypsin (24,000), and hemoglobin (65,000) have had their primary structures elucidated during the past two decades, and Frederick Sanger received the Nobel Prize in 1958 for determining the amino-acid sequence in beef insulin (5,734). The two-dimensional representation of the primary structure of lysozyme is shown in Fig. 14-2.

Primary structure of beef insulin

## Secondary Structure

Protein chains contain many polar functional groups. Although one might presuppose that these chains are free to assume any number of possible shapes, they actually exist in definite preferred conformations held together by intramolecular or intermolecular hydrogen bonds. The intramolecular H-bonds have severe

Hydrogen bond

geometric constraints and impose a limited number of possible configurations upon the protein chain, usually an $\alpha$-helix. A helix is inherently an asymmetric structure and both right-handed and left-handed helices are possible (Fig. 14-3). In the

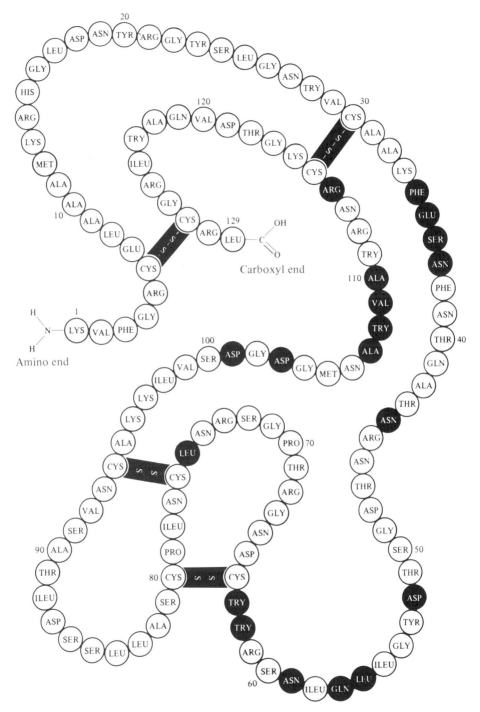

**Figure 14-2.** Two-dimensional representation of the primary structure of lysozyme. (*Redrawn from* David C. Phillips, "The Three-dimensional Structure of an Enzyme Molecule," *Scientific American*, November 1966, p. 79. Copyright ©1966 by Scientific American, Inc. All rights reserved.)

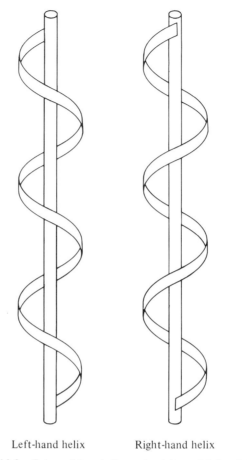

Left-hand helix          Right-hand helix

**Figure 14-3.** Polypeptide $\alpha$-helix can be right- or left-handed.

$\alpha$-helix, each carbonyl is hydrogen-bonded to an amide hydrogen which is three residues back in the chain. The helix has 3.6 amino-acid residues per turn, and the R groups extend outward from the helix backbone.

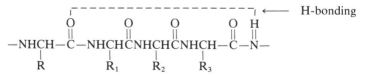

In the helix, H-bonding occurs between positions shown

Proteins can also exist in a second type of structure known as a *pleated sheet*. In this form, two proteins lie adjacent to one another and are held together by intermolecular hydrogen-bonding.

296

Peptide #1                    Peptide #2
Antiparallel (head-to-tail)
sheet

Parallel (head-to-head)
sheet

The sheets are pleated in order to reduce interaction between R groups, and this protein form is favored only when the R groups are rather small.

### Tertiary and Quaternary Structure

If all proteins were pure $\alpha$-helices, one would expect their overall gross structure to be rodlike. Although these fibrous proteins exist, most proteins are globular. This tertiary structure is the result of many factors in the overall protein makeup, several of which are outlined below:

1. If the protein chain contains proline, there is no N—H available for hydrogen-bonding. This disrupts the continuity of the helix.

2. When cysteine is present in the protein chain, disulfide bonds can occur which link two widely separated portions of the chain together in a "loop" (see, for example, the structure of insulin). Loop formation not only tends to disrupt the H-bonding pattern in the helix, but also enhances the tendency of the protein to be globular.

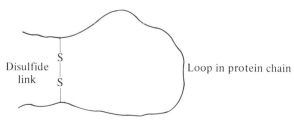

Disulfide
link                                    Loop in protein chain

3. There are electrostatic interactions between free —COOH and —NH$_2$ groups in the protein chains, forming —COO$^-$ or NH$_3^+$ and thus attracting those positions of the chain to one another.

4. In the R-group residues we may also have H-bonding between polar groups. Similarly, nonpolar R groups also prefer to orient themselves in proximity to one another (like attracts like).

All these factors compete with one another to determine the overall, or tertiary, protein structure.

Quaternary protein structure results when several proteins join together in some aggregate form. Several proteins may actually join together to form a specific unit; each individual participating protein is then referred to as a subunit. Hemoglobin contains four independent subunits, each having its own unique secondary and tertiary structure. Collogen has several proteins twisted together like a rope. Many enzymes are aggregations of several different kinds of proteins which yield a specific asymmetric reactive surface. When these aggregations are separated, the enzyme may lose its specificity. Thus quaternary structure, in its own right, can be responsible for quite specific biochemical transformations.

## PEPTIDE SYNTHESIS

The synthesis of complex polypeptides and naturally occurring proteins is an intriguing problem for the synthetic organic chemist. Amino acids are bifunctional molecules, and the main problem in peptide synthesis is being able to link the units together in a known, predictable fashion. Most ordinary methods of forming the amide bond (see Chapter I-12) yield mixtures of peptides when applied to syntheses with amino acids. For this reason, we must devise a strategy which allows for the reaction of one functional group at a time. This may be accomplished by using "protecting" or "blocking" groups.

---

**Exercise 14-6:**   Why will the following reactions yield mixtures of peptides instead of proceeding as written?

(a)

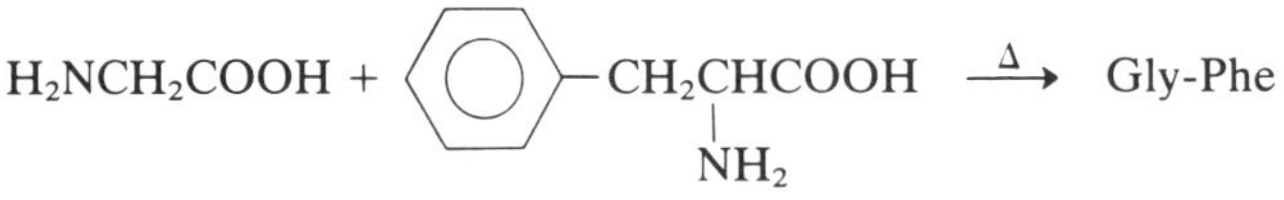

$$H_2NCH_2COOH + \text{(C}_6\text{H}_5\text{)}-CH_2CHCOOH \xrightarrow{\Delta} \text{Gly-Phe}$$
$$\qquad\qquad\qquad\qquad\qquad\qquad\quad | \atop NH_2$$

(b)

$$\overset{\displaystyle O}{\underset{\displaystyle NH_2}{CH_3CH\overset{\|}{C}Cl}} + H_2NCH_2COOH \xrightarrow{\Delta} \text{Ala-Gly}$$

---

One can protect or block either the amino or the carboxyl group. In practice, it is easier to block the amino group and operate on the carboxyl function. It is also desirable to convert the carboxyl function into a more reactive group, or in other words, to *activate* it. Then we can react this molecule with a second amino acid and obtain a dipeptide of *known* structure.

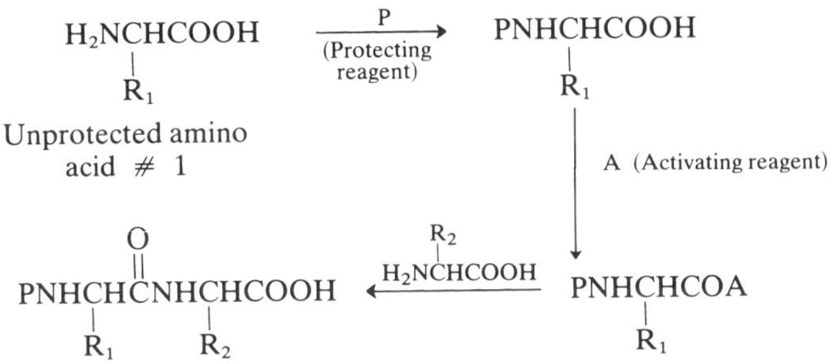


For a protecting group to be useful, it must be easily removed in order to obtain the free dipeptide. Some chemical reagents which have been utilized as protecting groups are shown below:

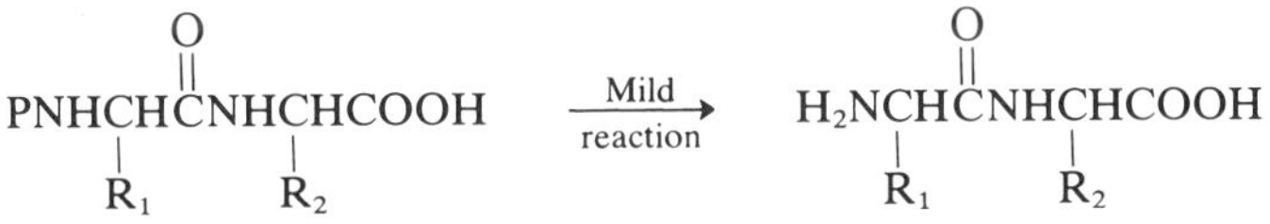


Finished dipeptide

Protecting Reagents:

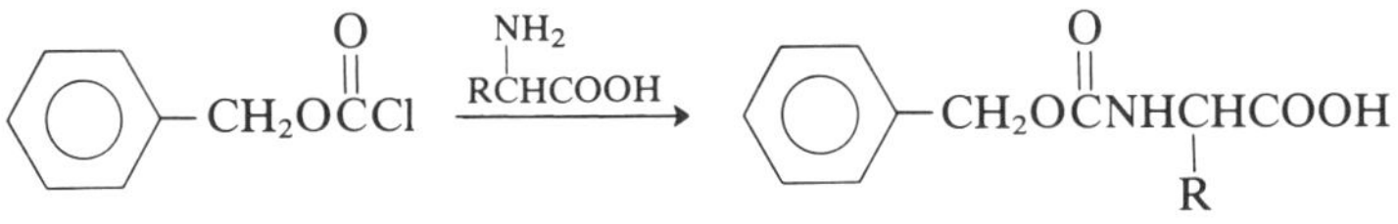


Carbobenzoxy chloride
CBO-Cl

CBO-amino acid

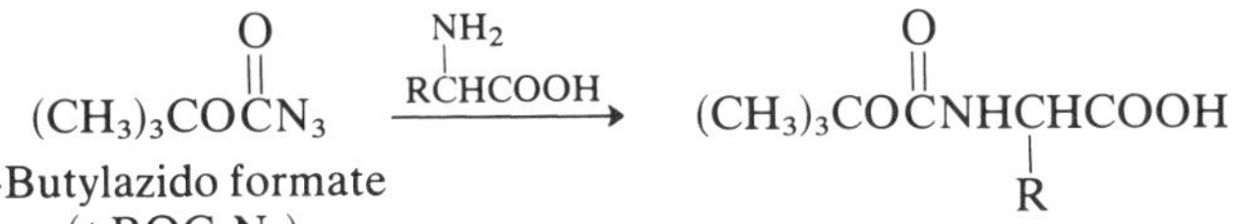


*t*-Butylazido formate
(*t*-BOC-N$_3$)

*t*-BOC-amino acid

Removal of Protecting Group:

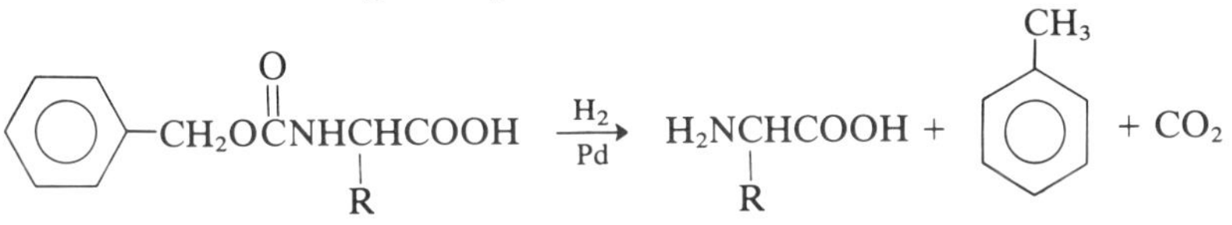

$$(CH_3)_3CO\overset{\overset{\displaystyle O}{\|}}{C}NHCHCOOH \xrightarrow{\quad CF_3COOH \quad} H_2NCHCOOH + (CH_3)_2C = CH_2 + CO_2$$
$$\qquad\qquad\qquad R \qquad\qquad\qquad\qquad R$$

Activating Groups:
The carbonyl group is usually converted to either an anhydride or ester to increase its reactivity.

$$PNHCH\overset{\overset{\displaystyle O}{\|}}{C}OH \xrightarrow{\quad O_2N-\text{\textcircled{}}-OH \quad} PNHCH\overset{\overset{\displaystyle O}{\|}}{C}-O-\text{\textcircled{}}-NO_2$$
$$\qquad R_1 \qquad\qquad\qquad\qquad\qquad R_1$$

$$\Big\downarrow \quad H_2N-CHCOOH$$
$$\qquad\qquad R_2$$

$$PNHCH\overset{\overset{\displaystyle O}{\|}}{C}-NHCHCOOH$$
$$\qquad R_1 \qquad\qquad R_2$$

**Exercise 14-7:**   Show how you would prepare the following tripeptide from the three amino acids and whatever protecting and activating reagents you desire:

Ala-Gly-Phe

## SOLID-PHASE PEPTIDE SYNTHESIS

Although the selective use of protecting and activating groups greatly simplifies peptide synthesis, isolation and purification of the peptide is time-consuming, particularly if your goal is a complex polypeptide with 50–100 amino-acid units. During the last decade Merrifield and co-workers at the Rockefeller Institute have developed a new synthetic method that eliminates these stepwise isolation and purification problems. In this new technique, an amino acid is bonded to a polymer

$$\boxed{\text{polymer}}-CH_2Cl + \overset{\ominus}{O}-\overset{\overset{\displaystyle O}{\|}}{C}-CHNHP \longrightarrow \boxed{\phantom{xxx}}-CH_2O\overset{\overset{\displaystyle O}{\|}}{C}CH-NHP$$
$$\qquad\qquad\qquad\qquad R_1 \qquad\qquad\qquad\qquad\qquad\qquad R_1$$

Amino acid bonded
to resin

support which contains reactive groups. The attachment is usually through the amino-acid carboxyl group by ester formation.

Once the amino acid is bonded to the resin, the protecting group may be removed, and a second amino acid reacted with the free amino group. The peptide

$$\boxed{\phantom{xxx}}-CH_2O\overset{\overset{\displaystyle O}{\|}}{C}\underset{\underset{\displaystyle R_1}{|}}{C}HNH_2 \;+\; A-\overset{\overset{\displaystyle O}{\|}}{C}-\underset{\underset{\displaystyle R_2}{|}}{C}HNHP$$

$$\downarrow$$

$$\boxed{\phantom{xxx}}-CH_2O\overset{\overset{\displaystyle O}{\|}}{C}\underset{\underset{\displaystyle R_1}{|}}{C}HNH\overset{\overset{\displaystyle O}{\|}}{C}\underset{\underset{\displaystyle R_2}{|}}{C}H-NHP \quad \text{Dipeptide bonded to resin}$$

chain may then be built by alternately removing the amino protecting group and reacting with a new amino acid. When the desired polypeptide chain has been constructed, the product may be removed from the resin by hydrolysis in $F_3CCOOH/HBr$. This procedure has now been automated, and the building of a

$$\boxed{\phantom{xxx}}-CH_2O\overset{\overset{\displaystyle O}{\|}}{C}\underset{\underset{\displaystyle R}{|}}{(CHNH)_n}\overset{\overset{\displaystyle O}{\|}}{C}\underset{\underset{\displaystyle R}{|}}{C}HNH_2 \quad \text{Bond polypeptide}$$

$$\downarrow \text{CF}_3\text{COOH/HBr}$$

$$\boxed{\phantom{xxx}}-CH_2OH \;+\; HOOC\underset{\underset{\displaystyle R}{|}}{(CH-NH)_n}\overset{\overset{\displaystyle O}{\|}}{C}-\underset{\underset{\displaystyle R}{|}}{C}H-NH_2 \quad \text{Free polypeptide}$$

polypeptide can be carefully programmed without isolating any intermediates. Several large naturally occurring proteins, such as ribonuclease (124 amino acid residues) and the human growth-factor hormone (188 residues), have been synthesized using this technique, and their physical and physiological properties have been shown to be similar to the natural substances.

**Exercise 14-8:**   Show all steps in the solid-phase synthesis of the following polypeptide from the component amino acids.

Ala-Phe-Ser-Phe

# PROTEIN METABOLISM

When proteins are digested, enzymes known as *proteases* break down the various polypeptide chains into small polypeptides and amino acids. These amino acids are then transported from the digestive system through the bloodstream to various parts of the body where protein *synthesis* is then carried out where needed. As pointed out earlier, the *essential* amino acids, which cannot be synthesized by the body, must be included in our protein intake in order for our internal protein synthesis to take place. The *nonessential* amino acids can be synthesized in the body from nonprotein sources. This constant intake and uptake requires that our bodies maintain an amino-acid reservoir in constant equilibrium with protein intake and use.

Proteins, via their constituent amino acids, also act as a nitrogen reservoir. They are normally broken down enzymatically to yield keto acids and ammonia which can then be utilized in other biosynthetic pathways. The resulting keto acids may then be utilized in the citric-acid cycle to produce energy in a manner similar to, but more efficient than, carbohydrates. Excess nitrogen is excreted in the form of urea.

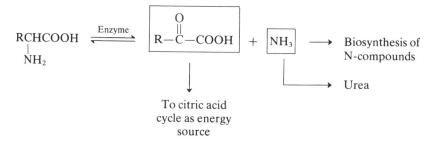

Finally, amino acids may serve as biosynthetic sources of other amino acids via keto acids and enzymes known as *transaminases*.

$$
\underset{\substack{\mid \\ NH_2 \\ \text{Amino acid 1}}}{R_1CHCOOH} + \underset{\text{Keto acid 2}}{R_2\overset{\overset{O}{\|}}{C}COOH} \xrightleftharpoons{\text{Enzyme}} \underset{\text{Keto acid 1}}{R_1\overset{\overset{O}{\|}}{C}COOH} + \underset{\substack{\mid \\ NH_2 \\ \text{Amino acid 2}}}{R_2CHCOOH}
$$

# REVIEW OF NEW TERMS AND CONCEPTS

Define each term and, if possible, give an example of each:

α-Amino acid                                   Neutral amino acids
Dipolar, or zwitterion                         Basic amino acids
Naturally occurring amino acids                Van Slyke determination

| | |
|---|---|
| Peptide bond | Acidic amino acids |
| Polypeptide | Strecker synthesis |
| Primary protein structure | Ninhydrin reaction (test) |
| Tertiary protein structure | Dipeptide |
| Protease | Protein |
| Sanger reagent | Secondary protein structure |
| Pleated-sheet structure | Quaternary protein structure |
| Activating group | End-group analysis |
| Transaminase reaction | α-Helix structure |
| Amino acid chirality | Blocking, or protective, group |
| Isoelectric point | Solid-phase synthesis |
| Essential amino acids | |

## PROBLEMS: CHAPTER I-14

1. Without consulting Table 14-1, write the structures for six common, naturally occurring α-amino acids.

2. Without consulting the chapter, write the structures for three essential amino acids.

3. For those amino acids you listed in (1) and (2), write the commonly accepted biochemical shorthand notation.

4. Determine whether the following acids are $R$ or $S$:

(a)

$$CH_3 \underset{COOH}{\overset{H}{|}} NH_2$$

(b)

$$HSCH_2 \underset{H}{\overset{NH_2}{|}} COOH$$

(c)

$$\bigcirc\!\!-CH_2 \underset{NH_2}{\overset{COOH}{|}} H$$

(d)

(indole)$\underset{H}{} \underset{H}{\overset{COOH}{|}} NH_2$

5. Which of the above acids have an L configuration?

6. Write the structure for phenylalanine present in each of the following solutions:

(a) acidic          (b) basic          (c) neutral

7. Carry out the following conversions using any reagents necessary:

(a) 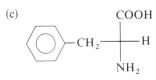 $\longrightarrow$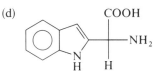

(b) $(CH_3)_2CHCH_2Br \longrightarrow (CH_3)_2CHCH_2\underset{\underset{NH_2}{|}}{C}HCOOH$

(c)

$HOOCCH_2CH_2COOH \longrightarrow H_2N\overset{\overset{O}{\|}}{C}CH_2\underset{\underset{NH_2}{|}}{C}HCOOH$

8. Devise a complete synthesis of the amino acid ornithine from any compounds containing three carbon atoms or fewer and any necessary inorganic reagents.

9. Draw the complete structure for glutathione, L-glutamyl-L-cysteinyl-glycine. Draw a structural isomer of glutathione.

10. Write complete equations for the reaction of phenylalanine with each of the following reagents

   (a) HONO           (b) acetyl chloride          (c) carbobenzoxychloride

11. The isolelectic point of phenylalanine is 6.1. For the following acids, predict whether you would expect their isoelectric points to be higher, lower, or about the same as alanine. Explain.

   (a) arginine          (b) aspartic acid          (c) valine

12. How would you expect the enzyme pepsin to react with gramicidin-S?

13. Show how pepsin, trypsin, and chymotrypsin would react with the following polypeptide:

   Gly-Tyr-Ala-Phe-Lys-Ala-Gly-Phe-Gly-His-Gly

14. How could the actions of carboxypeptidase and aminopeptidase differ in the hydrolysis of the following peptide:

   Ala-Phe-Phe-Ala-Gly-Gly

15. Draw the structure of a peptide containing two disulfide links.

PART **II**

# Core
# Enrichment

# Introduction to Reaction Mechanisms

When we write the equation for a typical reaction in organic chemistry, we are simply indicating that we know the reactants, products, and stoichiometry of the process.

$$\underset{\text{Reactants}}{A + B} \longrightarrow \underset{\text{Products}}{C + D} \qquad (1.1)$$
$$[?]$$

However, the equation tells us nothing concerning the "how" of the reaction process. Organic chemists refer to this "how" as the *reaction mechanism*. In this chapter we will explore methods of determining mechanisms of various organic transformations and explain why such topics as the nature of *intermediates*, the formation of *transition states*, and the study of *reaction kinetics* are vitally important to such determinations.

## ENERGY PROFILE DIAGRAMS AND THE REACTION COORDINATE

The primary goal of reaction mechanism studies is to obtain an intimate picture of the nature of *reaction intermediates* and *transition states*. In order to illustrate the essential differences between these two entities, *reaction profile diagrams* are quite

307

useful. Consider a simple one-step reaction between two molecules, A and B. If we were able to follow the reaction process in detail from beginning to end, plotting the total *potential* energy of the system as a function of the reaction progress, we could envision a plot such as Fig. 1-1. Both the energy and the reaction coordinate scales are dimensionally arbitrary; however, they do illustrate several points. The energy difference between the products and the reactants is represented by $\Delta H$, the enthalpy of reaction. The transition state, or activated complex $(AB)^{\ddagger}$, is represented as the arrangement of atoms at the point of *maximum* potential energy during the course of reaction. The activated complex is not a true chemical entity in that it has no finite existence; it is merely a state of maximum energy through which the reactants *must* pass before collapsing into the observed reaction products. The activation energy,

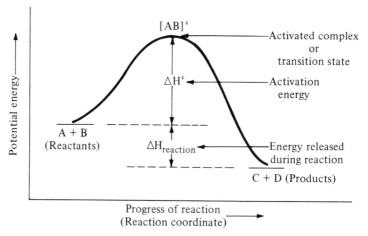

**Figure 1-1.** Energy profile diagram for a reaction proceeding through a simple transition state [AB]$^{\ddagger}$.

$\Delta H^{\ddagger}$, is a direct measure of the energy barrier that must be surpassed by the reactants in order to achieve chemical reaction. In its most elementary form, the activation energy may be envisioned as being derived from thermal energy, and is related to the amount of heating necessary to achieve a reasonable reaction rate.

Reactions proceeding through true chemical intermediates have quite different profile diagrams. Consider A and B reacting together to form an intermediate I, which subsequently forms products C and D.

$$A + B \longrightarrow [I] \longrightarrow C + D \tag{1.2}$$

A typical reaction profile diagram for this process is represented in Fig. 1-2. The intermediate I lies in a potential-energy well, and represents an energy minimum during the course of the reaction. As such, its existence is finite, and to a certain extent the depth of the well governs the lifetime of the intermediate and its relative concentration in the reaction medium. In some reactions, intermediates are stable

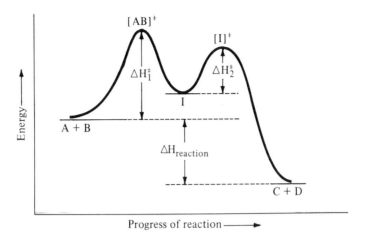

**Figure 1-2.** Energy profile diagram for a reaction proceeding through an intermediate.

enough to be isolated, and the reaction may be studied in two discrete stages. It should also be noted that there are *two* activated complexes involved whenever a true intermediate intervenes in the transformation of reactants to products.

**Exercise 1-1:** Draw a reaction profile diagram for a reaction which proceeds through two intermediates. How many transition states are involved in this type of reaction? How many separate activation steps are necessary? Which step in your scheme requires the *largest* activation energy?

## THE RATE-DETERMINING STEP

In any organic reaction proceeding through a number of consecutive steps and intermediates, the slowest, or *rate-determining*, step dominates all mechanistic considerations.

$$A + B \xrightarrow{\text{Slow}} [I] \xrightarrow{\text{Fast}} \text{Products} \qquad (1.3)$$

$$A + B \xrightarrow{\text{Fast}} [I] \xrightarrow{\text{Slow}} \text{Products} \qquad (1.4)$$

For example, in Eq. (1.3) the overall rate of the reaction is dependent on the rate of formation of [I], while in Eq. (1.4) product formation is limited by the rate of decomposition of [I]. In the first process, the concentration of the intermediate, I, may be so low that detection becomes a problem, while in the latter reaction the

309

concentration of I can become large enough to allow isolation and identification. Therefore, the nature of any organic reaction depends on the *relative rates* of the various steps in the overall scheme, and the task of determining which of these steps is rate determining is one of prime importance.

## REACTION KINETICS

The rate of chemical reaction is proportional to the concentration of those species directly involved in the rate-determining step. In Eq. (1.3), for example, the rate of disappearance of one of the reactants, A, would depend on the concentrations of both A and B. This relationship can be expressed in several ways. Since a reaction is normally followed by determining the change in concentration of one of the reactants as a function of time, we can write:

$$\frac{A_2 - A_1}{t_2 - t_1} = \frac{\Delta A}{\Delta t} \propto [A] \text{ and } [B] \tag{1.5}$$

where $A_2$ = quantity of A at time 2 $(t_2)$
$A_1$ = quantity of A at time 1 $(t_1)$.

$$-\frac{\Delta A}{\Delta t} = \text{rate of disappearance of A} \tag{1.6}$$

$$-\frac{d A}{dt} = \text{rate of disappearance of A} \atop \text{(calculus notation)} \tag{1.7}$$

$$-\frac{\Delta A}{\Delta t} \approx -\frac{d A}{dt} = k[A][B] \tag{1.8}$$

One can read the last expression, Eq. (1.8), as follows: "the rate of disappearance of A with respect to time is equal to the product of the proportionality constant, $k$, and the concentration of the reactants A and B." The proportionality constant, $k$, is referred to as the *rate constant* of the reaction, while the equation as a whole is termed the *rate expression*, or *rate law*.

The number of concentration terms that appear in the rate expression is referred to as the *reaction order*. For a general reaction involving A and B in the rate-determining step, the reaction order is the sum total of all exponents associated with each concentration term [Eqs. (1.9)–(1.10)].

$$n A + m B \xrightarrow{\text{Slow}} \text{Products}$$

$$-\frac{d A}{dt} = k[A]^n[B]^m \tag{1.9}$$

$$\boxed{\text{reaction order} = n + m} \tag{1.10}$$

By far the vast majority of organic reactions are either first- or second-order.

---

**Exercise 1-2:** Determine the kinetic order with respect to each reactant and also the *total* kinetic order for the reaction in each of the following:

(a) $-\dfrac{dA}{dt} = k[A][B]$

(b) $-\dfrac{dA}{dt} = k[A]^2$

(c) $-\dfrac{dA}{dt} = k[A][B]^2$

(d) $-\dfrac{dA}{dt} = k[A][B][H_3O^+]$

---

Reaction order does not tell us how many molecules are actually reacting at the molecular level in the rate-determining step, or in other words, the *molecularity*. Unimolecular reactions involve only one molecule, while bimolecular reactions involve two molecules in the rate-determining collision. However, we do not have any reasonable method of determining reaction molecularity, and thus tend to equate the kinetic order with the number of molecules actually reacting. It *is* true that all bimolecular reactions are second order, but the converse is not true. However, in the absence of any negative evidence, the experimental rate expression and total reaction orders give a good indication of the nature and number of reactants in the rate-determining step.

Let us consider a simple first-order reaction. In order to demonstrate the reaction order experimentally, the rate of disappearance of A is followed as a function of time. This is usually accomplished by removing small samples (aliquots) of reaction mixture maintained at constant temperature, at certain time intervals, quenching the reaction, and analyzing the mixture to determine the percentage of reactants and products. Mathematical solution of rate expressions tells us *how* the concentration of reactants should vary with time, and the process is outlined in Eqs. (1.11)–(1.14) for a typical first-order process. If we plot ln $A_0/A$ vs. $t$ for a first-order reaction, we should obtain a straight line whose slope is the rate constant, $k$ (Fig. 1-3). The mathematical manipulations involved for second-order reactions are similar, but more complicated and beyond the scope of this discussion.

$$-\frac{dA}{dt} = k[A] \qquad \text{First order in A} \qquad (1.11)$$

rearranging

$$-\frac{dA}{A} = k\,dt \qquad \text{to separate variables } dA \text{ and } dt \qquad (1.12)$$

and integrating

$$\ln A_0 - \ln A = kt \qquad (1.13)$$

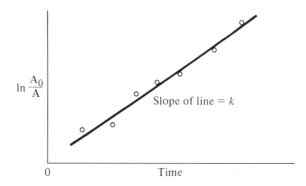

**Figure 1-3.** Plot of a typical first-order reaction A $\xrightarrow{k}$ product.

where   A = concentration of A at any time $t$
        $A_0$ = concentration of A at $t = 0$ (beginning)

and finally

$$\ln \frac{A_0}{A} = kt \qquad (1.14)$$

**Exercise 1-3:** (a) A useful concept used in the study of first-order reactions is the half-life, defined as the time necessary for exactly one-half of the original material to be converted to products. Prove that $t_{\frac{1}{2}} = \ln 2/k$ for any reaction. How many half-lives must pass before only $\frac{1}{16}$ of the original material remains? Does the half-life of a substance double if we double the original amount of material? (b) The half-life of $^{14}C$ is 5,700 years. If we assume that no new $^{14}C$ can be incorporated into wood once a tree is cut down, how much of the original $^{14}C$ would remain in the wood used to make a Roman chariot (100 A.D.)?

## THE EFFECT OF TEMPERATURE ON REACTION RATES

In any macroscopic sample containing a large number of molecules, a distribution of energy states exists (Maxwell–Boltzmann distribution). In Fig. 1-4 the distribution is shown for a "typical" sample. For any particular energy at a given temperature, the fraction of molecules having a particular energy $E_i$ is given by an exponential relationship, Eq. (1.15), where $N_i$ is the number of molecules having $E = E_i$, $N_0$ is the total number of molecules in the sample, and $k$ is Boltzmann's constant. As the temperature increases, two observations are generally true: (1) The sample has a

312

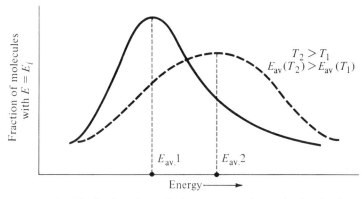

**Figure 1-4.** Distribution of energies in a macroscopic sample of molecules.

greater "spread" of energies, and (2) the average energy of the sample increases.

$$N_i = N_0 \, e^{-E_i/\mathscr{k}T} \tag{1.15}$$

$$\mathscr{k} = 1.3805 \times 10^{-16} \, \text{erg} \, {}^\circ\text{K}^{-1} \, \text{molecule}^{-1} \tag{1.16}$$

We have previously defined the reaction activation energy as the energy necessary to cause chemical reaction to occur when two molecules chemically react with one another. Thus the *rate* of reaction is increased by a rise in temperature, because of the increase in the number of molecules having $E_i \geqslant E_a$. The rate constant, $k$, is related to the activation energy by a similar exponential relationship, where A is a proportionality constant. Activation energies for a typical organic

$$k = \text{A} \, e^{-E_a/RT} \tag{1.17}$$

$$\ln k = \ln \text{A} - \frac{E_a}{RT} \tag{1.18}$$

$$[T \text{ in } {}^\circ\text{K} \, (273.2 + {}^\circ\text{C}); \, R = 1.98 \, \text{cal/mole} \, {}^\circ\text{K}]$$

reaction may be obtained then, by plotting the logarithms of a number of rate constants obtained at several different temperatures as a function of $1/T$ (Fig. 1-5).

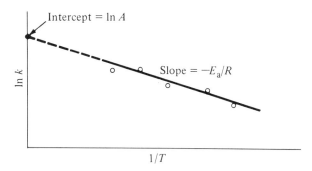

**Figure 1-5.** Arrhenius plot to determine activation energies.

313

Although this simple approach involves several assumptions, $E_a$'s obtained in this manner are approximately equal to $\Delta H^{\ddagger}$ from the mathematically more complicated transition theory. In essence, the activation energy is a measure of the energy necessary to break, bend, or stretch the bonds of the species involved in the rate-determining step. Thus, accurate measurements of $E_a$ can eliminate several possibilities in determining what is happening in the formation of transition states or intermediates.

## CATALYSIS

There are many reactions in organic chemistry that do not proceed at reasonable rates under conditions normally obtainable in the laboratory. These processes are dominated by high activation energies which cannot be overcome by simply raising the temperature. *Catalysts* accelerate the rates of these reactions by *lowering* the activation energy of the process, usually by converting one of the reactants to a *more reactive* intermediate, thus changing the pathway of the reaction to one of lower energy. A typical reaction profile diagram for this process is shown in Fig. 1-6. The most common catalysts in organic chemistry are acids and bases, while enzymes catalyze most biochemical reactions. Transition metals and their compounds are useful catalysts in chemical industry, particularly in catalytic cracking and reforming processes.

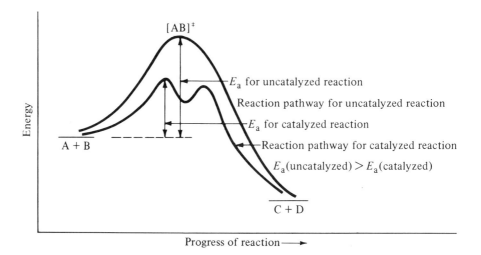

**Figure 1-6.** Comparison of reaction profile diagrams for catalyzed and uncatalyzed reactions.

A typical example of an acid-catalyzed reaction is the conversion of an alcohol to an olefin, Eq. (1.19), proceeding through an intermediary oxonium ion. The

$$
\underset{\underset{\text{RCH}_2\overset{\displaystyle |}{\text{C}}\text{HR}}{\overset{\text{OH}}{}}}{} \quad \xrightarrow[\Delta]{\text{H}_3\text{O}^+} \quad \text{RCH}{=}\text{CHR} + \text{H}_2\text{O} \tag{1.19}
$$

rate-determining step is the loss of water from the oxonium ion, and the overall rate of disappearance of the original alcohol can be written:

$$
\overset{\text{OH}}{\underset{|}{\text{RCH}_2\text{CHR}}} + \text{H}^+ \quad \underset{\text{Fast}}{\overset{K_{\text{eq.}}}{\rightleftharpoons}} \quad \overset{{}^+\text{OH}_2}{\underset{|}{\text{RCH}_2\text{CHR}}} \tag{1.20}
$$

$$
\overset{{}^+\text{OH}_2}{\underset{|}{\text{RCH}_2\text{CHR}}} \quad \xrightarrow[(-\text{H}_2\text{O})]{\text{Slow}} \quad \text{RCH}_2\overset{+}{\text{C}}\text{HR} + \text{H}_2\text{O} \tag{1.21}
$$

$$
\text{RCH}_2\overset{+}{\text{C}}\text{HR} \quad \xrightarrow[(-\text{H}^+)]{\text{Fast}} \quad \text{RCH}{=}\text{CHR} \tag{1.22}
$$

$$
-\frac{d(\overset{\text{OH}}{\underset{|}{\text{RCH}_2\text{CHR}}})}{dt} = k[\overset{{}^+\text{OH}_2}{\underset{|}{\text{RCH}_2\text{CHR}}}] \tag{1.23}
$$

The concentration of the intermediary oxonium ion is difficult to determine, however, since the concentration of the intermediate is determined by the position of the original equilibrium involving the catalyst $\text{H}^+$, the concentration may be expressed in terms of the original reactants. The catalyst concentration does appear in the overall kinetic expressions even though it does not participate directly in the rate-determining step and is not consumed in the reaction.

$$
K_{\text{eq.}} = \frac{[\overset{{}^+\text{OH}_2}{\underset{|}{\text{RCH}_2\text{CHR}}}]}{[\overset{\text{OH}}{\underset{|}{\text{RCH}_2\text{CHR}}}][\text{H}^+]} \tag{1.24}
$$

$$
[\overset{{}^+\text{OH}_2}{\underset{|}{\text{RCH}_2\text{CHR}}}] = K_{\text{eq.}}[\overset{\text{OH}}{\underset{|}{\text{RCH}_2\text{CHR}}}][\text{H}^+] \tag{1.25}
$$

$$
-\frac{d[\overset{\text{OH}}{\underset{|}{\text{RCH}_2\text{CHR}}}]}{dt} = \underbrace{k\,K_{\text{eq.}}}_{k_{\text{experimental}}}[\overset{\text{OH}}{\underset{|}{\text{RCH}_2\text{CHR}}}][\text{H}^+] \tag{1.26}
$$

**Exercise 1-4:**    Consider a base-catalyzed reaction proceeding through an intermediary

$$RH + B^- \rightleftharpoons R^- + BH$$

$$R^- + X \xrightarrow{\text{Slow}} \text{Products}$$

anion (by proton removal).  Show that the rate law for this reaction is of the form

$$-\frac{d[RH]}{dt} = k[RH][X][B] \quad \text{where } B^- \text{ is the solvent anion and BH the solvent.}$$

Is this an example of a trimolecular reaction?

## SOLVENT EFFECTS ON REACTION RATES

The role of the solvent in organic chemistry is complicated by the fact that we have such a variety of solvents available, ranging from nonpolar hydrocarbons to solvents of high polarity, both *protic* (having active hydrogen) and *aprotic* (no active hydrogens).  The study of elementary inorganic reactions is limited almost solely to aqueous solutions, but in organic chemistry we essentially have no such limitations.  The most intriguing question concerning the role played by the solvent in a typical organic reaction is the relative solvation energies of the reactants and of the transition state, or intermediate.  In many organic reactions, for example, neutral reactants are converted into charged intermediates.  Thus we would expect solvents

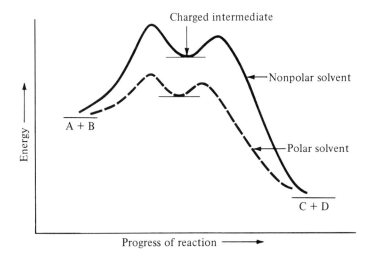

**Figure 1-7.**  Comparison of reaction profile diagrams for reactions proceeding through charged intermediates in solvents of differing polarity.

of high polarity to be more effective in promoting this transformation because of their greater ability to solvate the charged intermediate or transition state and thus lower the activation energy. If one were to study this type of reaction in a series of solvents of differing polarity, one would expect the rate of reaction to increase as the solvent polarity increased; this is in fact what is observed. The rationale behind this observation can be envisioned by looking at a reaction profile diagram for a reaction proceeding through charged intermediates (Fig. 1-7). Two factors are involved in determining the effectiveness of solvation—the relative solvation energies of the reactants and intermediates and the insulating ability of the solvent with respect to the solvated ions. This latter property can be approximated by the *dielectric constant* (D) of the solvent. In general, the higher the dielectric constant, the better the solvent will be for reactions requiring solvation of an ionic transition state. Some typical solvents and their dielectric constants are listed in Table 1-1.

Table 1-1.   Boiling Points and Dielectric Constants for Typical Organic Solvents.

| Solvent | B.p. (°C) | D (T, °C) | Solvent | B.p. (°C) | D (T, °C) |
|---------|-----------|-----------|---------|-----------|-----------|
| $H_2O$ | 100 | 78 (25) | MeCOMe | 56 | 21 (25) |
| $Me_2SO$ | 189 | 47 (25) | $C_5H_5N$ | 116 | 12 (25) |
| $MeNO_2$ | 101 | 36 (30) | $C_6H_6$ | 80 | 3.3 (25) |
| $MeC{\equiv}N$ | 82 | 36 (25) | $CCl_4$ | 77 | 2.2 (25) |
| MeOH | 65 | 33 (25) | $n\text{-}C_6H_{14}$ | 88 | 1.9 (25) |

**Exercise 1-5:**   In each of the following reactions, consider the type of transition state or intermediate and its charge distribution, and determine whether the rate would increase, decrease, or stay the same in switching from a nonpolar to a polar solvent.

(a) $EtO^{\ominus} + CH_3CH_2I$ $\xrightarrow{\text{Slow}}$ $[\overset{\delta-}{EtO}\text{---}CH_2\text{---}\overset{\delta-}{I}]$ $\xrightarrow{\text{Fast}}$ $EtOEt + I^{\ominus}$
     $\underset{CH_3}{|}$

(b) $CH_3CH_2\overset{\overset{CH_3}{|}}{\underset{\underset{CH_3}{|}}{C}}\!-\!Br$ $\xrightarrow{\text{Slow}}$ $CH_3CH_2\overset{\overset{CH_3}{|}}{\underset{\underset{CH_3}{|}}{C^{\oplus}}}, Br^{\ominus}$ $\xrightarrow[\text{Fast}]{OH^-}$ $CH_3CH_2\overset{\overset{CH_3}{|}}{\underset{\underset{CH_3}{|}}{C}}\!-\!OH + Br^{\ominus}$

## REVIEW OF NEW CONCEPTS AND TERMS

Define each term and, if possible, give an example of each:

Transition state                    Activated complex
Reaction coordinate              Reaction order

| | |
|---|---|
| Rate equation | Rate-determining step |
| Rate constant | Molecularity |
| Catalyst | Activation energy (enthalpy) |
| Dielectric constant | Arrhenius plot |
| Reaction intermediate | Solvent polarity |
| Reaction profile diagram | |

## PROBLEMS: CHAPTER II-1

**1.** Consider the reaction profile diagram below and answer the questions concerning the reaction it represents.

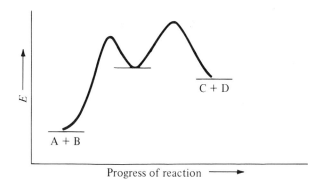

(a) Is the reaction exothermic or endothermic?

(b) How many transition states are involved in this reaction?

(c) How many intermediates?

(d) Which stage of the reaction sequence requires the greater activation energy, the first or second?

**2.** In the reaction $A + B \rightarrow C + D$ the following rate data were collected:

| Initial concentrations [moles/liter] | | |
|---|---|---|
| A | B | Relative rate |
| [0.100] | [0.100] | 1.00 |
| [0.200] | [0.100] | 2.00 |
| [0.400] | [0.100] | 4.00 |
| [0.100] | [0.200] | 1.00 |

Suggest a rate law $(-d\text{A}/dt = \ldots)$ which is consistent with these data.

**3.** (a) When 3-*tert*-butyl-1,3,5-hexatriene is heated at temperatures above 50°, ring closure to 2-*tert*-butyl-1,3-cyclohexadiene occurs. The progress of the reaction may be followed by taking samples of the reaction mixture at various times and measuring the quantities of each component by gas chromatography. The following data were collected at 85°C.

| | Mole fraction | |
|---|---|---|
| Time (hr) | Triene (A) | Diene (B) |
| 0 | 1.000 | 0.000 |
| 1 | 0.956 | 0.044 |
| 2 | 0.878 | 0.122 |
| 4 | 0.829 | 0.171 |
| 8 | 0.641 | 0.359 |
| 10.5 | 0.541 | 0.459 |
| 15 | 0.430 | 0.570 |
| 24.5 | 0.227 | 0.773 |

(b) In addition to the rate constant determined in part (a), the following rate constants were determined at various temperatures. Determine the activation energy for the ring closure reaction by plotting $\ln k$ vs. $1/T$.

| $k \ (\mathrm{sec}^{-1}) \times 10^5$ | Temp (°K) |
|---|---|
| 0.54 | 348 |
| 7.82 | 373 |
| 31.9 | 388 |
| 84.6 | 398 |

**4.** In the following reaction, would the rate be greater in highly polar or in nonpolar solvents?

$$R_3\overset{+}{N}-CH_2CH_3 + OH^- \xrightarrow{\Delta} R_3N + CH_2{=}CH_2 + H_2O$$

# Organic Reaction Mechanisms:
## Substitution, Elimination, and Addition

### NUCLEOPHILIC SUBSTITUTION REACTIONS

Nucleophilic substitution processes have perhaps been more widely studied than any other class of organic reactions.

$$Nu\colon + R{-}X \longrightarrow Nu{-}R + \colon X \tag{2.1}$$

where Nu: = attacking nucleophile
     X = leaving group
     R = alkyl group.

As shown in Eq. (2.1), a nucleophilic reagent displaces a leaving group X from an $sp^3$-hybridized carbon. Kinetically, nucleophilic substitution reactions fall into two distinct categories, which are described conventionally as $S_N1$ and $S_N2$ (Substitution, Nucleophilic, 1st or 2nd order). See Eq. (2.2) and Eq. (2.3). Thus

$$\frac{\Delta(RX)}{\Delta t} \approx \frac{-d(RX)}{dt} = k(RX) \quad S_N1 \text{ reaction} \tag{2.2}$$

$$\frac{\Delta(RX)}{\Delta t} \approx \frac{-d(RX)}{dt} = k(RX)(Nu) \quad S_N2 \text{ reaction} \tag{2.3}$$

the nucleophile does not participate in the rate-determining step in an $S_N1$ process and the rate is independent of nucleophile concentration. However, in an $S_N2$ process, the nucleophile *is* involved in the rate-determining step and the substitution rate is dependent on the nucleophile concentration. These two reaction pathways are usually envisioned as shown below.

$$\text{Nu:} + \text{R—X} \xrightarrow{\text{Slow}} [\text{Nu}\cdots\text{R}\cdots\text{X}] \longrightarrow \text{Nu—R} + \text{:X} \qquad (2.4)$$

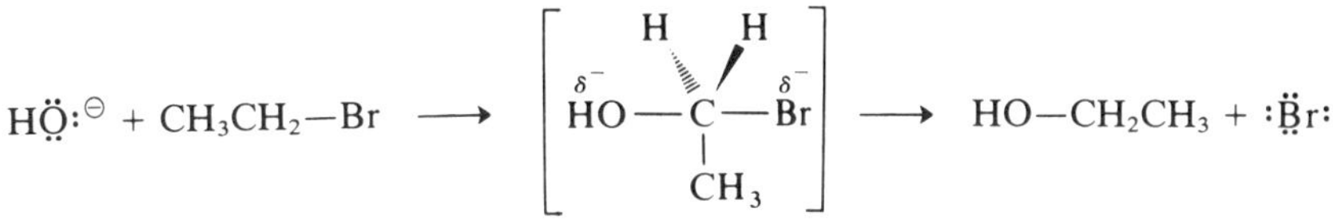

$$\text{HO:}^{\ominus} + \text{CH}_3\text{CH}_2\text{—Br} \longrightarrow \left[ \begin{array}{c} \text{H} \quad \text{H} \\ \overset{\delta^-}{\text{HO}} \text{—} \overset{\phantom{x}}{\text{C}} \text{—} \overset{\delta^-}{\text{Br}} \\ \text{CH}_3 \end{array} \right] \longrightarrow \text{HO—CH}_2\text{CH}_3 + \text{:Br:}^{\ominus} \qquad (2.5)$$

Typical $S_N2$ reaction scheme

$$\text{R—X} \underset{}{\overset{\text{Slow}}{\rightleftharpoons}} \text{R}^{\oplus} + \text{X}^{\ominus} \qquad (2.6)$$

$$\text{R}^{\oplus} + \text{:Nu} \xrightarrow{\text{Fast}} \text{R—Nu} \qquad (2.7)$$

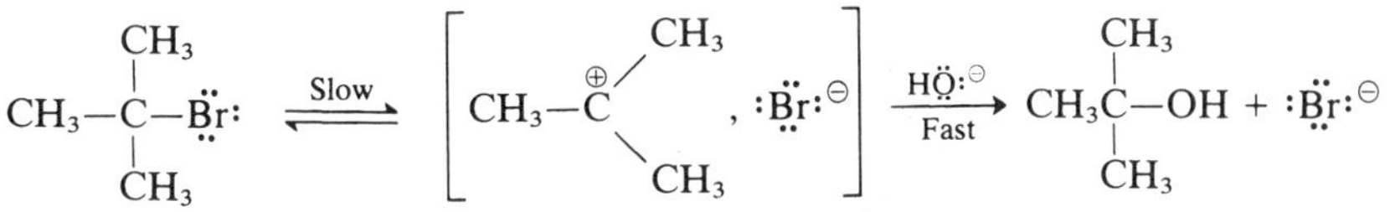

$$\begin{array}{c} \text{CH}_3 \\ | \\ \text{CH}_3\text{—C—Br:} \\ | \\ \text{CH}_3 \end{array} \overset{\text{Slow}}{\rightleftharpoons} \left[ \begin{array}{c} \text{CH}_3 \\ \overset{\oplus}{\diagup} \\ \text{CH}_3\text{—C} \quad , \text{:Br:}^{\ominus} \\ \diagdown \\ \text{CH}_3 \end{array} \right] \xrightarrow[\text{Fast}]{\text{HO:}^{\ominus}} \begin{array}{c} \text{CH}_3 \\ | \\ \text{CH}_3\text{C—OH} + \text{:Br:}^{\ominus} \\ | \\ \text{CH}_3 \end{array} \quad (2.8)$$

Typical $S_N1$ reaction scheme

In $S_N2$ reactions, the strength of the incoming nucleophile is a dominating aspect of the overall reaction; the better the nucleophile, the faster the reaction. Good nucleophiles are good bases, but nucleophilic character is also enhanced by the ability to participate in the transition state by becoming polarized and dispersing the charge distribution. Thus large, easily polarized groups or atoms are often better nucleophiles than smaller ions of greater basicity. Polarizability is also important in

$$I^{\ominus} > Br^{\ominus} > Cl^{\ominus} > F^{\ominus}$$
$$RS^{\ominus} > RO^{\ominus} > HO^{\ominus}$$

Relative nucleophilic strengths of typical ions

determining the effectiveness of a group's leaving ability, since it must also be able to disperse charge in the transition state. Similarly, less basic groups are easily displaced by incoming nucleophiles or by the solvent, as are groups which form particularly resonance-stabilized ions. Table 2-1 lists some typical nucleophiles and leaving groups.

Table 2-1. Typical Nucleophiles and Leaving Groups
        in Substitution Reactions

| Typical nucleophiles | Typical leaving groups |
|---|---|
| $I^-$, $Br^-$, $Cl^-$, $CN^-$ | $I^-$, $Br^-$, $Cl^-$, $F^-$ |
| $HO^-$, $RO^-$, $ArO^-$, $RCOO^-$ | $ArSO_3^-$, $RCOO^-$ |
| $H_2N^-$, $R_2N^-$, $R_3N$, $H_3N$ | |
| $H_2O$, $ROH$, $ROR$ | $-\overset{+}{\underset{H}{O}}H$, $-\overset{+}{\underset{H}{O}}R$, $-\overset{+}{S}R_2$, $-N\equiv N^+$ |
| $RS^-$, $H^-$ | |

## STEREOCHEMISTRY AND SUBSTITUTIONS

The differences between the $S_N1$ and $S_N2$ transition states are exemplified by considering the stereochemical fate of a typically active chiral center. In an $S_N2$ process, the incoming nucleophile must approach from the *back side* in a direction colinear with the leaving group. By displacing the leaving group in this fashion, the *absolute configuration* of the chiral center *inverts*. In an $S_N1$ substitution, however,

$$ \text{(2.9)} $$

Absolute configuration inverts in $S_N2$ substitution

the initial ionization of a chiral bromide produces a *planar intermediary carbonium ion*, thus destroying the asymmetry of the chiral center. When the nucleophile attacks the carbonium ion, the probability of attack from either above or below the plane is identical and a racemic product results. The $S_N1$ substitution of an optically active chiral bromide typically results in a loss of optical activity, while $S_N2$ substitution results in retention of optical activity but inversion of absolute configuration.

Optically active bromide

Planar carbonium ion (nonchiral)

Racemic mixture (optically inactive)

$$ \text{(2.10)} $$

322

**Exercise 2-1:** Consider the structure of $S_N1$ and $S_N2$ transition states. What is the hybridization of the C atom in each case? *Hint:* Consider the geometry of the intermediate and match with the geometries of the various possible carbon hybridizations.

## EFFECT OF STRUCTURE ON REACTION PATHWAY

Carbonium-ion stability and ease of formation both follow a predictable pattern: $3° > 2° > 1° > CH_3^{\oplus}$. One would expect, then, that tertiary halides might prefer to react via $S_N1$ pathways. The $S_N2$ transition state requires that five groups surround a carbon atom, creating a highly hindered arrangement. In this situation, methyl, or primary, halides might be expected to accommodate this crowding, but certainly not tertiary groupings. This leads to the prediction that reactivity orders for these two mechanisms should be opposite:

$$3° > 2° > 1° > CH_3X \qquad S_N1 \text{ reactivities}$$

$$3° < 2° < 1° < CH_3X \qquad S_N2 \text{ reactivities}$$

Alkyl groups such as allyl or benzyl are special cases in that they are primary halides (can participate in $S_N2$ transition states) and yet form extremely stable carbonium ions (can form $S_N1$ transition states). Both transition states are accessible for these halides, and they are quite reactive compared with most other halides.

$$
\begin{array}{c}
CH_2=CHCH_2Cl \\
\text{Allyl chloride}
\end{array}
\quad
\begin{array}{c}
\xrightarrow{S_N1} \left[ H_2C \cdots CH \cdots CH_2 \right]^{\oplus} \xrightarrow{Z:} \\[2mm]
\xrightarrow[S_N2]{Z:} \left[ \begin{array}{c} CH=CH_2 \\ | \\ \overset{\delta^-}{Z} \cdots \overset{}{C} \cdots \overset{\delta^-}{Cl} \\ / \quad \backslash \\ H \quad\quad H \end{array} \right]
\end{array}
\longrightarrow CH_2=CHCH_2Z \qquad (2.11)
$$

Vinylic and aromatic halides, on the other hand, are quite resistant toward either $S_N1$ or $S_N2$ displacement. The potential leaving group is bonded in both of these to an $sp^2$-hybridized carbon and the bond is much more difficult to break. In addition, the negative $\pi$-electron cloud tends to repel the incoming nucleophile. Under normal circumstances neither class of halides will react under conditions favorable to $S_N1$ displacement of tertiary halides or $S_N2$ displacement of primary halides.

**Exercise 2-2:**    Arrange the following halides according to their reactivities with dilute sodium hydroxide under $S_N2$ conditions:

$$CH_3CH=CHCH_2Cl, \qquad CH_3CH=CHCl, \qquad CH_3CH_2CH_2CH_2Cl$$

$$\begin{array}{c} CH_3 \\ | \\ CH_3CH_2C-Cl, \\ | \\ CH_3 \end{array} \qquad \begin{array}{c} CH_3CH_2CHCH_3, \\ | \\ Cl \end{array} \qquad H_3C-\!\!\!\left\langle \bigcirc \right\rangle\!\!\!-CH_2Cl$$

## ELIMINATION REACTIONS

Elimination processes have a great deal in common with substitution reactions. Actually, either may be considered a side reaction of the other, with reaction conditions dictating the major pathway. By far the most common elimination process is beta, or 1,2-elimination initiated by a strong nucleophile.

$$Nu\!:^{\ominus} + \ -\overset{|}{\underset{|}{C_\beta}}-\overset{X}{\underset{|}{C_\alpha}}- \ \longrightarrow \ Nu\!:\!H + \ \overset{\diagdown}{\diagup}C_\beta\!=\!C_\alpha\overset{\diagup}{\diagdown} + X\!:^{\ominus} \qquad (2.12)$$
$$\overset{|}{H}$$

where:  Nu: = attacking nucleophile
          X = leaving group on $\alpha$-carbon
          H = proton on $\beta$-carbon.

However, acid-catalyzed eliminations, particularly in alcohols, are also quite com-

$$\begin{array}{c} OH \\ | \\ RCHCH_2R \end{array} \xrightarrow[\substack{\text{or} \\ \text{Lewis acid} \\ \text{catalyst}}]{H_3O^+} RCH=CHR + H_2O \qquad (2.13)$$

mon, and eliminations from "onium" type intermediates are synthetically quite useful, although less common.

$$RCH_2CH_2\overset{\oplus}{N}(CH_3)_3, OH^{\ominus} \xrightarrow{\Delta} RCH=CH_2 + H_2O + (CH_3)_3N \qquad (2.14)$$

At least three possible mechanisms may be envisioned for 1,2-elimination, depending on the relative timing of the bond-breaking sequences in the transition state:

(A)

$$-\overset{\underset{|}{X}}{\underset{\underset{|}{H}}{\overset{|}{C}}}-\overset{|}{\underset{|}{C}}- \xrightarrow{\text{Slow}} \left[ -\overset{|}{\underset{|}{C}}-\overset{\oplus}{\underset{|}{C}}- \right], X^{\ominus} \xrightarrow[\substack{(-H^+) \\ \text{Fast}}]{\text{Nu:}} \overset{\diagdown}{\phantom{}}C=C\overset{\diagup}{\phantom{}} + H:Nu \qquad (2.15)$$

(B)

$$\underset{\text{Nu:}\curvearrowleft}{-\underset{\underset{|}{\overset{}{H}}}{\overset{\overset{X\curvearrowright}{}}{C}}\overset{|}{C}-} \xrightarrow{\text{Slow}} \left[ -\overset{\overset{X}{|}}{C}\!\!=\!\!\overset{|}{\underset{\underset{\ddots}{\overset{}{H}}}{C}}- \right] \xrightarrow{\text{Fast}} \underset{\text{Nu:H}}{\overset{\diagdown}{\phantom{}}C=C\overset{\diagup}{\phantom{}}} \qquad (2.16)$$

Conjugate base

(C)

$$\underset{\text{Nu:}\curvearrowleft}{-\underset{\underset{|}{\overset{}{H}}}{\overset{}{C}}-\overset{\overset{X}{|}}{\underset{|}{C}}-} \xrightarrow{\text{Fast}} \left[ \underset{\text{Nu:H}}{\overset{\ominus}{\underset{|}{C}}-\overset{\overset{X}{|}}{\underset{|}{C}}-} \right] \xrightarrow{\text{Slow}} \overset{\diagdown}{\phantom{}}C=C\overset{\diagup}{\phantom{}} + X^{\ominus} \qquad (2.17)$$

In Mechanism A, the slow step is an initial ionization of the leaving group–$\alpha$-carbon bond to yield an intermediary carbonium ion, followed by rapid proton removal. This is a first-order process, independent of nucleophilic concentration, and is designated as an $E_1$ reaction. The first step of this sequence is identical to the first step in an $S_N1$ reaction.

In mechanism B, loss of H and X is synchronous. As the nucleophile removes the proton from the beta carbon, an electron redistribution occurs smoothly, leading to observed products without extensive movement or rearrangement of the various nuclei. The overall process is second order (first order in both substrate and nucleophile), and is designated as an $E_2$ reaction. As in the case of $E_1/S_N1$ processes, the temptation is great to relate $E_2$ and $S_N2$ reactions. However, the transition states are *not* related and in fact are competing directly with one another.

Mechanism C is less common than either A or B, and being second order is difficult to distinguish from typical $E_2$ reactions. In general, the process is limited to structures in which the *carbanion* intermediate is stabilized by electron-withdrawing substituents on the beta carbon (such as halogen of phenylsulfonate). These reactions are designated as $E_1cB$ (elimination–first-order–conjugate base).

Acid-catalyzed dehydrations are most closely related to $E_1$-type reactions in that they generally proceed through carbonium-ion transition states after initial oxonium ion formation.

$$\underset{}{\overset{\overset{\text{OH}}{|}}{\text{RCH}_2\text{CHR}}} \xrightarrow{H^{\oplus}} \underset{\substack{\text{Oxonium} \\ \text{ion}}}{[\overset{\overset{\oplus}{\text{OH}_2}}{\text{RCH}_2\text{CHR}}]} \rightleftharpoons \underset{\substack{\text{Carbonium} \\ \text{ion}}}{\overset{\text{H}_2\text{O} +}{[\text{RCH}_2\overset{\oplus}{\text{CHR}}]}} \rightleftharpoons \underset{+ H^{\oplus}}{\text{RCH}=\text{CHR}}$$

In substitution reactions we were concerned with the fate of chiral reaction centers, as they helped us to determine the nature of the reaction transition state. In elimination reactions, our concern is directed toward two different stereochemical questions: product geometry, and the required orientations of leaving groups, if any.

In $E_1$ and acid-catalyzed eliminations, the most stable olefin is formed from the intermediary carbonium ion. Thus in the dehydration of 2-butanol, not only is the major product 2-butene, but the geometry of the incipient product is predominantly *trans*. The identity of the leaving group prior to the ionization step, and its orientation with respect to the $\beta$-proton, is relatively unimportant in acyclic molecules. However, the vast bulk of experimental evidence for $E_2$ eliminations shows that the leaving groups must have a *trans* or *anti* orientation. This requirement can

*anti* orientation of leaving groups in $E_2$ transition state

determine geometric identity of the product. For example, consider the $E_2$ elimination of HX from the diastereomeric compounds A and B. Since there is only *one*

Compound A

Compound B

possible orientation in each compound which places X and H *anti* to each other, the product geometry is easily predicted. If more than one orientation is possible (say, in a compound with two beta-hydrogens) then elimination will occur from the more *stable anti* configuration.

More-stable *anti* configuration

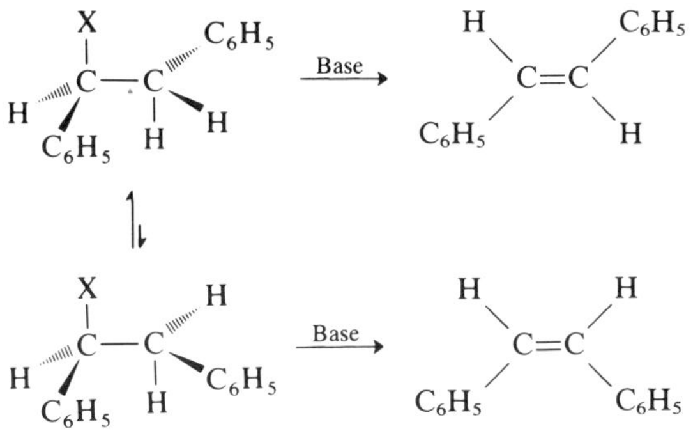

More-favored product

Less-favored product

Less-stable *anti* configuration

In cyclic compounds, the leaving groups can only achieve the required *anti* configuration when they occupy *trans-diaxial* positions, even though this implies that reaction occurs from a less-stable conformation. This observation is strikingly confirmed in the classical study of HCl eliminations of the various isomers of

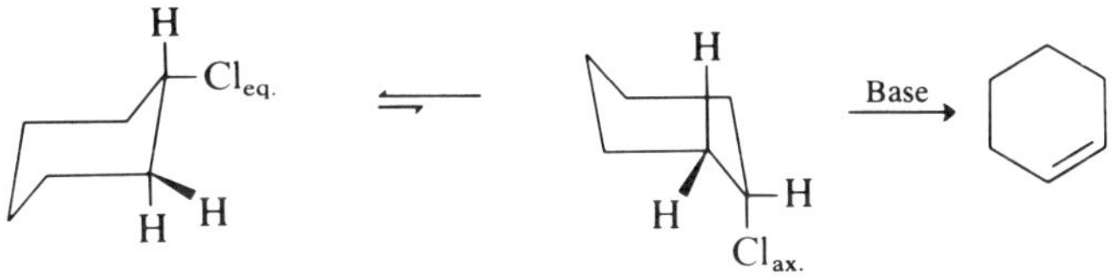

More-stable
conformation

Cl in Equatorial Position

Less-stable
conformation

Cl in Axial Position

1,2,3,4,5,6-hexachlorocyclohexane. Only one isomer, in which the chlorine atoms alternate around the ring in *trans* orientation, cannot achieve an *anti* orientation of H and Cl. This particular isomer undergoes HCl elimination *several thousand times slower* than any of the other $C_6H_6Cl_6$ isomers.

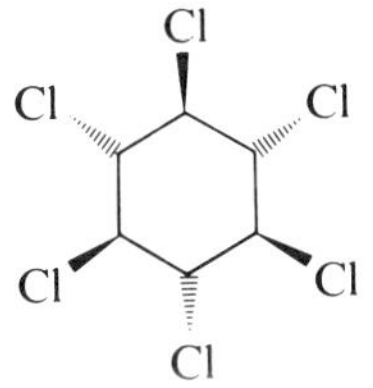

cannot achieve a 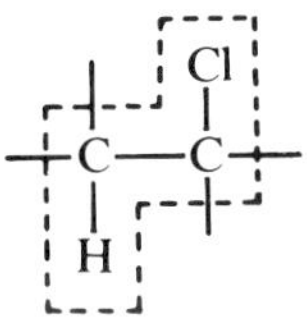                    orientation

*Syn elimination* has been observed for certain classes of organic reactions involving a cyclic transition state, for example, in acetate pyrolysis or amine pyrolysis (Cope reaction).

*syn* orientation of leaving groups in some reactions

Acetate pyrolysis or amine oxide pyrolysis (Cope reaction)

## COMPETITION BETWEEN $S_N$ AND E PATHWAYS

It was previously indicated that substitution and elimination accompany one another in any reaction. This is easily seen in the case of $S_N1$ and $E_1$ reactions, since they have common intermediates. However, the competition between $S_N2$ and $E_2$ is more subtle because of the different transition states.  By far the most common method of shifting the ratio of substitution to elimination is by raising the temperature,

$$RCH \overset{\curvearrowright}{\underset{\underset{\diagdown \!\! :Z}{H \curvearrowleft}}{-}} CH_2 \overset{\nearrow}{-} \overset{\cdot}{X} \quad \text{vs.} \quad RCH_2 - \overset{\overset{X^{\delta^-}}{|}}{\underset{\underset{Z^{\delta^-}}{|}}{CH_2}}$$

$$E_2 \text{ TS} \qquad\qquad\qquad S_N2 \text{ TS}$$

elimination being favored by higher temperatures.  Elimination is also more readily accomplished with tertiary leaving groups, substitution being favored for primary leaving groups.  Since elimination is related to the removal of a proton, the *base strength* of the attacking group is of more importance than its nucleophilicity.  An ideal combination of these effects is found in high-boiling tertiary amines, such as N,N-diethylaniline, which make excellent dehydrohalogenating media, with little if any substitution.

## ADDITION REACTIONS OF C=C

When one considers the structure of the double bond in three-dimensional space, it is not surprising that its most characteristic reaction is addition initiated by attack of an electrophile on the diffuse, negatively-charged $\pi$-electron cloud.  One can envisage

$$X \overset{\oplus}{} \longrightarrow$$

the electrophile actually forming a weak bond with the $\pi$ cloud prior to further reaction, and such complexes are referred to as $\pi$ *complexes.*  Such complexes with

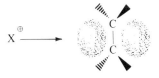

$\pi$ complex
(X buried in
$\pi$ cloud)

$\pi$ complex

good electrophiles like silver ion are well known, and have been shown to be formed reversibly.  Once the electrophile is held loosely within bond-forming distance of the carbon framework of the original double bond, $\sigma$-bond formation becomes feasible, resulting in carbonium-ion formation. The carbonium ion then can undergo rapid attack by whatever nucleophiles are present. The entire scheme for electrophilic addition can be represented as a two-stage reaction. For all practical purposes

Reaction scheme for electrophilic addition to C=C

$\pi$-complex formation may be ignored in considering product formation, as carbonium-ion stability is the determining factor in the addition of unsymmetrical reagents:

*In electrophilic addition to a carbon–carbon double bond, the most stable carbonium ion is formed as an intermediate.*

Electrophilic addition of an unsymmetrical reagent

Of course, this principle of preferential addition of unsymmetrical reagents to unsymmetrical olefins was first recognized in 1869 by Russian chemist Vladimir Markownikoff, who studied the addition of various acids, $H^+Y^-$, to various unsymmetrical olefins. The results of his study prompted the formulation of the famous "Markownikoff's rule" memorized by generations of organic-chemistry students:

*In the addition of an acid to a carbon–carbon double bond, the proton attaches itself to the carbon atom containing the greatest number of hydrogens.*

One of the early triumphs of carbonium-ion theory was the successful interpretation of Markownikoff's rule in terms of modern organic theory.

The addition of symmetrical reagents, such as $Br_2$ and $Cl_2$, also follows two-step ionic pathways involving carbonium-ion intermediates. The presence of

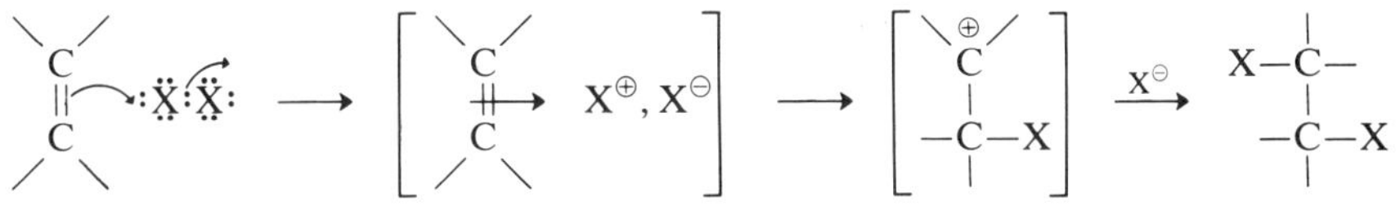

X = Cl, Br

the intermediary carbonium ion was inferred from the fact that nucleophiles other than $X^-$ can "trap" the intermediate by chemical reaction. For example, if the bromination is carried out in aqueous solution with dissolved sodium salts, such as $Na^+$ and $Cl^-$, several products are observed—one for *each* nucleophile present. It is

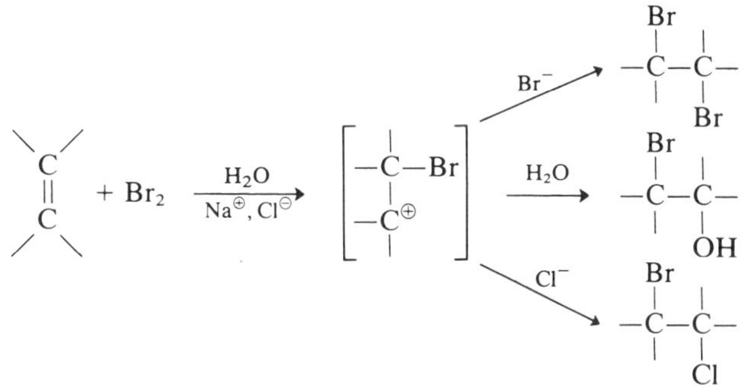

interesting to note that there is good experimental evidence for the intermediary carbonium ion in the preceding reaction to be written as a "cyclic bromonium" structure; however, for purposes of determining the products of addition, the more common carbonium-ion form will suffice for our present discussion. Cyclic "onium" structures in organic chemistry are far less common than the corresponding acyclic "carbonium" forms. Interestingly enough, there is currently very little evidence for the existence of other cyclic "halonium" ions.

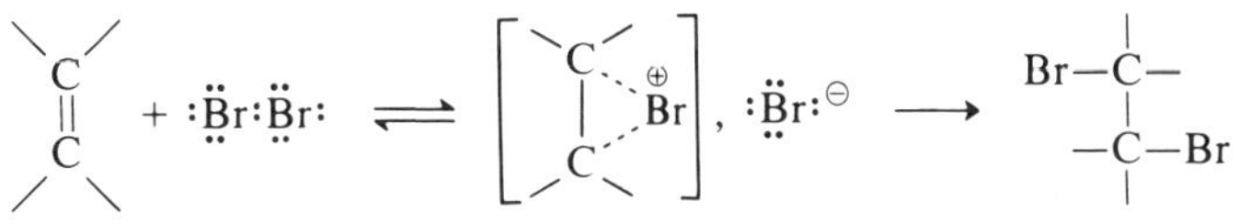

Cyclic bromonium ion

Under normal conditions, nucleophilic attack on a carbon–carbon double bond is suppressed by the repulsion between the $\pi$-electron cloud and the nucleophilic electron pair. However, if electron-withdrawing groups are attached to the double bond, thus reducing electron density about the $\pi$ bond, nucleophilic attack becomes feasible. Strong electron-withdrawing groups such as $-C\equiv N$ or $-NO_2$ facilitate nucleophilic addition. One of the more important reactions of this class is a process known as *cyanoethylation*, leading to difunctional synthetic intermediates.  It is

$$CH_2=CHC\equiv N \xrightarrow[ROH]{RO^-} [ROCH_2\overset{\ominus}{C}HC\equiv N] \xrightarrow{ROH} ROCH_2CH_2C\equiv N$$

"Cyanoethylation" reaction

important to note that in nucleophilic addition, the more stable *carbanion* is formed as an intermediate.  This leads to the opposite of Markownikoff addition.

$$CH_2=CHC\equiv N$$

Cyanoethylene
acrylonitrile

$$\xrightarrow{RO^\ominus} \quad \overset{\quad\quad OR}{\underset{}{\overset{\ominus}{\ddot{C}}H_2-CH-C\equiv N}} \quad \text{Unstable, not formed}$$

$$\xrightarrow{RO^\ominus} \quad \left[ \underset{OR}{CH_2\overset{\ominus}{C}H-C\equiv N} \quad \longleftrightarrow \quad \underset{OR}{CH_2-CH=C=\ddot{N}:^\ominus} \right]$$

Resonance-stabilized

## REVIEW OF NEW CONCEPTS AND TERMS

Define each of these terms and, if possible, give specific examples of each:

| | |
|---|---|
| Nucleophilic substitution reaction | $S_N1$ reaction |
| $S_N2$ reaction | Polarizable nucleophile |
| Stereochemistry of $S_N1$ and $S_N2$ processes | 1,2-elimination reaction |
| *Anti* elimination | *Syn* elimination |
| $E_1$ reaction | $E_2$ reaction |
| $S_N/E$ ratio in organic reactions | Electrophilic addition |
| Nucleophilic addition | $\pi$ complex |
| Markownikoff's rule | Markownikoff's rule stated in terms of modern theory |

## PROBLEMS: CHAPTER II-2

1. Give structures for the products expected for dehydrobromination of each of the following compounds.  If more than one product is formed, indicate major and minor products.

(a)

$$\underset{CH_3CH_2CH_2\overset{|}{C}HCH_3}{\overset{Br}{}}$$

(b)

(c)

Br

(d)   $\overset{\displaystyle Br}{\underset{\displaystyle CH_3}{CH_3\overset{|}{\underset{|}{C}}CH_2CH_2CH_3}}$

(e) $CH_3CH_2\underset{\underset{\displaystyle Br}{|}}{CH}CH_2CH_3$

(f) $CH_3CH_2\underset{\underset{\displaystyle Br}{|}}{CH}CH_2CH_2CH_2CH_3$

**2.** For each of the following olefins, indicate which carbonium ion(s) would be formed by addition of a proton ($H^+$). Explain your choice if more than one ion is possible.

(a) $CH_3CH_2CH{=}CH_2$

(b)

$-CH_3$

(c) $CH_2{=}CHCH{=}CH_2$

(d)

$-CH{=}CH_2$

(e) $CH_3CH{=}C(CH_3)_2$

(f)

$\underset{\displaystyle CH_3}{\overset{\displaystyle CH_3}{C}}{=}CH-$

**3.** When chlorobenzene is treated with $KNH_2$, aniline is formed in an apparent nucleophilic substitution. However, if chlorobenzene is labeled with carbon 14 (radioactive tracer), the following distribution is obtained:

$* = {}^{14}C$ tracer                    $1:1$

Suggest a mechanism that would account for the observed isotope distribution. (*Hint:* Consider what other well-known reaction can be initiated by a strong base.)

**4.** Explain the relative $S_N2$ reaction rates for the following series of halides with sodium ethoxide in ethanol at $55°$:

| $CH_3CH_2Br$ | $CH_3CH_2CH_2Br$ | $CH_3\overset{\displaystyle CH_3}{\overset{|}{C}H}CH_2Br$ | $(CH_3)_3CCH_2Br$ |
|---|---|---|---|
| $1.0$ | $2.8 \times 10^{-1}$ | $3.0 \times 10^{-2}$ | $4.2 \times 10^{-6}$ |

**5.** Consider the nucleophilic displacement of bromide from optically active α-phenylethylbromide by a nucleophilic solvent such as methanol. What would the stereochemistry of the product be if the reaction proceeded via an $S_N2$ pathway?

$(R)$ configuration

An $S_N1$ pathway? If only 20–30% of the *optical activity* were *retained* in this reaction, what conclusions could be drawn from this result? (*Hint:* Consider the *timing* of the displacement process.)

6. When β-phenethyl bromide is treated with sodium ethoxide in ethanol, styrene is the product. In order to determine if the reaction follows $E_2$ or $E_1cB$ mechanisms, the same reaction was run in EtOD and the reaction allowed to proceed to 50% completion. No deuterium was found in either the starting material or the product. What conclusions concerning the mechanism ($E_2$ or $E_1cB$) can be drawn from the experiment? If deuterium *had* been found in either product or reactant what conclusion could be drawn?

7. When isotopically labeled allyl chloride is allowed to equilibrate in a polar solvent, a scrambling of the label is observed. Explain.

$$CH_2{=}CHCD_2Cl \longrightarrow CH_2{=}CHCD_2Cl + CD_2{=}CHCH_2Cl$$

$${}^*CH_2{=}CHCH_2Cl \longrightarrow CH_2{=}CH\overset{*}{C}H_2Cl + \overset{*}{C}H_2{=}CHCH_2Cl$$

$${}^* = {}^{14}C \text{ tracer}$$

8. When optically active 2-iodooctane is equilibrated with radioactive iodide ion, optical activity is lost and radioactive iodide is incorporated in the product. Suggest a mechanism that would account for this observation. If your mechanism is correct, would the rate of incorporation of radioactive iodide be less than, equal to, or greater than the rate of racemization?

$R$-2-Iodooctane
(optically active)

9. Would you expect allyl chloride and benzyl chloride to undergo nucleophilic substitution via $S_N1$ or $S_N2$ mechanisms? Explain your choice.

10. In many reactions in which carbonium ions are possible reaction intermediates (acid-catalyzed dehydration of primary alcohols, for example), a mixture of products is obtained in which rearrangement has occurred. Suggest a mechanistic explanation for each of the following reactions. (*Hint:* Consider how you would get a *more* stable carbonium ion as an intermediate.)

(a) $CH_3CH_2CH_2CH_2OH \xrightarrow[\Delta]{H_2SO_4} CH_3CH_2CH{=}CH_2 + CH_3CH{=}CHCH_3$
   20%                80%

(b)
$$\underset{\displaystyle CH_3CH_2CH_2\overset{\displaystyle OH}{\overset{|}{C}}HCH_3}{} \xrightarrow{HBr} \underset{\displaystyle CH_3CH_2CH_2\overset{\displaystyle Br}{\overset{|}{C}}HCH_3}{} + \underset{\displaystyle CH_3CH_2\overset{\displaystyle Br}{\overset{|}{C}}HCH_2CH_3}{}$$
   86%                14%

(c)
$$CH_3{-}\overset{\displaystyle CH_3}{\underset{\displaystyle CH_3}{\overset{|}{\underset{|}{C}}}}{-}CH_2OH \xrightarrow[\Delta]{H_2SO_4} CH_3\overset{\displaystyle }{\underset{\displaystyle CH_3}{\overset{}{\underset{|}{C}}}}{=}CHCH_3 + CH_2{=}\overset{\displaystyle }{\underset{\displaystyle CH_3}{\overset{}{\underset{|}{C}}}}{-}CH_2CH_3$$

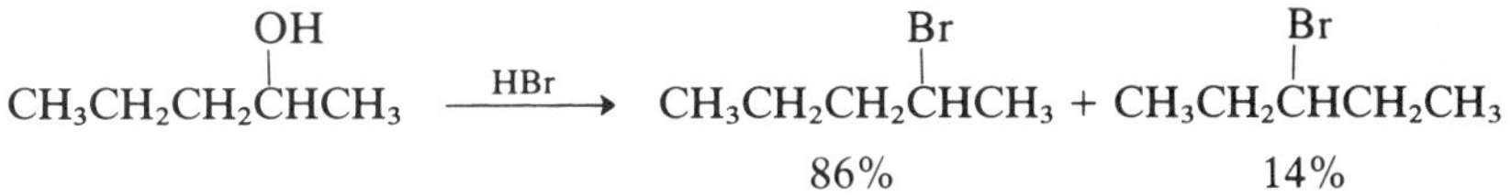

   Major              Minor
   product            product

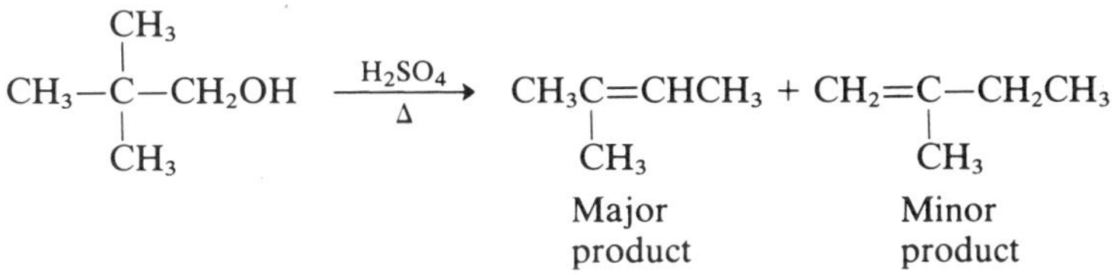

11. Arrange the following compounds in order of their reactivity toward nucleophilic displacement with $I^-$.

(a)
$$CH_3CH_2CH_2\overset{\displaystyle CH_3}{\underset{\displaystyle CH_3}{\overset{|}{\underset{|}{C}}}}{-}Br$$

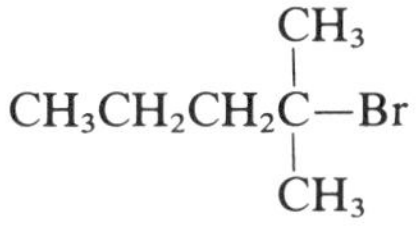

(b) $CH_3(CH_2)_5Br$

(c) $-CH_2Br$

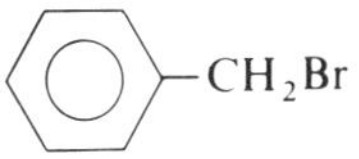

(d)
   Br

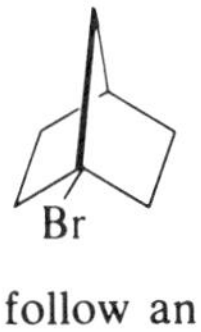

12. Carbonium-ion stability has been shown to follow an allyl, benzyl $> 3° > 2° > 1° > CH_3^+$ order. Would you expect free radicals, $R\cdot$, to follow the *same* order? Explain.

# 3

# Determination of Structure

Before World War II, the determination of the structure of organic molecules was limited to time-consuming, and often inaccurate, chemical methods. With the development of modern-day electronics, sophisticated instruments are now available which allow us to investigate molecular structure. Organic molecules can absorb electromagnetic radiation, and the type and energy of the absorbed radiation yields an amazing variety of information concerning the molecule's structure. In this chapter we will learn how these absorption processes are related to molecular structure, and how structure may be deduced by examining several different types of absorption spectra.

## MOLECULAR STRUCTURE AND THE ABSORPTION OF LIGHT

Light has both wave and particle properties, and electrons in atoms behave in similar fashion. You will recall that electron motion in atoms is *quantized* and that only *certain* energy levels may be occupied. Molecular structure is much more complicated than atomic structure, yet many of the principles governing the absorption of light by atoms also govern light absorption by molecules.

When light, or other electromagnetic radiation, comes in contact with an organic molecule, its energy determines whether it will be absorbed or not. If absorption takes place, the energy corresponds to the difference between two quantum levels. Energy having wavelengths either greater or smaller than this difference will not be absorbed. Thus the wavelength of absorption is a direct measure of quantum-level difference in a molecule. This absorption of energy is illustrated in Fig. 3-1.

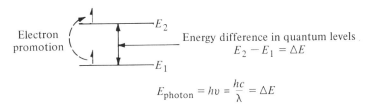

**Figure 3-1.**   Energy absorption process in atoms and molecules.

In order for a photon to be absorbed, its energy *must* correspond to the energy difference between two quantum levels. When the energy is absorbed, an electron is *promoted* from a lower energy level to a higher energy level.

## ENERGY LEVELS IN MOLECULES— VARIETY IN MOTION

Molecules are different from atoms in that many more degrees of freedom of motion are possible, and this diversity results in many different kinds of quantum levels. The total energy of a molecule may be roughly broken down into several discrete types:

$$E_{total} = E_{electronic} + E_{vibrational} + E_{rotational} + E_{translational}$$

$$\xleftarrow{\hspace{4cm}}$$
Increasing energy

Electronic energy levels are associated with the $\sigma$, $\pi$, and nonbonded electron pairs in the molecule. Absorption of light in this region of the spectrum usually breaks the chemical bond. Vibrational and rotational excitation are more or less self-descriptive, and differ from electronic excitation in that the bonds are merely deformed rather than broken. Translational energies are associated with how fast a molecule travels through space and are of little interest for the purposes of structure determination. Most chemical bonds in organic molecules are in the range of 50–150 kcal/mole, which corresponds to the ultraviolet (UV) and visible regions of the electromagnetic spectrum. Vibrational energies correspond to the infrared (IR) region.

UV-visible and IR spectrophotometers are the workhorses of organic chemistry, and operate basically on the same principles. A light source produces a broad spectrum of light in a given spectral region, and the particular wavelength desired for irradiation of the sample is chosen by passing the light through a rotating prism or diffraction grating. Light of any particular wavelength, with initial intensity $I_0$, is then passed through the sample. If any light is absorbed, then the intensity is diminished to a new value, $I$. The difference between these two intensities is usually measured as a ratio, which is related to the solution concentration, path length of the sample cell, and a characteristic of each compound known as the *absorptivity*, or probability of absorbing a photon. This is usually formulated as the *Beer–Lambert law*, which is illustrated in Fig. 3-2.

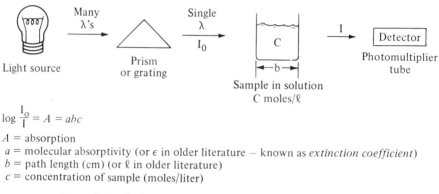

$$\log \frac{I_0}{I} = A = abc$$

$A$ = absorption
$a$ = molecular absorptivity (or $\epsilon$ in older literature – known as *extinction coefficient*)
$b$ = path length (cm) (or $\ell$ in older literature)
$c$ = concentration of sample (moles/liter)

**Figure 3-2.** Basic spectrophotometer operation (IR and UV-visible)

Example:  In the ultraviolet spectrum of Fig. 3-3 the absorption of a $5 \times 10^{-5}$ M sample in a 1 cm cell yields the following absorptivity:

$$A = abc; \qquad a = \frac{A}{bc} = \frac{(0.80 - 0.10)}{(1)(5 \times 10^{-5})}$$

$$\boxed{A = 14,000}$$

The spectrum would be recorded as:

$$\lambda_{max}(\varepsilon_{max})\ 242\ nm\ (14{,}000)$$

Both the wavelength of absorption (242 nm) and the absorptivity (14,000) are characteristic of the compound. The former is related to the energy necessary to raise a particular electron in a chemical bond to a higher energy level, while the latter is a measure of the efficiency of the process.

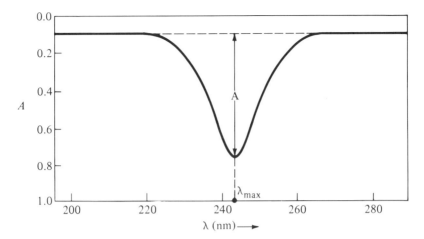

**Figure 3-3.** A typical UV spectrum.

## ULTRAVIOLET—VISIBLE SPECTRA
## AND ORGANIC STRUCTURE

The *ultraviolet* region of the spectrum is generally regarded as including the range 200–400 nm. In this region, both bonding and nonbonding electrons are directly excited in the molecule. The *visible* region extends from 400 to 750 nm, and derives its name from the fact that these wavelengths are perceived by the human eye and are responsible for objects being "colored." However, there is really no difference between the mechanism of light absorption in the UV and visible regions other than photon energy (wavelength). In the visible region, blue-violet corresponds to the more energetic photons (about 400 nm) while the red end is of lower energy (about 750 nm).

In organic molecules one finds sigma ($\sigma$), pi ($\pi$), and nonbonded electrons. When ultraviolet or visible light is absorbed, an electron is *promoted* or *excited* to a higher energy level, which is normally represented as an *anti* bonding orbital $\sigma^*$ or $\pi^*$. For reasons of symmetry, electron transitions between $\pi \rightarrow \sigma^*$ or $\sigma \rightarrow \pi^*$ are forbidden (have a very low probability), thus there are four possible transitions:

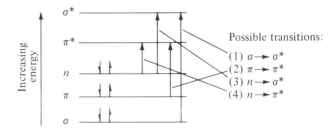

339

For all practical purposes, the $\sigma \rightarrow \sigma^*$ transitions require so much energy (less than 200 nm) that they are observable only under vacuum, since air and solvents would also absorb light in this spectral region. This greatly reduces their usefulness for studying organic structure. However, the $\pi \rightarrow \pi^*$ transition ($C=C$, $C\equiv C$, conjugated $C=C-C=C$, and aromatic structures ⬡) is easily studied (200–800 nm). The $n \rightarrow \pi^*$ transition is observed for compounds containing oxygen ($>C=\ddot{O}:$). Some typical ultraviolet spectra are listed in Table 3-1.

Table 3-1.   Typical Organic Absorption Spectra

| Compound | Absorbing unit | Transition | $\lambda_{max}$ (nm) | $a_{max}$ |
|---|---|---|---|---|
| $CH_3CH=CH_2$ | $>C=C<$ | $\pi \rightarrow \pi^*$ | 185 | 10,000 |
| $CH_2=CHCH=CH_2$ | $>C=C-C=C<$ (Acyclic) | $\pi \rightarrow \pi^*$ | 217 | 21,000 |
| ⬡ | $>C=C-C=C<$ (Cyclic) | $\pi \rightarrow \pi^*$ | 256 | 8,000 |
| $CH_2=CHCH=CHCH=CH_2$ | | | | |
| | $+C=C+_3$ | $\pi \rightarrow \pi^*$ | 258 | 50,000 (*E*-isomer) |
| ⬡ | Aromatic ring | $\pi \rightarrow \pi^*$ | 180 200 260 | 47,000 7,000 200 |
| $CH_3\overset{\overset{O}{\|\|}}{C}CH_3$ | $>C=\ddot{O}:$ | $n \rightarrow \pi^*$ | 276 | 12 |
| $CH_2=CH\overset{\overset{O}{\|\|}}{C}CH_3$ | $>C=C-C=O$ $>C=O$ | $\pi \rightarrow \pi^*$ $n \rightarrow \pi^*$ | 219 324 | 3,600 24 |

The absorbing units are sometimes referred to as *chromophores*. To a large extent they are independent of the rest of the molecule, but attached groups can shift the absorption to either a higher or a lower wavelength by increasing or decreasing the electron density. Solvents also may shift the position of absorption. Note that as conjugation increases ($C=C \rightarrow C=C-C=C \rightarrow C=C-C=C-C=C$), the wavelength of absorption increases dramatically, as does the absorptivity. Thus ultraviolet spectroscopy is particularly useful in detecting the presence of conjugation in a molecule, and in distinguishing between conjugated and isolated double bonds.

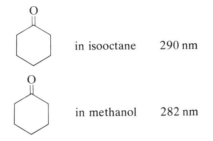

258 nm          263 nm

Alkyl group shifts $\lambda_{max}$

| | | |
|---|---|---|
| $CH_2{=}CHCH_2CH{=}CH_2$ | 187 nm | Nonconjugated |
| $CH_3CH{=}CHCH{=}CH_2$ | 224 nm | Conjugated |

Conjugation shifts $\lambda_{max}$

in isooctane     290 nm

in methanol     282 nm

Solvent interaction shifts $\lambda_{max}$

---

**Exercise 3-1:** Explain how you might use ultraviolet spectroscopy to distinguish between the following pairs of compounds:

(a) 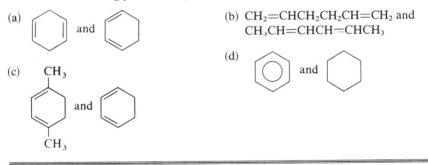 and

(b) $CH_2{=}CHCH_2CH_2CH{=}CH_2$ and
$CH_3CH{=}CHCH{=}CHCH_3$

(d) and

(c) $CH_3$ ... $CH_3$ and

---

# INFRARED SPECTRA
# AND ORGANIC STRUCTURE

The next region of the electromagnetic spectrum is the infrared. In this portion of the spectrum photons are no longer energetic enough to break even weak $\pi$ bonds. Instead, the photon causes the bond to *vibrate* with a characteristic frequency. In order to appreciate why infrared spectra are so complex, one must appreciate what kinds of vibrating motion are possible for even simple molecular ensembles. Examples of typical fundamental vibrations are shown below:

1. Stretching vibrations: $-\overset{\leftrightarrow}{C}-\overset{\leftrightarrow}{H}$

(a) 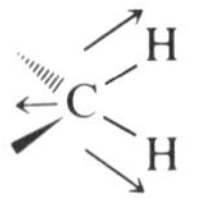      *Symmetrical* stretching—hydrogens vibrate in phase with one another ($2950 \, cm^{-1}$)

(b) 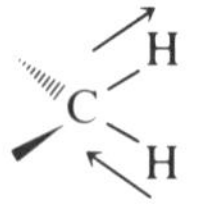      *Asymmetrical* stretching—hydrogens vibrate out of phase with one another ($2850 \, cm^{-1}$)

2. Bending vibrations—there are several different types of bending motions, many of which have special designations.

(a) 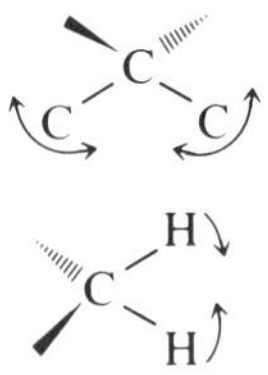      *Scissoring* (in plane of paper) ($1450 \, cm^{-1}$)

(b) 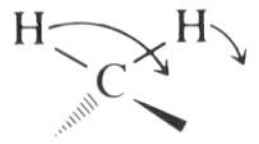      *Wagging* (out of plane of paper) ($1250 \, cm^{-1}$)

(c) 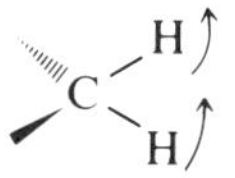      *Rocking* (in plane of paper) ($750 \, cm^{-1}$)

(d) 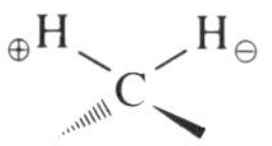      *Twisting* (one hydrogen moves behind the plane of the paper while the other moves toward you) ($1250 \, cm^{-1}$)

Infrared spectra may be recorded in terms of wavelength (cm) or reciprocal wavelength $(1/\lambda)$ ($cm^{-1}$). The latter is really an energy term, and to calculate the energy required to excite a particular vibration, one has only to multiply by the constants $h$ and $c$:

$$E_{vibration} = \frac{hc}{\lambda} = \underset{\substack{\uparrow \\ Constants}}{(hc)} \times \boxed{\frac{1}{\lambda}}$$

Units of $cm^{-1}$ referred to as *wave numbers*

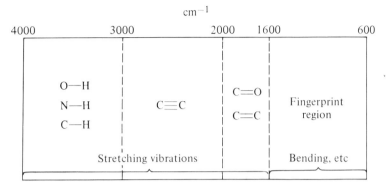

$$cm^{-1}$$

| 4000 | 3000 | 2000 | 1600 | 600 |

O—H
N—H
C—H

C≡C

C=O
C=C

Fingerprint region

Stretching vibrations

Bending, etc

**Figure 3-4.** Approximate group positions in a typical infrared spectrum.

Simple group vibrations, such as a C—H stretch or a C=O stretch, are referred to as *characteristic frequencies*. Although their values, in $cm^{-1}$, differ slightly from compound to compound, they always occur in the same place in the spectrum (Fig. 3-4). Stretching frequencies are extremely characteristic and are most useful in identifying functional groups such as OH, $NH_2$, and C=O. These vibrations may be found at the high-energy end of the infrared (1600–3600 $cm^{-1}$). Most molecules have very complicated bending regions (1600–600 $cm^{-1}$) which are difficult to analyze; however, no two molecules have exactly the same spectrum in this portion of the infrared and for that reason it is referred to as the *fingerprint region*. A short list of characteristic group absorptions is found in Table 3-2.

When you attempt to analyze an unknown spectrum, you should first scan the stretching region for the following easily identifiable bands: OH, NH, CH, C≡C, C≡N, C=C, and C=O. Once the functional groups are identified, you can make some reasonable predictions concerning the structure of the molecule. In Fig. 3-5 several infrared spectra are analyzed in this fashion.

**Exercise 3-2:** For each of the following compounds identify the functional groups that would give rise to IR absorption bands, and give approximate absorption frequencies (see Table 3-2).

(a)
$$\underset{\underset{}{OH}}{CH_2=CHCH_2\overset{|}{C}HCH_3}$$

(b) ⟨benzene ring⟩—$CH_2C≡CH$

(c) $CH_3$—⟨benzene ring⟩—$\overset{\overset{O}{\|}}{C}OCH_3$

(d) $CH_2=CHCOOH$

(e) ⟨cyclohexene ring⟩=O

(f) $CH_3\underset{\underset{}{OH}}{\overset{|}{C}H}$—⟨benzene ring⟩—$\overset{\overset{O}{\|}}{C}-H$

343

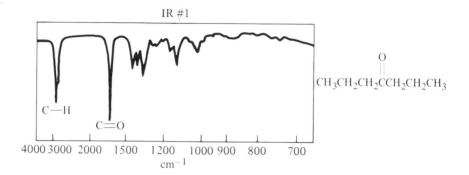

IR #1

C—H

C=O

4000 3000 2000    1500    1200   1000 900   800        700
                              cm$^{-1}$

$$CH_3CH_2CH_2\overset{\overset{\displaystyle O}{\|}}{C}CH_2CH_2CH_3$$

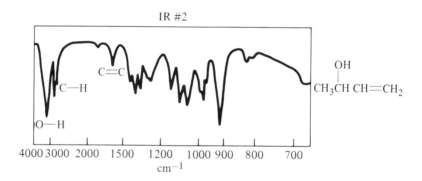

IR #2

C=C

C—H

O—H

4000 3000 2000    1500    1200   1000 900   800        700
                              cm$^{-1}$

$$CH_3\overset{\overset{\displaystyle OH}{|}}{CH}\ CH{=}CH_2$$

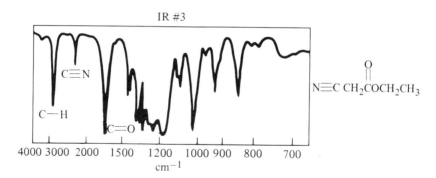

IR #3

C≡N

C—H

C=O

4000 3000 2000    1500    1200   1000 900   800        700
                              cm$^{-1}$

$$N{\equiv}C\ CH_2\overset{\overset{\displaystyle O}{\|}}{C}OCH_2CH_3$$

**Figure 3-5.** Analysis of infrared spectra by the group frequency procedure can identify the functional groups in the molecule.

Table 3-2. Common Characteristic IR Group Stretching
          Vibrations

| Group | Type | Group frequency range (cm$^{-1}$) |
|-------|------|-----------------------------------|
| O—H | Alcohols[a] | 3200–3600 |
| | Acids[b] | 3000–3500 |
| N—H | Amines | 3300–3500 |
| C—H | Alkane | 2850–3000 |
| | Alkene | 3020–3080 |
| | Aromatic | 3000–3100 |
| | Alkyne | 3300 |
| | Aldehyde | 2700–2800 |
| C≡N[c] | Nitrile | 2210–2260 |
| C≡C[c] | Alkyne | 2100–2250 |
| C=C | Alkene | 1600–1680 |
| | Aromatic | 1400–1600 |
| C=O[d] | Aldehyde | 1720–1740 |
| | Ketone | 1705–1725 |
| | Carboxylic acid | 1700–1725 |
| | Ester | 1730–1750 |
| | Amide | 1640–1700 |
| C—X | Halogen | 600–800 (Cl) |
| | | <600 (Br) |

[a] If hydrogen-bonded, these peaks will be broad.

[b] Acids are hydrogen bonded and the absorption will be broad.

[c] When present, these absorptions are weak.

[d] When present, these absorptions will be the strongest stretching frequencies in the spectrum.

# NUCLEAR MAGNETIC RESONANCE AND ORGANIC STRUCTURE

Nuclear magnetic resonance (nmr) spectroscopy is the organic chemist's best friend. For years, chemists have dreamed of a technique that would enable them to examine the structure of a molecule in intimate detail. While UV and IR spectra provide some of this information, the detailed analyses involved with these techniques are usually too difficult except for the identification of the simple functional groups described above.

The nuclei of many, but not all, elements spin like electrons and generate a nuclear magnetic field. When these nuclei, in a compound, are placed in a strong external field, they align themselves either with or against the field. When the sample is then irradiated with radio-frequency energy, usually 60 or 100 MHz, absorption takes place at certain values of the irradiating frequency and is a measure of the energy difference between the two different magnetic alignments.

$$\Delta E = \frac{\gamma h H}{2\pi}$$

where $\Delta E$ = energy necessary to realign a nucleus in a magnetic field
   $\gamma$ = gyromagnetic constant—different for each nucleus
   $H$ = strength of the external magnetic field (about 14,000 Gauss)

While all this is very interesting, it would be of little value to an organic chemist if all the nuclei commonly found in organic compounds reacted to an external field. However, only protons ($^1H_1$ nuclei) absorb in this fashion, $^{12}C$ and $^{16}O$ do not. Other nuclei which will show nmr behavior include $^{13}C$, $^{15}N$, $^{19}F$, $^{31}P$, and $^2H$ (deuterium). Thus it is possible to study nmr of other nuclei by isotope substitution, for example $^{13}C$ for $^{12}C$, in organic molecules. Nmr spectra contain a wealth of information which may be characterized as follows:

1. The number of signals
2. The relative areas of each signal
3. The position of the signal with respect to an internal standard—called the *chemical shift.*
4. The influence of one proton on adjacent protons—magnetic coupling or spin–spin splitting.

We will discuss each of these in turn.

## Number of Signals

If each proton in an organic molecule felt the full effect of the external applied field, then all protons in a given sample would absorb radio-frequency energy at the same frequency. However, the differing electron densities throughout the molecule can shield the protons from the magnetic field. Since each proton therefore senses a different external field, they absorb at different frequencies, and we see a distinct signal for each different kind of proton in the molecule. Fortunately, each *chemically* different proton is also *magnetically* different, and we can predict how many signals should be present in an nmr spectrum by identifying the number of chemically distinct protons. Several examples are shown below:

$$\underset{a}{CH_3}-\underset{\underset{\underset{a}{CH_3}}{|}}{\overset{\overset{a}{CH_3}}{\overset{|}{C}}}-\underset{b}{H}$$

Two signals: C—H and —CH$_3$
(all 9 methyl H's are equivalent)

$$\underset{a}{CH_3}\underset{b}{CH_2}\underset{c}{OH}$$
Three signals: CH$_3$—, —CH$_2$—, and —OH

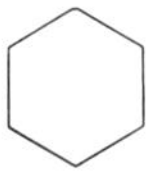

One signal: All —$CH_2$— are equivalent

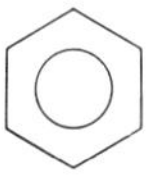

$$\overset{H}{\underset{}{|}}$$
One signal: All 6 —C= are equivalent

$$\underset{a\quad b\quad c\quad b\quad a}{CH_3CH_2\overset{\overset{Cl}{|}}{C}HCH_2CH_3}$$   Three signals: (a) Both —$CH_3$'s are equivalent
                                            (b) Both —$CH_2$—'s are equivalent
                                            (c) C—H

---

**Exercise 3-3:**   Predict how many signals you would expect for each of the following structures:

(a) 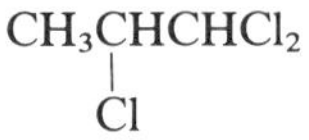 $CH_3\overset{\overset{Cl}{|}}{C}HCHCl_2$

(b) (structure)

(c) $CH_3\overset{\overset{CH_3}{|}}{\underset{\underset{CH_3}{|}}{C}}CH_2OH$

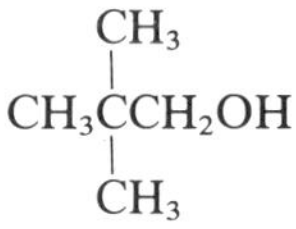

(d) 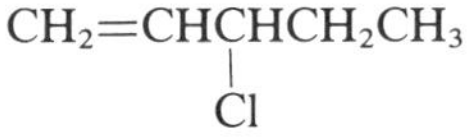 $CH_2{=}CH\overset{\overset{Cl}{|}}{C}HCH_2CH_3$

(e) 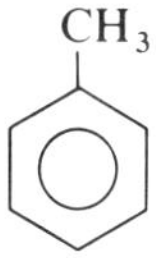 $CH_3$ (structure)

(f) 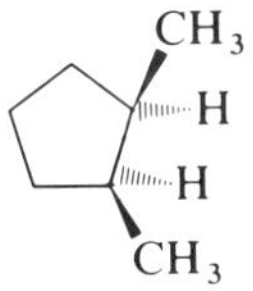 (structure) $CH_3$, H, H, $CH_3$

---

## Signal Area

The strength of the nmr signal is proportional to the number of protons giving rise to that signal. This is normally measured as the area under the absorption curve. Thus if we have three peaks in an nmr spectrum and their respective areas are

$$\text{Signal area} \propto \text{Number of protons}$$

$45:30:15$ (arbitrary units), then we can assume that the numbers of protons giving rise to these signals are in the ratio $3:2:1$. You must keep in mind, however, that these are *ratios*, and not the actual numbers of protons. For example, the above ratio would also be observed for $6:4:2$ or $12:8:4$ proton samples.

## The Chemical Shift

It is very difficult to measure the exact position of a proton signal in absolute frequency units. However, it is relatively easy to measure signal position with respect to an internal standard which does not react with the sample. The most common standard used today is tetramethylsilane, $(CH_3)_4S$, (TMS). Almost all proton signals appear *downfield* from the TMS signal. We can define a scale to measure the position of the proton with respect to TMS, and the numerical value of the signal on this scale is called the *chemical shift*. Two related scales are utilized: the $\delta$ (delta) or the $\tau$ (tau) scale. If we are utilizing a 60 MHz instrument, the most common instrument in use

$$\delta = \frac{\Delta\nu \text{ (Hz)} \times 10^6}{\text{Frequency (Hz)}} \quad \text{Defined } \delta \text{ scale}$$

$$\Delta\nu = \text{shift in Hz from TMS}$$

$$\tau = 10 - \delta$$

today, a shift $(\Delta\nu)$ of 15 Hz downfield from TMS would give rise to a $\delta$ value of 4.0. Since a shift of 15 Hz represents 4 parts per million of the 60 MHz frequency, each $\delta$

$$\delta = \frac{15 \times 10^6}{60 \times 10^6} = 4.0 \quad (\tau = 10.0 - 4.0 = 6.0)$$

or $\tau$ unit is 1 ppm and chemical shifts are thus expressed as ppm from TMS. Table 3-3 lists chemical shifts for the more common types of protons. Note that proton signals affected by highly electronegative elements are shifted downfield. Figure 3-6

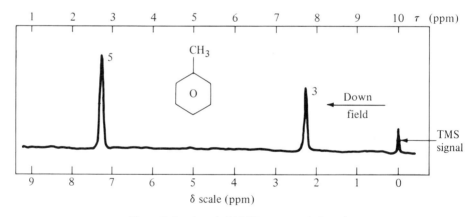

**Figure 3-6.**   A typical NMR spectrum (toluene).

shows a typical simple nmr spectrum for toluene. There are two signals—one for the methyl protons at $\delta$ 2.3 (relative area 3) and one for the ring protons at $\delta$ 7.2 (relative area 5).

The values in Table 3-3 are derived from a large number of samples and represent "normal" ranges. Abnormal shielding or deshielding in a molecule may shift signals outside these ranges.

Table 3-3. Common Proton Chemical Shifts

| Proton type | | Chemical shift | |
|---|---|---|---|
| | | $\delta$ (ppm) | $\tau$ (ppm) |
| $RCH_3$ | (Primary) | 0.9 | 9.1 |
| $R_2CH_2$ | (Secondary) | 1.3 | 8.7 |
| $R_3CH$ | (Tertiary) | 1.5 | 8.5 |
| ⬡—H | (Aromatic) | 6–8.5 | 1.5–4 |
| C=C—H | (Vinylic) | 4.6–6.0 | 4.0–5.4 |
| C≡C—H | (Acetylene) | 2–3 | 7–8 |
| C=C—$CH_2$— | (Allylic) | 1.7–2.2 | 7.8–8.3 |
| C—OH | (Alcohols) | 3.4–4 | 6–6.6 |
| —C—H ‖ O | (Aldehyde) | 9–10 | 0–(+1) |
| —C—OH ‖ O | (Acid) | 10.5–12 | (−2)–(−0.5) |
| R—$NH_2$ | (Amine) | 1–5 | 4–9 |
| $CH_3F$ | | 4.0 | 6.0 |
| $CH_3Cl$ | | 3.5 | 6.5 |
| $CH_3Br$ | | 2.8 | 7.2 |

## Magnetic Coupling (Spin–Spin Splitting)

The magnetic fields generated by each spinning hydrogen nucleus are not only affected by an external applied magnetic field but also by the fields generated by their near neighbors (adjacent protons). These neighboring protons may either reinforce or subtract from the effective force of the applied field. The net effect of this interaction is to "split" the proton signal, and the two interacting nuclei are said to be *coupled*. The amount of the shift, both upfield and downfield, is referred to as the *coupling constant, J* (in Hz).

Consider two adjacent protons, $H_a$ and $H_b$, which are magnetically nonequivalent. $H_a$ will be aligned either *with* the external field $H_0$ or *against* the external field. Thus $H_b$ will feel the effect of $H_0 + H_a$ or $H_0 - H_a$ and the $H_b$ signal

$$-\overset{|}{\underset{\underset{H_a}{|}}{C}}-\overset{|}{\underset{\underset{H_b}{|}}{C}}- \quad \uparrow H_0$$

$H_a$ will align with this field
$H_a\uparrow$
or against the field
$H_a\downarrow$

will be split into *two* signals corresponding to these two combinations. Similarly $H_a$ will be split into two signals by $H_b$, and the total spectrum of this coupled system

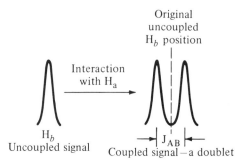

will appear as two doublets. In general, a proton signal will be split into $(n + 1)$ peaks by $(n)$ adjacent protons. Figure 3-7 outlines some similar splittings, and Fig. 3-8 illustrates two typical nmr spectra.

$$CH_3-\overset{|}{\underset{|}{C}}-$$
$$\phantom{CH_3-C}H$$
$$\phantom{CH_3}a\quad\phantom{-}b$$

The methyl signal (*a*) will appear as a doublet
$(n + 1) = 1 + 1 = 2$—split by $H_b$

The $-\overset{|}{\underset{|}{C}}-H$ signal (*b*) will appear as a quartet

$(n + 1) = 3 + 1 = 4$—split by $CH_3$

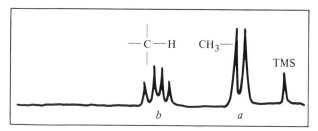

$$CH_3CH_2-$$
$$a\quad\ b$$

The methyl signal (*a*) will appear as a triplet—split by $-CH_2-$

The methylene signal (*b*) will appear as a quartet—split by $CH_3$

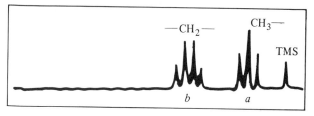

**Figure 3-7.** Some typical splitting patterns in NMR spectra.

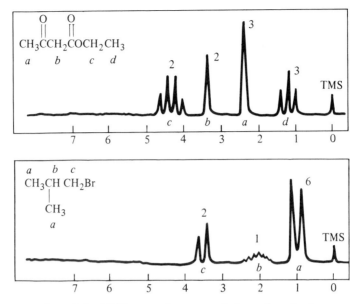

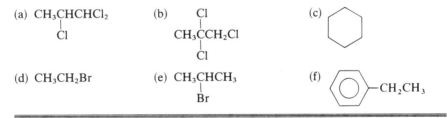

**Figure 3-8.**   NMR spectra for some typical organic molecules.

---

**Exercise 3-4:**   Predict the nmr spectra for the following simple compounds. Estimate the signal positions from Table 3-3 and calculate the splitting (if any) from the $(n + 1)$ formula. Assume that only *adjacent* protons contribute to signal splitting (identical protons do not split one another).

(a)  $CH_3CHCHCl_2$
         |
         $Cl$

(b)        $Cl$
              |
         $CH_3CCH_2Cl$
              |
              $Cl$

(c)

(d)  $CH_3CH_2Br$

(e)  $CH_3CHCH_3$
              |
              $Br$

(f)

---

## MASS SPECTROMETRY
## AND ORGANIC STRUCTURE

As the name implies, mass spectrometry is a technique by which the mass, or molecular weight, of an organic compound may be determined. A stream of highly energetic electrons is allowed to bombard a molecule in the gas phase at reduced pressure, and these electrons cause the molecule to fragment. The manner in which the breakup occurs is related to the molecular structure. To determine the molecular

weight, the *parent ion* peak, M⁺, must be identified.  Usually this peak is the highest
mass peak in the spectrum.

$$M \xrightarrow{e} M^\oplus + 2\,e$$

Molecule         Parent
in gas           molecular
phase            ion

$$M^\oplus \longrightarrow \text{Fragment ions}$$

The mass spectrometer actually measures the mass-to-charge ratio (m/e) of
each fragment ion by accelerating them through a magnetic field, the heavier
fragments being deflected less than lighter ions.  A plot of the relative intensities of
the various m/e peaks is referred to as the *mass spectrum*.  A typical mass spectrum is
shown in Fig. 3.9.  The M⁺ is prominent in this spectrum (m/e 58).  However, this is

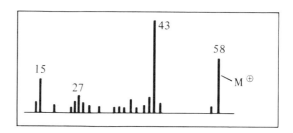

**Figure 3-9.**   Mass spectrum of acetone.

only one piece of information concerning the structure that can be deduced from the
spectrum.  An equally if not more important aspect is the *ion abundance*, or the
relative intensities of the various peaks.  Ion abundance is related to preferred
fragmentation patterns associated with the breakup of the molecule.  We have
previously learned that some organic ions are more stable than others, for example,
3° carbonium ions are more stable than 1° carbonium ions, and that benzylic and
allylic ions are particularly stable because of electron delocalization.  A molecular
ion, M⁺, tends to dissociate into its most stable fragments, then give rise to the most
abundant ions.  In the spectrum shown in Fig. 3-9, the most intense peak (m/e 43)
may be accounted for by loss of one methyl group, the main pathway by which the
molecular ion fragments.

$$[CH_3\overset{\overset{\displaystyle O}{\|}}{C}CH_3]^{\oplus} \longrightarrow [CH_3\cdot] + [CH_3\overset{\overset{\displaystyle O}{\|}}{C}{}^{\oplus} \longleftrightarrow CH_3C{\equiv}\overset{\oplus}{O}]$$

M⁺                                   m/e 43
m/e 58                          Fragment is
                          resonance-stabilized

**Exercise 3-5:**   In the following compounds, the $M^+$ peak and the most abundant ion(s) of the mass spectrum are given.  Postulate *reasonable* fragmentation patterns for each example:

(a)  $(CH_3CH_2)_3N$ $\qquad\qquad$ $M^+$ (m/e 101), m/e 86

(b)  $(CH_3)_3CCH{=}CH_2$ $\qquad$ $M^+$ (m/e 84), m/e 69, m/e 41

(c) $\qquad\quad$ O
$\qquad\qquad\quad$ ‖
$\qquad$ $(CH_3)_2CHCCH_3$ $\qquad$ $M^+$ (m/e 86), m/e 43

(d)  $CH_3OCH_2CH_2OCH_3$ $\qquad$ $M^+$ (m/e 90), m/e 45

(e)  $(CH_3)_3CCH_2NH_2$ $\qquad\quad$ $M^+$ (m/e 87), m/e 30

Many different fragmentation patterns are possible for large organic molecules, including more complicated reaction paths such as group eliminations and molecular rearrangements.  By comparing a large number of mass spectra for compounds in a series, or having similar structural features, an organic chemist may discern fragmentation patterns which allow him to interpret mass spectra of totally unknown compounds.  Many structural features may be determined from these fragmentation patterns that are difficult to distinguish on the basis of absorption spectroscopy (IR, UV, nmr) alone.

## REVIEW OF NEW TERMS AND CONCEPTS

Define each term and, if possible, give an example of each:

| | |
|---|---|
| Ultraviolet excitation and absorption | Infrared excitation and absorption |
| Absorptivity | Beer–Lambert law |
| $n \to \pi^*$ transition | $\pi \to \pi^*$ transition |
| Stretching frequency | Bending frequency |
| Characteristic frequency | Fingerprint region |
| Nuclear magnetic resonance | Chemical Shift |
| $\delta$ and $\tau$ scales | Spin–spin coupling |
| Mass spectrometry | Molecular peak |
| Fragmentation pattern | |

## PROBLEMS: CHAPTER II-3

**1.** One might expect the absorption of light in the ultraviolet region of the spectrum to yield very sharp line spectra corresponding to the $\Delta E$ between the two states (say,

$\pi \rightarrow \pi^*$); yet the recorded spectra are usually very broad peaks. Why?

    instead of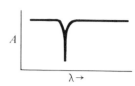

2. A vibrational stretching frequency is related to both the force constant (strength) of the bond and mass of the atoms involved. Would you expect a C—H bond to have a higher or lower stretching frequency than a C—D bond, assuming that the force constants were nearly the same?

$$\nu_{vib} = \frac{1}{2\pi} \sqrt{\frac{k}{\mu}}$$

$$\overset{\longleftrightarrow}{Ⓐ} \underline{\quad\quad} \overset{\longleftrightarrow}{Ⓑ}$$

where  $k$ = force constant
       $\mu$ = reduced mass

$$\mu = \frac{M_A M_B}{M_A + M_B}.$$

3. (a) Would you expect the O—H stretching frequency in pure water to be narrow or broad? Explain.

   (b) Would there be any difference in the appearance of the O—H frequency in a pure alcohol and one which was wet (say, 5% $H_2O$)? Explain.

4. Compare the UV spectra of 1,3-butadiene and 1,3,5-hexatriene (Table 3-1). Predict what the spectrum of 1,3,5,7-octatetraene might be like ($\lambda_{max} + a_{max}$).

5. Dehydration of $CH_3CH{=}CH{-}CH_2\overset{\overset{\displaystyle OH}{|}}{C}HCH_2CH_3$ yields two different products. What might they be and how would you distinguish between them using ultraviolet spectroscopy?

6. Two compounds having the same molecular formula, $C_2H_6O$, have quite different IR spectra. Compound A has a strong absorption at 3400 cm$^{-1}$, while compound B does not. Suggest formulas for A and B which account for this difference.

7. Suggest structures for the following compounds based on the available spectral data:

   (a) $C_3H_6O$    UV:  $\lambda_{max}$ 280 ($a_{max}$ 20)
                    IR:  3000, 1750 cm$^{-1}$
                    nmr:  $\delta$ 2.1 (singlet)

   (b) $C_3H_8O$    UV:  transparent, 200–360 nm
                    IR:  3400, 2980, 1465 cm$^{-1}$
                    nmr:  $\delta$ 1.8 (doublet, 6H), $\delta$ 3.6–4.2 (septet, 1H),
                          $\delta$ 4.8 (singlet, 1H)

(c) $C_6H_{15}N$     UV:   transparent
                 IR:   2980, 2900, 1465 cm$^{-1}$
            nmr:   $\delta$ 1.0 (triplet, 3H), $\delta$ 2.4 (quartet, 2H)

8. In the
$$\begin{array}{cc} H_a & H_b \\ | & | \\ -C & -C- \\ | & | \end{array}$$
system we showed that each proton signal will split into a doublet.

However, in
$$\begin{array}{c} | \\ -C-CH_3, \\ | \\ H_a \end{array}$$
$H_a$ is split into a quartet whose peak heights are in the ratio 1:3:3:1. Explain this ratio. (*Hint:* Consider all possible alignments of the methyl protons with $H_a$.)

9. Predict the appearance of $H_b$ proton signal for each of the following:

$$\begin{array}{ccc} H_a & H_b & H_c \\ | & | & | \\ -C & -C & -C- \\ | & | & | \end{array}$$

(a) $J_{ab} = J_{bc} = 5$ Hz

(b) $J_{ab} = 5$ Hz, $J_{bc} = 10$ Hz

(c) $J_{ab} = 2$ Hz, $J_{bc} = 20$ Hz

Remember that $J$ values tell you how *much* a signal has been split by the adjacent proton, and will be additive only if identical.

10. Predict what information the UV, IR, and nmr spectra of each of the following might contain. Draw the expected nmr spectrum.

(a) $CH_3CH_2CHCl_2$

(b)
$$\begin{array}{c} O \\ || \\ CH_3CH_2CCH_2CH_3 \end{array}$$

(c)
$CH_3$
$CH_3$

(d)
$$\begin{array}{c} CH_3 \\ | \\ CH_3-C-OH \\ | \\ CH_3 \end{array}$$

(e)
$CH_3$
$CH_3$

(f)
$$\begin{array}{c} O \\ || \\ CH_3CH_2COCH_3 \end{array}$$

(g) $(CH_3)_2CHCHO$

11. Cyclohexane, $C_6H_{12}$, has two different kinds of protons—6 axial and 6 equatorial—yet the nmr spectrum shows only one signal (singlet) at room temperature. Build a model of cyclohexane and identify the two different kinds of hydrogen. (If you have difficulty, consult Chapter II-4.) Suggest a reason or mechanism which would explain the room-temperature nmr.

**12.** Predict the most abundant ion or ions that you might expect in the mass spectra of the following compounds:

(a)

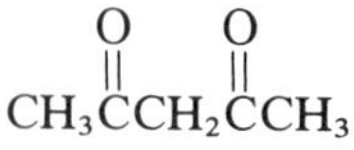

$$CH_3\overset{O}{\overset{\|}{C}}CH_2\overset{O}{\overset{\|}{C}}CH_3$$

(b)

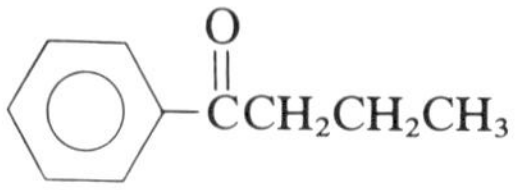

(c)

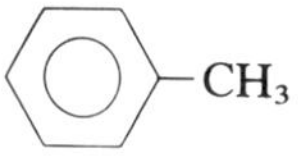

(d) $(CH_3)_4C$

(e)

# 4

# Conformation and Molecular Structure

At ordinary temperatures, molecules are in constant motion. If we neglect translational movement through space, molecular motion may be considered as a combination of various *vibrations* and *rotations*. Vibronic motion is relatively localized and rarely involves more than a few atoms at a time, however, rotational changes may involve the whole molecule. Although many *rotational isomers* may be envisioned, only a relatively few stable *conformations* are favored in which internal repulsive forces are minimized. In this chapter we discuss rotational isomerism, and how molecular conformation influences chemical stability.

## ROTATION ABOUT THE C—C BOND IN SIMPLE MOLECULES

In the early days of organic chemistry, rotation about the carbon–carbon single bond was considered to be "free." In other words, it was assumed that there were no interactions between groups on adjacent carbon atoms as they rotated past each other. As long as we think in terms of nuclei, this is probably true, but the electron clouds associated with the nuclei definitely do affect one another. These repulsive forces give rise to small energy barriers to rotation about single bonds, and the net

As Ⓑ rotates past Ⓐ, their electron clouds repel each other, creating an energy barrier to the rotation.

result of all internal repulsions is the tendency for molecules to assume preferred *molecular conformations* in which the repulsive forces are minimized. Thus at any particular time at a given temperature, an equilibrium between favorable and unfavorable conformations exists, with the former predominating. As we lower the temperature, the most stable conformation will increasingly predominate.

Unfavorable conformation
repulsive forces maximized

⇌

Favorable conformation
repulsive forces minimized

## REPRESENTATION OF CONFORMATION— NEWMAN PROJECTIONS

In our previous discussions of organic stereochemistry, the usefulness of projections for describing three-dimensional formulas has been emphasized. Conformational interactions can best be demonstrated by *Newman projections* in which the observer looks down the carbon–carbon bond at the projection of the six groups attached directly to the two carbons in question.

Look down C—C axis

≡

Front carbon atom
Rear carbon atom

⇌

120° rotation

# CONFORMATIONS OF ACYCLIC
HYDROCARBONS: ETHANE AND BUTANE

If one imagines the rotation of one carbon atom with respect to the other in ethane, it should be obvious that two extreme *conformational isomers* (*conformers*) may be drawn. In the *staggered form* (I), groups on adjacent carbons are *anti* to one another,

| Staggered form | Eclipsed form | Staggered form |
| :---: | :---: | :---: |
| I | II | III |

and thus intermolecular repulsions are minimized. As we rotate one carbon with respect to the other from the staggered form, however, the groups move closer together and repulsive forces increase until they reach a maximum in the *eclipsed form* (II). Continual rotation past the eclipsed configuration lowers the internal molecular energy by decreasing the repulsive interactions until we again reach an energy minimum with a new staggered form (III). The amount of energy required to carry out this rotational sequence (I → II → III) may be termed the *rotational energy barrier*. In ethane this barrier is about 3 kcal/mole, and at room temperature this means that most ethane molecules (>95%) are in the staggered conformation at any particular time. However, rotation *is* occurring in *all* ethane molecules continuously, with the molecules spending the majority of their time in the more stable conformation. This *dynamic behavior* may be illustrated by plotting the potential energy of a molecule versus its internal conformation (Fig. 4-1).

---

**Exercise 4-1:** The energy barrier to rotation between two conformations may be expressed as $\Delta G = -RT \ln K$ where $\Delta G$ is the free-energy difference, $T$ is the temperature (°K), and $K$ is the equilibrium constant $K = [B]/[A]$ for $A \rightleftharpoons B$. Calculate what the ratio of staggered and eclipsed forms must be for ethane at room temperature (298°K) and a $\Delta G$ value of 3 kcal/mole. Is the staggered form favored by at least 95%?

---

The rotational energy changes in butane and larger molecules are much more complex. In ethane only hydrogens were required to pass each other during each rotation. In butane, there are methyl–hydrogen and methyl–methyl interactions as well, as illustrated in Fig. 4-2.

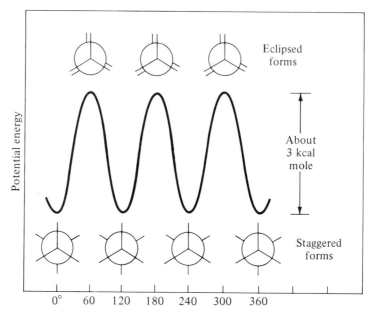

**Figure 4-1.** Rotational energy profile for ethane.

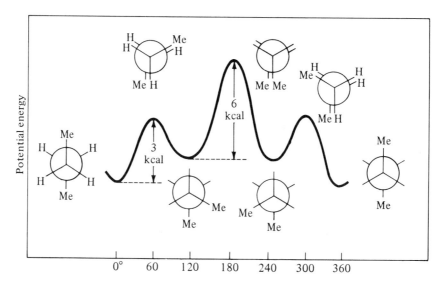

**Figure 4-2.** Rotational energy profile for butane.

360

It can immediately be seen that butane can be drawn in two different staggered forms, and that two different rotational energy barriers exist within the molecule. The completely staggered, or *anti*, form has the methyl groups separated as much as possible. In the second staggered form the methyl groups bear a *skew*, or *gauche*, relationship to one another. The *anti* form is about 0.8–1.0 kcal more stable than the *skew* form. At room temperature, butane may be considered to be oscillating

Anti form
(more stable)

Skew form

between the two staggered forms, with the *anti* form predominating at any particular instant.

**Exercise 4-2:**   Draw Newman projections for pentane, looking down the $C_2$—$C_3$ bond. How many staggered forms are possible? How many eclipsed forms? What is the most stable conformation? Draw a rough plot of how the potential energy of the molecule might change as a function of rotation about the $C_2$—$C_3$ bond (see Fig. 4-2).

## CONFORMATIONS OF HYDROCARBONS: CYCLOHEXANE

In the early days of organic chemistry all carbon rings were thought to be planar, or nearly so. Smaller rings, such as cyclopropane or cyclobutane are strained because of the large deviations of the C—C—C bond angles from the desired 109°28′. They are also relatively inflexible. Larger rings, however, not only are much more flexible, but can also accomodate tetrahedral bond angles without strain. The cyclohexane ring seems to occupy a particularly secure niche in nature's hierarchy in that it outnumbers all other-sized rings in naturally-occurring compounds, and hence deserves our close study. Build a model of cyclohexane and examine the carbon skeleton closely. Even though motion in a ring is restricted compared to acyclic structures, note that rotation of bonds in the cyclohexane ring is still possible, and that with care you can interchange two different rotational forms without breaking the ring. These conformations are referred to as *chair* and *boat* forms because of the peculiar shapes of the carbon skeletons.

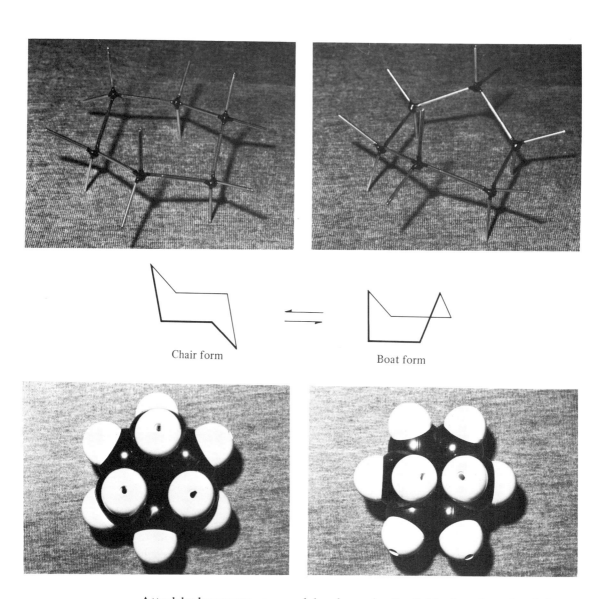

Chair form ⇌ Boat form

Attach hydrogens to your model and examine the finished product. Look down one of the C—C bonds. You should find that the chair form of cyclohexane has all bonds in staggered array, while the boat form has its bonds in eclipsed conformation. It should come as no great surprise, then, that the chair conformation of cyclohexane is the more stable and preferred conformation. Fig. 4-3 illustrates a common method of writing a projection formula for the chair form of cyclohexane and is referred to as a "Newman Projection."

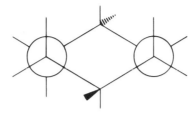

**Figure 4-3.** Newman projection of cyclohexane.

If you examine cyclohexane models quite closely, you will soon see that there appear to be two different kinds of hydrogens—a set of six that exist around the molecule's *equator*, and a set of six that are parallel to an *axis* drawn through the center of the molecule, as shown in Fig. 4-4. Build a model of a monosubstituted cyclohexane (say, 1-methyl or 1-chloro). Place the substituent in either an equatorial or an axial position. Now carry out any possible rotations of the ring and you will discover that it is possible to interconvert axial and equatorial isomers by "flipping" the ring (Fig. 4-5). During this process, all axial bonds in the one isomer are converted to equatorial bonds in the other (and vice versa). If we examine Newman

Six equatorial bonds                    Six axial bonds

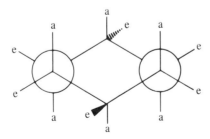

**Figure 4-4.** Cyclohexane has six equatorial and six axial bonds.

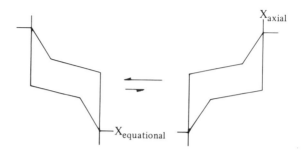

**Figure 4-5.** Rotational isomers are interconvertible by internal rotation.

projections, we can also determine which of the above two isomers is the more stable. When the substituent group is in an axial position, it is *skew* to two of the ring C—C bonds, while in an equatorial position it is not. We have previously seen in butane that *skew* or *gauche* conformations are 0.8–1.0 kcal/mole less stable than *anti* conformations. Thus equatorial isomers are more stable than axial isomers, and will predominate in any equilibrium.

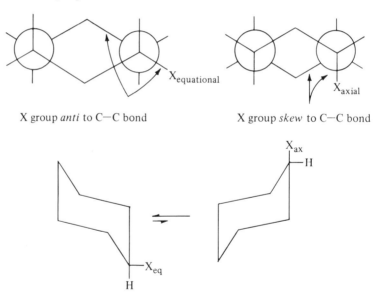

X group *anti* to C—C bond        X group *skew* to C—C bond

Equatorial isomer predominates at ordinary temperatures

## CONFORMATION AND RING
## GEOMETRIC ISOMERISM

Disubstituted cyclohexanes can exist as either *cis* or *trans* isomers. Our original definitions of *cis* (on the same side) and *trans* (across from) are more difficult to apply to ring systems undergoing constant rotational isomerization than they are for

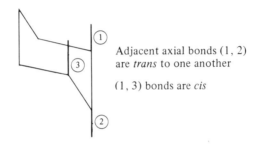

trans-1,2-Disubstituted
cyclohexane

cis-1,2-Disubstituted
cyclohexane

olefins. There are three possible conformational combinations for 1,2-disub-stitution: equatorial–equatorial (ee), equatorial–axial (ea), and diaxial (aa), and the question we must answer is which of these combinations can be matched with the appropriate geometric isomer.

Consider your model of cyclohexane. If you examine adjacent axial bonds, it is easy to see that they bear a *trans* relationship. Therefore, a 1,2-diaxial disubstituted

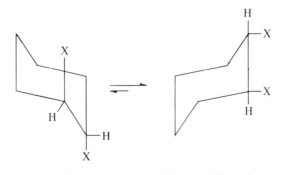

Adjacent axial bonds (1, 2)
are *trans* to one another

(1, 3) bonds are *cis*

cyclohexane must be a *trans* isomer. However, we also know that the diaxial compound is in conformational equilibrium with its 1,2-diequatorial counterpart, and furthermore, that the diequatorial conformer will predominate. If the ee and aa

Diaxial conformer                Diequatorial conformer

conformations are both *trans*, then the ea conformation *must* be *cis*. Projection formulas for these two isomers are shown below, along with a summary of conformations for *cis* and *trans* isomers for 1,2-, 1,3-, and 1,4-disubstituted cyclohexanes.

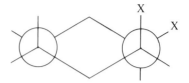

trans-1,2-Disubstituted
cyclohexane

cis-1,2-Disubstituted
cyclohexane

Table 4-1.   Conformations of Various
           Disubstituted Cyclohexanes

| Substitution | trans | cis |
|---|---|---|
| 1,2 | ee | ea |
| 1,3 | ea | ee |
| 1,4 | ee | ea |

**Exercise 4-2:**   Draw the *most stable* conformations for each of the following.  Refer to Table 4-1 if necessary.

(a)

Cl

Cl

(b)

Me

Me

(c)

Me

t-Bu

(d)

Me

t-Bu

(e)

Me

Cl

Cl

(f)

O

# CONFORMATIONAL CHANGES
# IN MACROMOLECULES

Large molecules of biological interest (macromolecules) can assume an almost infinite variety of conformations, but, like the small organic molecules discussed in this chapter, there are definite *preferred conformations*. Thus peptide chains and proteins can assume coiled, helical, and sheet conformations, depending on a variety of chemical and physical factors. Macromolecules are peculiarly susceptible to changes in their chemical or thermodynamic environment. Although the study of conformational changes in macromolecules is of extreme importance in biochemistry, a detailed discussion is beyond the scope of this text. The following examples will give you some indication of why biochemists must consider conformation and conformational changes in their attempts to understand protein structures and function.

Many polypeptides exist in helical conformations under normal conditions (see Chapter I-14). When these peptides are subjected to a change in either pH (hydrogen-ion concentration) or temperature, a change in conformation occurs which has been named the *helix-coil* transformation. The well-defined conformation of the α-helix is converted to a *random coil* whose conformation, as its name implies, lacks structural definition. In the pH-induced transformation, the conversion of ionizable groups from the ionized to the neutral state, or vice versa, is generally regarded as being responsible for the structural change. However, in the thermal conversion, conformation changes because of the increased vibrations of the atoms in the more rigid helix. When the vibrational energy becomes large, the rigid conformation converts to the random coil, which has more degrees of freedom.

Other more complicated examples of conformational changes include the *triple helix-random coil* conversion in collagen. In the basic collagen molecule, three left-handed helices form a super helix by wrapping around one another, and are held together by interchain hydrogen bonds. This unique system is converted to random chains by increasing temperature. Obviously, the manner in which the super helix unwraps itself is quite complicated and is still not totally understood. One thing is certain—the study of conformational changes in macromolecules is a fascinating subject, and one which will engage the minds of biochemists for many years to come.

## REVIEW OF NEW TERMS AND CONCEPTS

Define each term and, if possible, give an example of each:

| | |
|---|---|
| Rotational isomerism | Conformation |
| Conformer | Newman projection |
| Staggered conformation | Eclipsed conformation |
| *Skew* conformation | Rotational energy barrier |
| *Anti* configuration | Chair conformation |
| Boat conformation | Axial bond |
| Equatorial bond | Helix-coil transformation |

1. Answer the following questions by building and examining molecular models.

    (a) Do the following molecules have different axial and equatorial positions to which the X group might be bonded?

    (b) For the molecules in part (a) for which you found different axial and equatorial isomers, predict whether the axial isomer would be more or less stable than the corresponding axial isomers of cyclohexane:

2. Decalin, $C_{10}H_{18}$, consists of two cyclohexane rings fused together. The hydrogens at the ring junction may be either *cis* or *trans* to one another. Build models of both isomers and determine their most stable conformations. Draw them.

3. (a) Would you expect the percentage of the less stable conformations in any conformational equilibrium to increase or decrease with increasing temperature? Explain.

    (b) Axial and equatorial hydrogens are chemically different, and thus magnetically different. Yet the nmr of cyclohexane shows only one peak at room temperature. Why? Propose an experiment that might allow you to "see" both protons in the nmr spectrum.

4. Build a model of cycloheptane. Examine the possible molecular rotations and see how many *different* conformations you can discover. Draw them and predict which might be the most and least stable.

5. Would you expect the helix to random coil transition to be reversible? In other words, does the helix reassemble spontaneously when a solution containing only random coils is cooled?

# 5

# Aromatic Substitution

In Chapter I-6 we learned that aromatic molecules such as benzene react primarily by *electrophilic substitution*. In this chapter we will see how substituted groups attached to the aromatic ring affect the substitution pathway, directing the incoming electrophile to the *ortho, para*, or *meta* positions. We will also see how this behavior may be correlated with the ability of these substituents to either *donate* or *withdraw* electron density from the ring. Finally, we will also learn that certain groups can promote *nucleophilic aromatic substitution*.

## ELECTROPHILIC SUBSTITUTION REVISITED

We have previously seen that most substitution reactions of benzene and other aromatic molecules involve an initial attack of an electrophile. The mechanism of this attack is normally envisioned as occurring by discrete steps involving both $\pi$ and $\sigma$ complexes (Fig. 5-1):

Since the initial electrophilic attack involves the $\pi$-electron cloud of the aromatic substrate, it should not be surprising that substituent groups which affect the total $\pi$-electron density in the ring have a pronounced effect on substitution rate. Thus we would expect groups that *increase* electron density to promote substitution, and we refer to this class of substituents as *activating groups*. Such behavior is

369

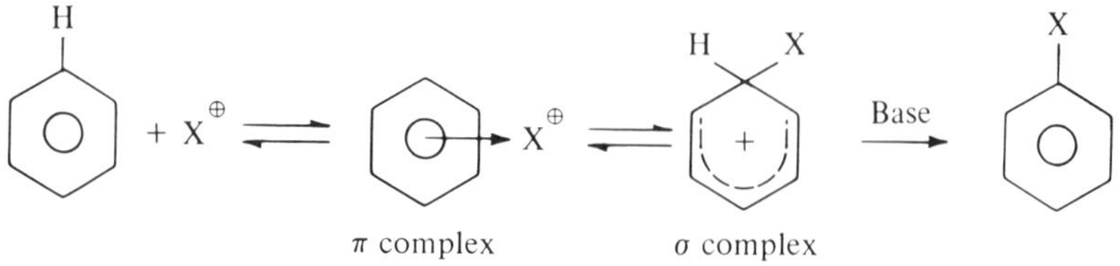

H
X
H
X
Base
π complex
σ complex


**Figure 5-1.** Generalized mechanism of electrophilic aromatic substitution.

manifested in studies of the relative substitution rates for a series of monosubstituted benzenes. Substituted benzenes containing an activating substituent will react *faster* than benzene, while those with a deactivating group will react more slowly than benzene.

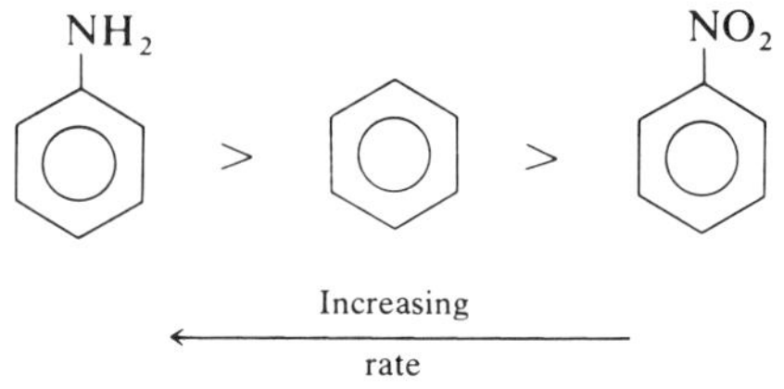

NH₂
NO₂
Increasing
rate


—$NH_2$ is an activating group
—$NO_2$ is a deactivating group

Relative rates of electrophilic substitution

## ORIENTATION OF SUBSTITUTION

In monosubstituted benzenes there are three possible positions that an incoming electrophile might attack: *ortho*, *meta*, or *para*. If the attack were random in nature,

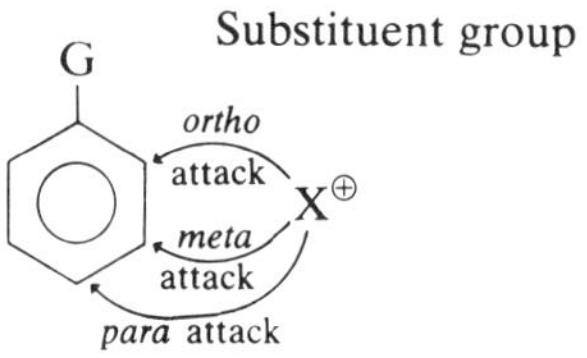

Substituent group
G
ortho
attack
X
meta
attack
para attack


we might expect a 2 : 2 : 1 product ratio, since there are two *ortho* and *meta* positions, but only one *para*. However, it has been found experimentally that substitution is *not* random, but falls into two distinct classes. Substituent G groups direct an incoming

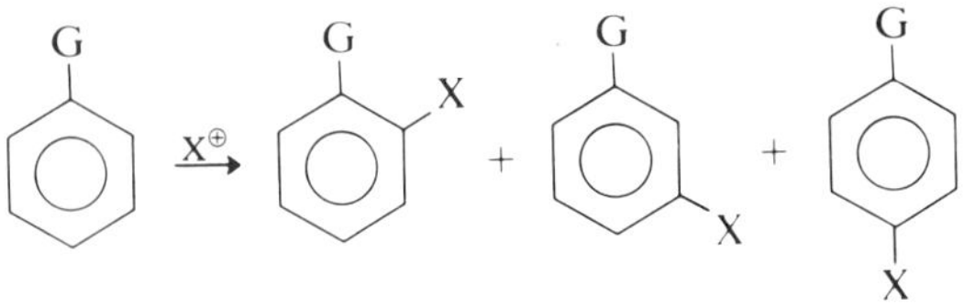

2   :   2   :   1

Random electrophilic substitution is never observed

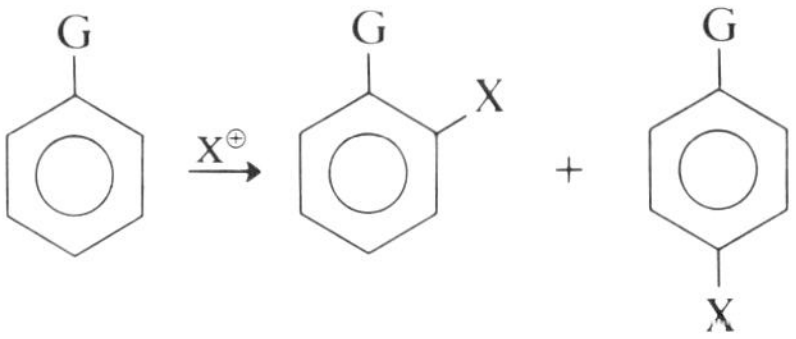

G is an *o,p*-director

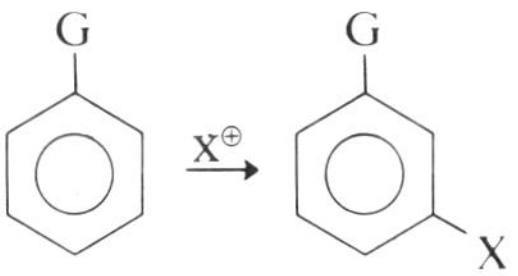

G is an *m*-director

electrophile either to the *ortho* and *para* positions (*o,p*-directors), or to the *meta* positions (*m*-directors). We have previously seen that substituent groups also may be divided into two groups (activating and deactivating), based on their effects on relative substitution rates. What is most interesting about these two phenomena is

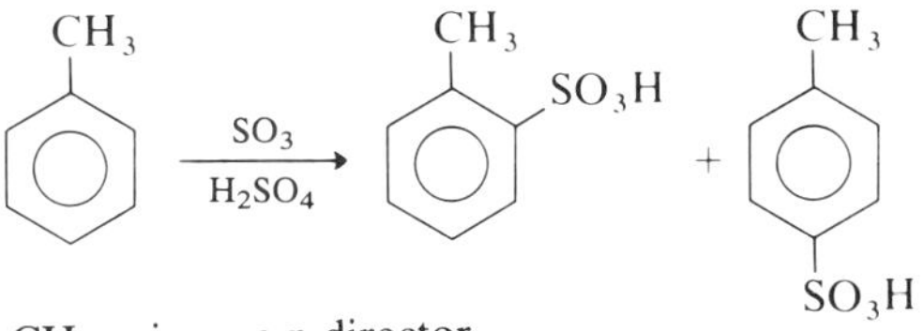

CH₃— is an *o,p*-director

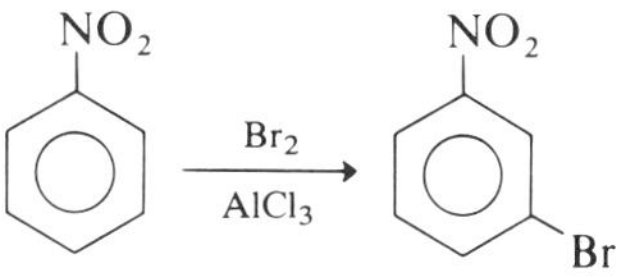

NO₂— is an *m*-director

that they seem to be related: Activating groups are also *o,p*-directors, while deactivating groups are also *m*-directors. Table 5-2 lists the more common groups and their direct influence. In comparing this list with Table 5-1, note that the only exception is halogen (see Problem 7 at the end of this chapter).

Table 5-1.   Substituent Effect on Electrophilic Substitution
              Reactivity in Aromatic Molecules

| Activating | Deactivating |
|---|---|
| $-NH_2$, $-NHR$, $-NR_2$<br>$-OH$, $-OR$<br><br>$\overset{O}{\overset{\|}{-O-C-R}}$,  $\overset{O}{\overset{\|}{-NH-C-R}}$<br><br>$-R$, | $-\overset{+}{N}R_3$<br>$-NO_2$, $-SO_3H$, $-COOH$<br>$-COR$, $-CHO$, $-COOR$<br>$-C\equiv N$<br>$-$Halogen |

Table 5-2.   Substituent Effect on Orientation
              in Electrophilic Substitution

| *Ortho,Para*-Directors | *Meta*-Directors |
|---|---|
| $-NH_2$, $-NHR$, $-NR_2$<br>$-OH$, $-OR$<br><br>$\overset{O}{\overset{\|}{-O-C-R}}$,  $\overset{O}{\overset{\|}{-NH-C-R}}$<br><br>$-R$, <br><br>$-$Halogen | $-\overset{+}{N}R_3$<br>$-NO_2$, $-SO_3H$, $-COOH$<br>$\overset{O}{\overset{\|}{-C-R}}$, $\overset{O}{\overset{\|}{-C-H}}$, $\overset{O}{\overset{\|}{-C-OR}}$<br>$-C\equiv N$ |

**Exercise 5-1:**   Determine the substitution products for each of the following reactions. Refer to the listing of substituent directing ability in Table 5-2.

(a)

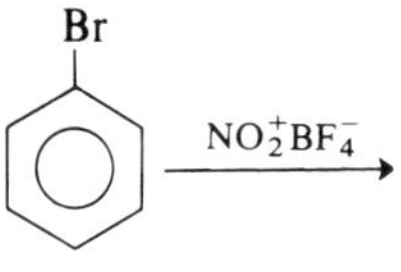

(b)

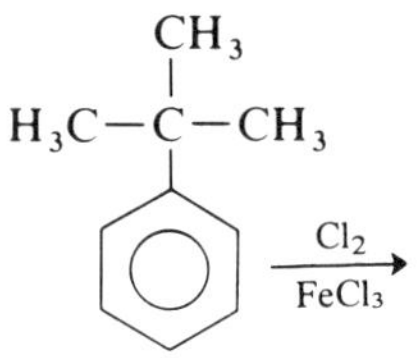

(c)

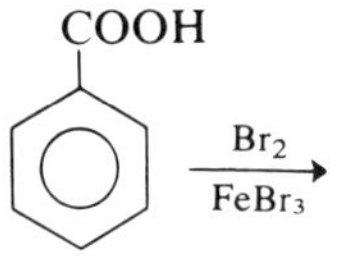

(d)

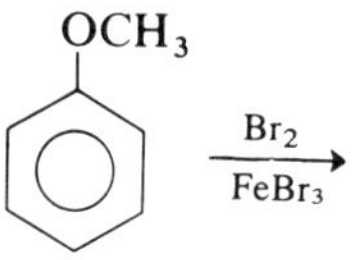

In substituted benzenes containing more than one substituent group, the substituents may either reinforce each other's directive ability, or they may oppose one another. In the case of an *m*-director versus an *o,p*-director, remember that the *meta* substituent is also a deactivating group, while the *o,p*-director activates the *ortho* and *para* positions with respect to the *meta* position. Thus the *o,p*-director will dominate the orientation of the attacking electrophile.

CH₃ / CH₂CH₃ structure

Both are *o,p*-directors. Attack will occur at both 1 and 2

CH₃ / NO₂ structure

CH₃ is an *o,p*-director; NO₂ is a *m*-director. Attack will occur at positions 1 and 2 and to a lesser extent at 3. Attack will *not* occur at 4

OCH₃ / NO₂ structure

OCH₃ is an *o,p*-director; NO₂ is a *m*-director. Attack will occur at position 1 only.

---

**Exercise 5-2:**  Predict the most probable nitration site(s) in each of the following molecules. State your reasons for choosing these attack positions.

(a)    OH ... NO₂

(b)    OH ... NO₂

(c)    NO₂ ... NO₂

(d)    COOH, Br

(e)    biphenyl

(f)    biphenyl with NO₂, NO₂

---

There are two possible explanations for the unique directive ability of each different substituent group in electrophilic aromatic substitution. Since the reaction involves an electrophilic attack on the $\pi$-electron cloud of the substituted benzene, one might imagine that the substituent group somehow reduces or increases the electron density at either the *o,p* or *m* positions, thus allowing attack at the remaining position(s) with the higher electron concentrations. An alternative postulate involves the stabilization of the intermediary $\sigma$ complex. A substituent group may either stabilize or destabilize the various possible $\sigma$ complexes derived from the attack at *o,p* or *m* positions, and the reaction pathway would proceed via whichever intermediate achieved the greater degree of stabilization. Let us consider how these mechanisms might operate.

### Electron Donation and Withdrawal

The general mechanism for electrophilic substitution involves an intermediary $\sigma$ complex which has a delocalized positive charge. Groups that *donate* electron density stabilize carbonium ions, while those that *withdraw* electron density destabilize such ions. Compare the structures of two typical substituent groups, $-NH_2$ and $-NO_2$.

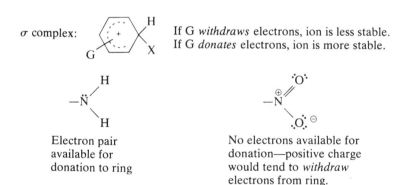

$\sigma$ complex:

If G *withdraws* electrons, ion is less stable.
If G *donates* electrons, ion is more stable.

Electron pair available for donation to ring

No electrons available for donation—positive charge would tend to *withdraw* electrons from ring.

The amino group ($-NH_2$) has an electron pair which may readily be donated to the ring, thus stabilizing any developing positive charge as the $\sigma$ complex is formed. However, the nitro group is adjacent to the ring and would only destabilize a positive charge in the ring. An amino group is therefore an activating group, while nitro is a deactivating group.

In general, atoms or groups capable of donating an electron pair to the aromatic ring are activating groups.

374

Deactivating groups usually are attached to the ring by an atom which is multiply-bonded to other atoms of greater electronegativity.  Since this results in a full or partial positive charge on these atoms, they are incapable of electron donation to the ring, and in fact withdraw electron density.

## Stabilization of the $\sigma$ Complex

In order to assess the ability of a group to either stabilize of destabilize an intermediary $\sigma$ complex, let us examine the result of a typical electrophilic attack at the three possible positions for both amino and nitro substituents:

Attack at *ortho* position

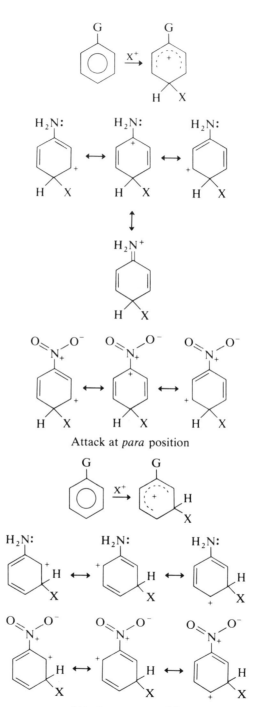

Attack at *para* position

Attack at *meta* position

376

Several structural features of the $\sigma$ complexes produced by $o$, $m$, or $p$ attack may be seen immediately:

1. Attack at either the *ortho* or *para* position to an amino group produces a $\sigma$ complex which is stabilized by electron donation of the unshared pair on the amino nitrogen. This additional delocalization thus favors attack at these positions over *meta* attack where such delocalization is impossible.
2. Attack at either the *ortho* or *para* position to a nitro group produces a $\sigma$ complex which is *destabilized* by having two positive charges adjacent to one another—a highly unstable electrostatic interaction. This destabilization thus forces the incoming electrophile to attack the *meta* position *by default*.
3. An amino group donates electron density to the ring, but only when attack is at the *ortho* or *para* positions. Thus it is an activating group as well as an $o,p$-director.
4. A nitro group withdraws electron density from the ring, but this withdrawal tends to be greater from the *ortho* and *para* positions, thus favoring attack only at the *meta* position.

The preceding arguments are emphasized by consideration of the resonance structure of aniline and nitrobenzene:

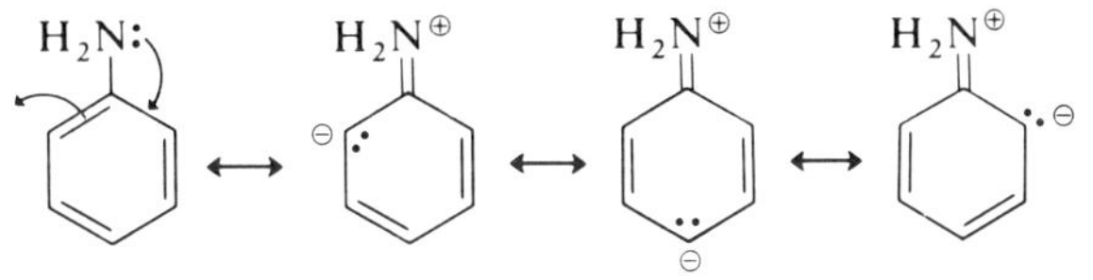

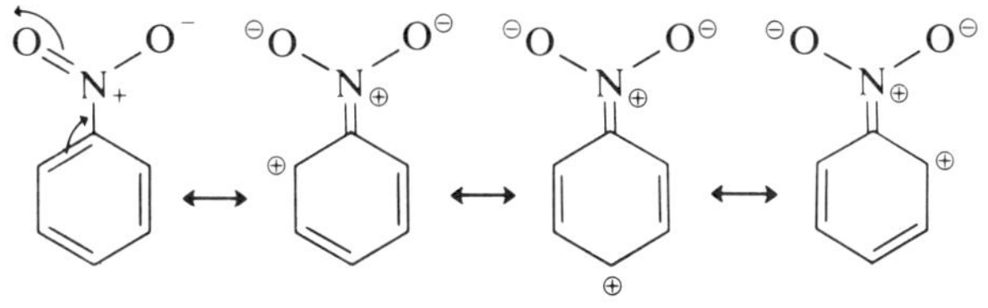

**Exercise 5-5:**   Write out the equations for *ortho*, *meta*, and *para* attack of $NO_2^+$ on toluene. Explain why the $CH_3$— group is an *ortho-*,*para*-director based on the structure of the $\sigma$ complexes formed by $o$, $m$, and $p$ attack.

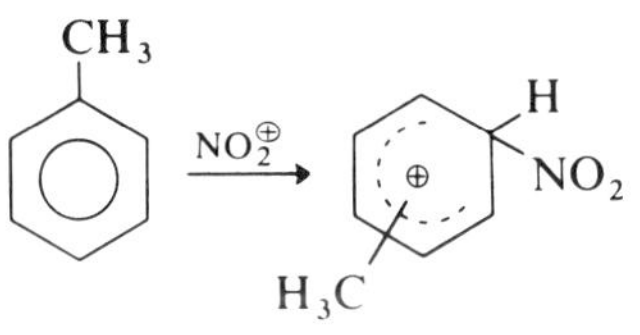

# NUCLEOPHILIC AROMATIC SUBSTITUTION

Aromatic and vinylic halides are much less reactive than the corresponding alkyl halides. In part this is due to the repulsion between the electron-rich nucleophile and the $\pi$-electron clouds. We have seen however, that substituents attached to an aromatic ring can markedly affect the ring's $\pi$-electron density. In order for nucleophilic substitution to become possible, several strong electron-withdrawing groups must be present to lower the electron density significantly and allow the approach of the nucleophile. The greater the number of electron-withdrawing groups on the ring, the milder the reaction conditions needed for substitution to occur, as seen below.

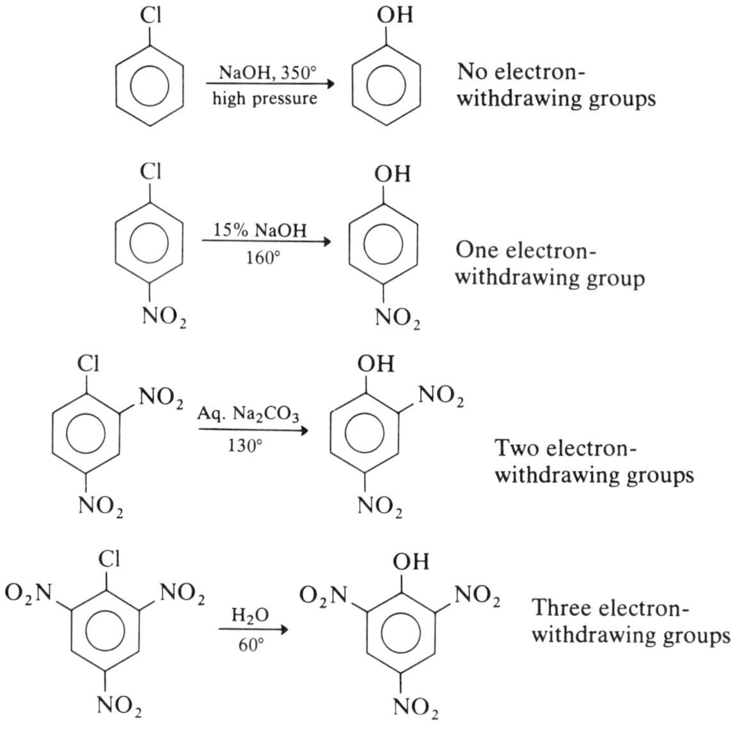

It can readily be seen that as the number of electron-withdrawing groups increases, the substitution of HO for Cl becomes easier, which is reflected both in the reaction conditions (temperature) and in the strength of the attacking nucleophile ($HO^- > H_2O$). Other substitution reactions occur with similar ease if the ring contains at least two deactivating groups.

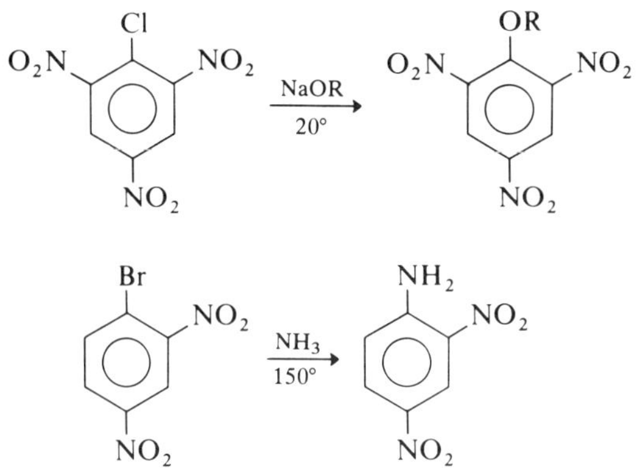

The mechanism for aromatic nucleophilic substitution is usually represented as an addition–elimination reaction. The intermediate involved is the negatively charged counterpart of the $\sigma$ complex we discussed previously for electrophilic substitution, and is a highly stabilized carbanion.

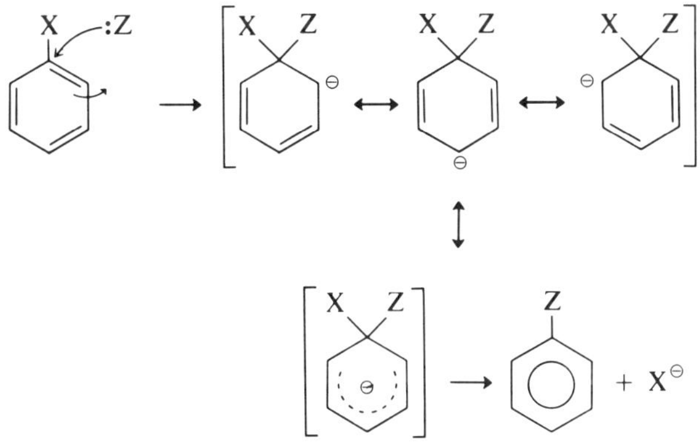

$\sigma$ complex

---

**Exercise 5-6:** In the examples of nucleophilic aromatic substitution shown in this chapter, the electron-withdrawing groups are *ortho* and/or *para* to the halide. Would they be equally effective if they were *meta* to the leaving group? (*Hint:* Consider the stabilities of the intermediary $\sigma$ complexes.)

# REVIEW OF NEW TERMS AND CONCEPTS

Define each term and, if possible, give an example of each:

Activating group                          Deactivating group
Electron-donor group                      Electron-withdrawing group
*o,p*-directing group                     *m*-directing group
Nucleophilic aromatic substitution

## PROBLEMS: CHAPTER II-5

**1.** Predict the products of the following reactions; pay particular attention to the *relative* directing abilities of the substituent groups.

(a)

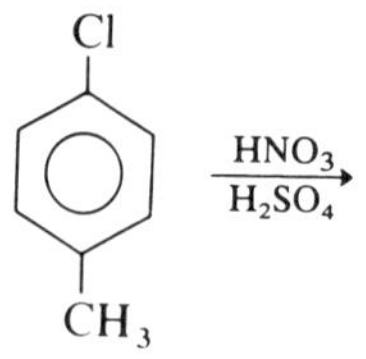

(b)

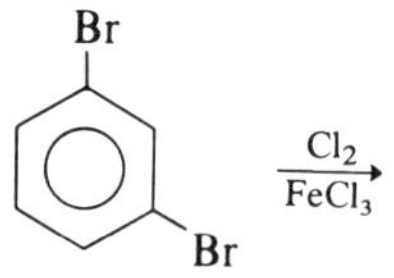

(c)

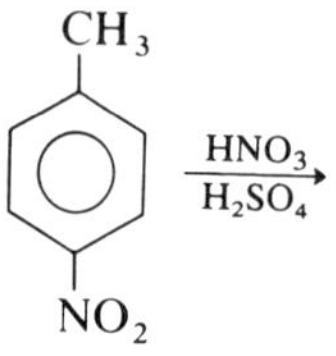

(d)

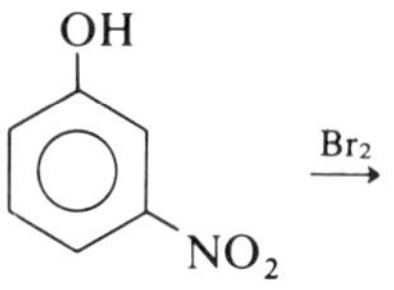

(e)

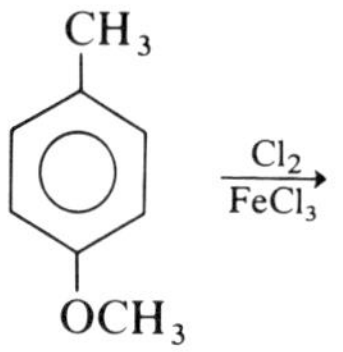

(f)

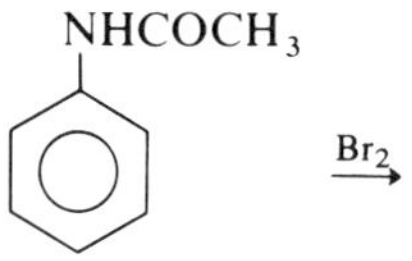

(g)

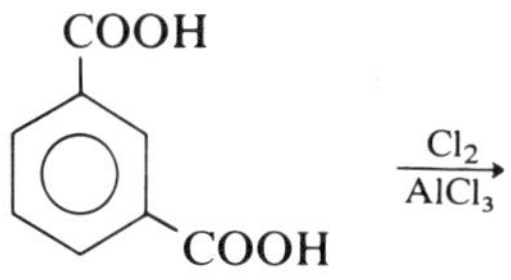

**2.** In each of the following, predict the more reactive compound under the given reaction conditions. Explain your choices.

(a)

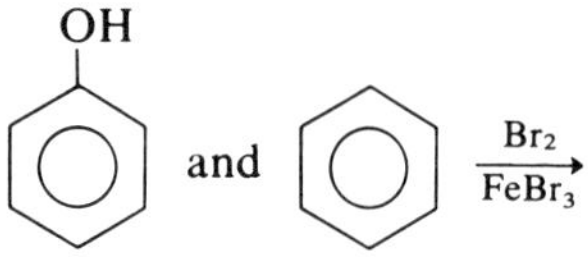

(b)

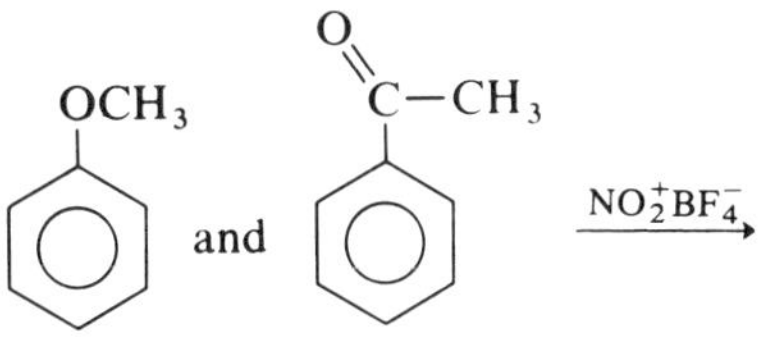

(c)

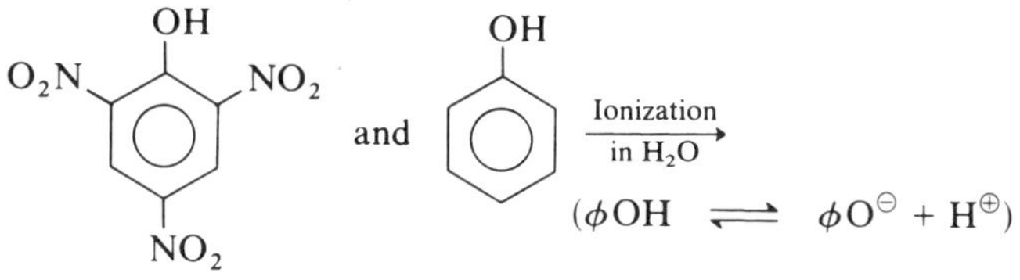

$$(\phi OH \rightleftharpoons \phi O^{\ominus} + H^{\oplus})$$

(d)

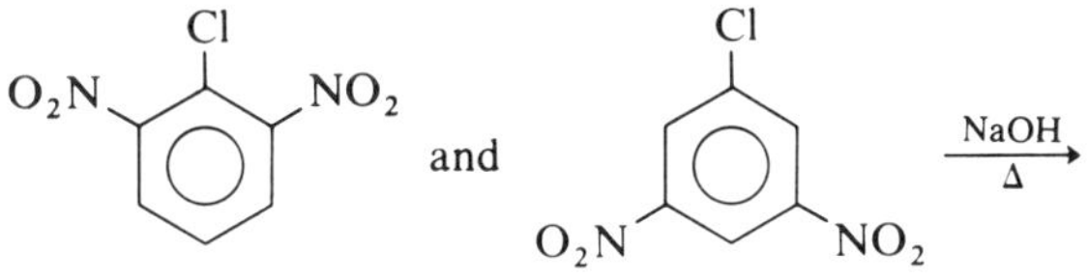

(e)

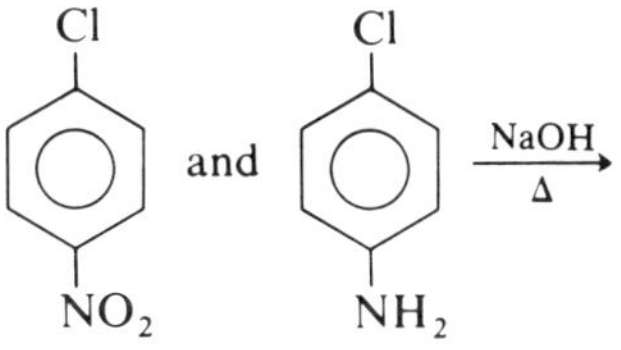

3. Give detailed structures for the $\sigma$ complexes derived from attack at *ortho*, *meta*, and *para* positions for each of the following reactions. Look up the directive ability of each substituent group in Table 5-2. Rationalize, in each case, why the group directs the incoming electrophile to the experimentally observed positions.

(a)

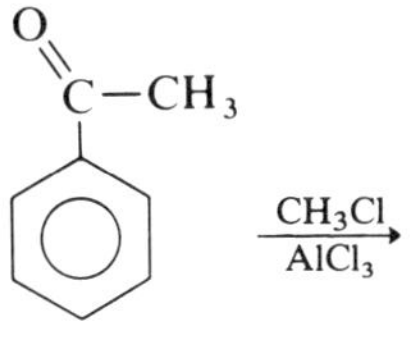

(b)   t-Bu

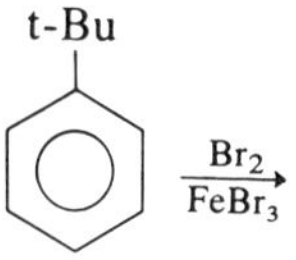

(c)   OCH$_2$CH$_3$

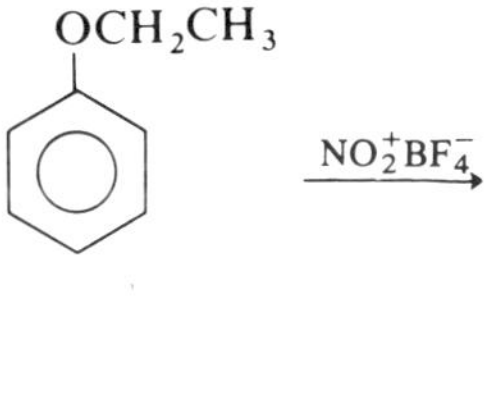

(d)   SO$_3$H

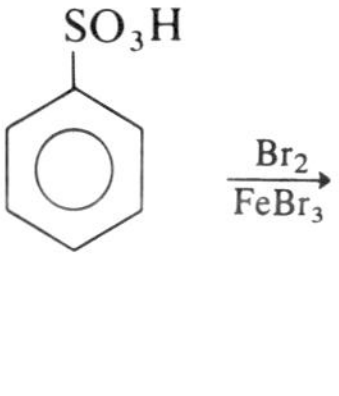

**4.** The amino group, $NH_2$, is one of the most powerful *o,p*-directing groups. Yet when aniline is nitrated with $H_2SO_4$, a significant quantity of *meta* isomer is obtained. Why?

**5.** Nitrobenzene does not undergo Friedel–Crafts alkylation. Can you suggest a reason?

**6.** Would you expect 4-chloro and 4-bromo pyridine to be more or less reactive toward nucleophilic substitution than bromo or chlorobenzene? Explain.

**7.** Now that you have an understanding of the mechanisms of electron withdrawal and donation, can you postulate why halogens are *ortho-*,*para*-directors and yet deactivate aromatic rings? (*Hint:* Consider other possible electronic effects in addition to resonance.)

# 6

# Fats, Oils, Waxes, Detergents, and Other Lipids

In Chapter I-12 we discussed how large straight-chain carboxylic acids occurred in nature as triglycerides, which are referred to as *fats*. Actually, fats, oils, waxes, soaps, and many other related substances may all be classified in a much broader classification known as *lipids*. Fats are energy reservoirs in the bodies of animals and they also function as a transport vehicle for other nonpolar substances. Cell walls contain *phospholipids* whose unique physical and chemical properties regulate membrane permeability. In this chapter we will review the nature of triglycerides and fatty acids, and how the components are related to function in nature. We will also learn how *soaps* may be obtained from fats, the "mechanism" of cleansing, and how synthetic *detergents* may be developed to imitate and improve soap behavior. Finally we will briefly examine the more complex lipids such as *steroids* and *prostaglandins* and why they are important to life in general.

## GENERAL CLASSIFICATION OF LIPIDS

Lipids may be divided into several distinct classes. Fats, oils, and waxes are *simple lipids* composed of esters of long-chain carboxylic acids. Fats and oils are triglycerides, while waxes are esters formed from long-chain monohydric (one OH group) alcohols. *Complex lipids* include both phospholipids, composed of glycerides in which one phosphate group has replaced a fatty acid, and *glycolipids*, glycerides containing a sugar unit. There are also several classes of compounds referred to as *derived lipids* which include terpenes, steroids, fat-soluble vitamins, and prostaglandins. In the following sections we will discuss several of these classes individually.

## FATTY ACIDS IN FATS, OILS, AND WAXES

Each fat and oil present in either plants or animals has a characteristic fatty-acid composition. In general, even-numbered, saturated, straight-chain acids are found together with certain unsaturated acids containing 16 or 18 carbons and one, two, or three double bonds. The most common of these acids are listed in Table 11-6 (Part 1, page 227), and include lauric ($C_{12}$), myristic ($C_{14}$), palmitic ($C_{16}$), and stearic ($C_{18}$) saturated acids; and oleic ($C_{18}$, 1 C=C), linoleic ($C_{18}$, 2 C=C), and linolenic ($C_{18}$, 3 C=C) as the major unsaturated acids.

Although the individual fatty-acid composition in a particular fat or oil may vary over a narrow range, in general the composition is quite characteristic. Thus beef fat is quite different from lard, and peanut oil quite different from cottonseed oil, even though the same fatty acids are present. A comparison of some familiar fats and oils is given in Table 6-1.

Table 6-1.   Fatty-Acid Composition of Typical Fats and Oils

| Fat or oil | Saturated Fatty Acids (%) | | | | Unsaturated Fatty Acids (%) | | |
|---|---|---|---|---|---|---|---|
| | $C_{12}$ | $C_{14}$ | $C_{16}$ | $C_{18}$ | $C_{18}$ (1) | $C_{18}$ (2) | $C_{18}$ (3) |
| Beef tallow | 0.2 | 2–3 | 25–30 | 21–26 | 39–42 | 2 | — |
| Lard | — | 1 | 25–30 | 12–16 | 41–51 | 3–8 | — |
| Butter | 1–4 | 8–13 | 25–32 | 8–13 | 22–29 | 3 | — |
| Cottonseed oil | — | 0–3 | 17–23 | 1–3 | 23–44 | 34–55 | — |
| Peanut oil | — | 0.5 | 6–11 | 3–6 | 39–66 | 17–38 | — |
| Linseed oil | — | 0.2 | 5–9 | 4–7 | 9–29 | 8–29 | 45–67 |

For example, a "typical" sample of peanut oil might contain 9% palmitic acid, 59% oleic acid, and 21% linoleic acid, while a "typical" cottonseed oil sample would have 20% palmitic, 30% oleic, and 45% linoleic oils. Thus even though they have the same components, the differing percentage compositions give each particular fat or oil its own peculiar characteristics.

The blend of fatty acids in the triglyceride has a profound effect on its physical characteristics. The main distinguishing difference between a fat and an oil is the relative percentage of saturated versus unsaturated acids. Fat glycerides are composed mainly of saturated fatty acids, while oil glycerides have high percentages of unsaturated fatty acids. The presence of unsaturation, in essence, lowers the melting points of the triglyceride. A striking example of this relationship is found in the manufacture of Crisco, which is made primarily from cottonseed oil, a liquid oil. In order to make the product convenient to use, however, the cottonseed oil is partially hydrogenated to convert the oil into a low-melting solid fat ($>C=C< + H_2 \rightarrow >CH-CH<$).

## SAPONIFICATION OF FATS AND OILS: SOAP FORMATION

When glycerides are hydrolyzed with strong base, the process is known as *saponification*. Glycerol is released and alkali metal salts of the component fatty acids are formed.

$$
\begin{array}{c}
\overset{\displaystyle O}{\underset{\displaystyle \|}{\phantom{.}}}\\
CH_2OCR_1\\
\overset{\displaystyle O}{\underset{\displaystyle \|}{\phantom{.}}}\\
CHOCR_2 \quad \xrightarrow[\Delta]{NaOH}\\
\overset{\displaystyle O}{\underset{\displaystyle \|}{\phantom{.}}}\\
CH_2OCR_3
\end{array}
\qquad
\begin{array}{c}
Na^{\oplus}\overset{\ominus}{O}-\overset{\overset{\displaystyle O}{\displaystyle \|}}{C}-R_1\\
Na^{\oplus}\overset{\ominus}{O}-\overset{\overset{\displaystyle O}{\displaystyle \|}}{C}-R_2 + \begin{array}{c}CH_2OH\\ CHOH\\ CH_2OH\end{array}\\
Na^{\oplus}\overset{\ominus}{O}-\overset{\overset{\displaystyle O}{\displaystyle \|}}{C}-R_3
\end{array}
$$

This reaction has been known since antiquity, when primitive man rendered animal fat with potash obtained from the burnt wood of campfires. The sodium salts of fatty acids are *soaps*, which have changed very little from ancient times, with the possible exception of increased product purity.

The mechanism of soap action has intrigued man since its discovery. We know that individual sodium alkanoates associate in solution to form conglomerates called *micelles* in which like groups associate. The long hydrocarbon chains dissolve each

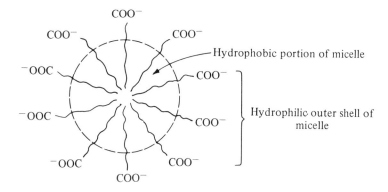

Representation of soap micelle

other to form a spherical globule whose outer circumference contains the ionic carboxyl groups. This outer polar shell is hydrophilic (water-loving), while the interior of the micelle is hydrophobic (water-fearing). As might be expected, individual micelles repel each other through like-charge forces. Soap's cleansing action results when the hydrophobic hydrocarbon chain dissolves in oil, fat, or grease adhering to clothes or skin. The hydrophilic carboxyl end of the soap enables the oil, fat, or grease to be dispersed in solution, a process known as *emulsification*. Thus the offending dirt is prevented from recollecting and can be rinsed away with water.

| Grease on cloth | Soap micelle | Clean cloth | Emulsified grease dispersed in solution |

## PROBLEMS WITH SOAP AND THE ADVENT OF DETERGENTS

Much of the world's water contains dissolved minerals, particularly magnesium and calcium salts (such water is commonly called *hard water*). These minerals precipitate soap from solution, forming the familiar soap scum (bathtub "ring"). To alleviate this problem, synthetic detergents (*syndets*) were developed. The ideal syndet retains the basic properties of soap (micelle formation), but will not precipitate in hard water. Long-chain alkyl sulfonates were found to be ideal candidates for soap

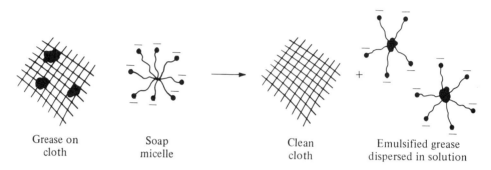

$$CH_3(CH_2)_nCH_2OH \xrightarrow{\text{H}_2\text{SO}_4} CH_3(CH_2)_nCH_2OSO_3H$$

$$n = 10, 12, 14, \ldots$$
Alkyl hydrogen sulfate

$$\downarrow \text{NaOH}$$

R⎍⎍⎍⎍⎍⎍⎍SO$_3^{\ominus}$Na$^{\oplus}$     $CH_3(CH_2)_nCH_2OSO_3^{\ominus}Na^{\oplus}$

Hydrophobic      Hydrophilic      Sodium alkyl sulfonate
end                   end              (syndet)

substitutes. Another class of detergents also widely used today are the alkyl benzene sulfonates (**ABS**).

R⎍⎍⎍⎍⎍⎍⎍⎍⎍⎍⎍⎍〈◯〉—SO$_3^{\ominus}$ Na$^{\oplus}$
Long straight alkyl chain

**Exercise 6-1:** Show how the following syndets might be prepared from naturally-occurring fatty acids:

(a) $CH_3(CH_2)_{10}CH_2OSO_3^{\ominus}Na^{\oplus}$ ("Dreft")

(b)

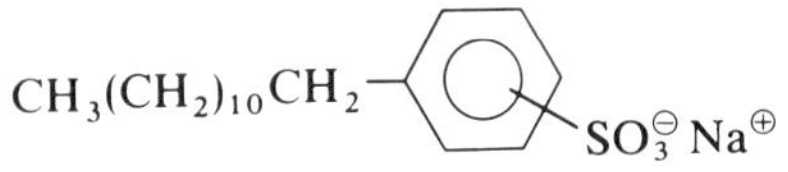

$$CH_3(CH_2)_{10}CH_2 - \hspace{-1em}\bigcirc\hspace{-1em} - SO_3^{\ominus}\ Na^{\oplus}$$

## WAXES

Waxes are esters formed from long-chain carboxylic acids, but instead of being bound in a glyceride, the alcohol portion is a long-chain monohydric alcohol. Waxes are not energy reservoirs like fats or oils, but serve instead as protective coatings in plants. Both the acid and the alcohol portions normally contain more carbons in the chain than are found in most fats and oils, typically $C_{24}$–$C_{36}$. Examples of three common waxes found in plants are shown below.

$$CH_3(CH_2)_{24}\overset{\overset{\displaystyle O}{\|}}{C}OCH_2(CH_2)_{28}CH_3 \quad \text{Myricyl cerotate}$$

$$CH_3(CH_2)_{14}\overset{\overset{\displaystyle O}{\|}}{C}OCH_2(CH_2)_{24}CH_3 \quad \text{Ceryl palmitate}$$

$$CH_3(CH_2)_{16}\overset{\overset{\displaystyle O}{\|}}{C}OCH_2(CH_2)_{24}CH_3 \quad \text{Ceryl stearate}$$

## PHOSPHOLIPIDS

Phosphoglycerides are similar to normal glycerides except that one carboxylic-acid residue has been replaced with a phosphoric-acid group. The parent structure for this family is referred to as a *phosphatidic acid*. Phospholipids contain not only the

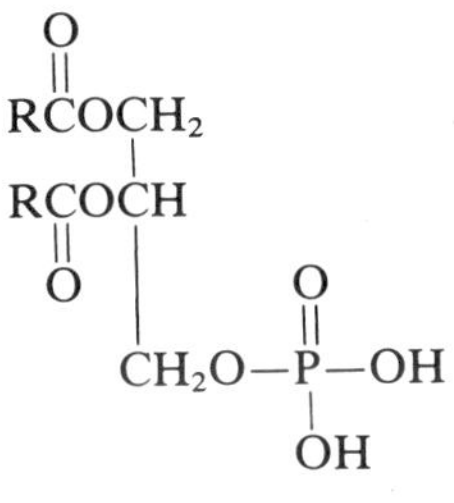

$$\begin{array}{l} \overset{\overset{\displaystyle O}{\|}}{R}COCH_2 \\[4pt] RCOCH \\[2pt] \overset{\displaystyle \|}{O}\ \ \ \ \ \ \ \ \ \overset{\displaystyle O}{\|} \\[2pt] \ \ \ \ \ CH_2O-P-OH \\[2pt] \ \ \ \ \ \ \ \ \ \ \ \ \overset{}{O}H \end{array} \qquad \text{Phosphatidic acid}$$

nonpolar ester links, but also a polar phosphoric-acid end-group. The two remaining acidic OH groups are still free for ester formation. Two such classes are formed from ethanolamine and serine (cephalins) and choline (lecithin).

$$\begin{array}{c}
RCOOCH_2 \\
| \\
RCOOCH \\
| \quad\quad O \\
\quad\quad || \\
CH_2OP(OH)_2
\end{array}
\quad + CH_2CH_2NH_2 \longrightarrow
\begin{array}{c}
RCOOCH_2 \\
| \\
RCOOCH \\
| \quad\quad O \\
\quad\quad || \\
CH_2OPOCH_2CH_2NH_2 \\
\quad\quad OH
\end{array}$$

Phosphatidylethanol amine

$$\begin{array}{c}
RCOOCH_2 \\
| \\
RCOOCH \\
| \quad\quad O \\
\quad\quad || \\
CH_2OP-OH \\
\quad\quad OH
\end{array}
\quad + CH_2CH_2\overset{\oplus}{N}(CH_3)_3 \longrightarrow
\begin{array}{c}
RCOOCH_2 \\
| \\
RCOOCH \\
| \quad\quad O \\
\quad\quad || \\
CH_2OPOCH_2CH_2\overset{\oplus}{N}Me_3 \\
\quad\quad OH
\end{array}$$

Phosphatidyl choline

$$\begin{array}{c}
HOCH_2CHCOOH \\
\quad NH_2
\end{array}
\longrightarrow
\begin{array}{c}
RCOOCH_2 \\
| \\
RCOOCH \\
| \quad\quad O \\
\quad\quad || \\
CH_2OPOCH_2CHCOOH \\
\quad\quad OH \quad NH_2
\end{array}$$

Phosphatidyl serine

Phospholipids are formed in cell membranes, and are particularly abundant in nerve and brain tissue. Since they contain both polar and nonpolar groupings, they are in many respects like soaps. They are extremely good emulsifying agents, having both hydrophobic and hydrophilic components. In the cell membrane they can form a bilayer, with the hydrophilic portions forming barriers to the passage of polar or ionic material into or out of the cell interior. Yet material does enter and leave the

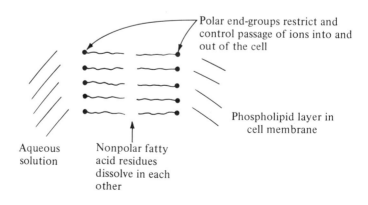

Polar end-groups restrict and control passage of ions into and out of the cell

Phospholipid layer in cell membrane

Aqueous solution

Nonpolar fatty acid residues dissolve in each other

cell. The mechanism of this movement across the bilayer is not yet well understood but seems to involve proteins which are imbedded in the membrane and aid the transport of polar or ionic material.

## COMPOUNDS DERIVED FROM LIPIDS

Many important classes of compounds in both plant and animal systems are derived from lipids, such as terpenes, steroids, certain vitamins, and prostaglandins. Although the structures of many of these compounds seem unrelated, they all depend on acetyl coenzyme-A (acetyl CoA). This complex molecule contains a terminal SH group which reacts with acetic acid to form an acetylated group. By a complex series of biochemical reactions involving, among other things, NADPH

$$\text{Coenzyme A} \quad -SH + CH_3\overset{O}{\overset{\|}{C}}OH \longrightarrow -S-\overset{O}{\overset{\|}{C}}CH_3 \equiv CH_3\overset{O}{\overset{\|}{C}}S-CoA \quad \text{Acetyl CoA}$$

(nicotinamide adenine dinucleotide), long hydrocarbon chains are built enzymatically, adding a two-carbon unit each time through the cycle. For this reason, most naturally occurring fatty acids contain an even number of carbon atoms. Terpenes, steroids, and many other lipids depend on this cycle for their existence in living systems. These pathways are briefly summarized in Fig. 6-1.

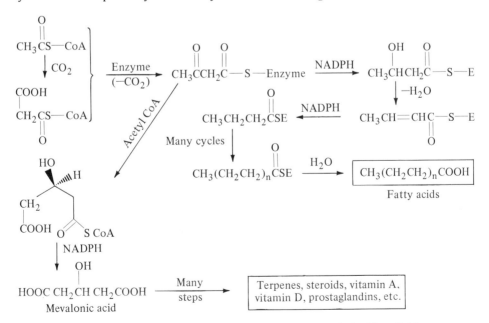

**Figure 6-1.**  Biosynthetic pathways leading to compounds derived from lipids.

## TERPENES

Terpenes may be structurally described as hydrocarbon skeletons derived from individual *isoprene* units. Monoterpenes are composed of two such isoprene units and contain 10 carbons. Other classes of terpenes contain higher multiples of isoprene units.

Isoprene

Terpene structure based on combining isoprene units

| Class | Number of carbons | Representative example |
|---|---|---|
| Monoterpenes | 10 | Ocimene |
| Sesquiterpenes | 15 | β-Cadinene |
| Diterpenes | 20 | Phytol |
| Triterpenes | 30 | Squalene |

**Exercise 6-2:** In the preceding examples show how molecules are constructed from simple isoprene units. For example, limonene has two identifiable isoprene units.

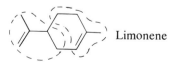

Limonene

Terpenes exist in a number of different forms. They may be open-chain or cyclic, and may contain a diverse number of functional groups including double bonds, alcohols, aldehydes, ketones, and carboxylic acids. Terpenes are responsible, to a large extent, for the odors of various plants and are exuded through the leaves. In fact, the morning haze and odor associated with pine forests such as those in the Great Smoky Mountains actually are caused by condensing clouds of terpenes and isoprene. Terpenes have been used by perfumists for years under the name of "essential oils," because of their frequently pleasant odors.

Tetraterpenes ($C_{40}$) such as lycoprene and $\beta$-carotene are important plant pigments. $\beta$-Carotene is also a precursor in the human body to vitamin A, which is intimately involved in the mechanism of sight.

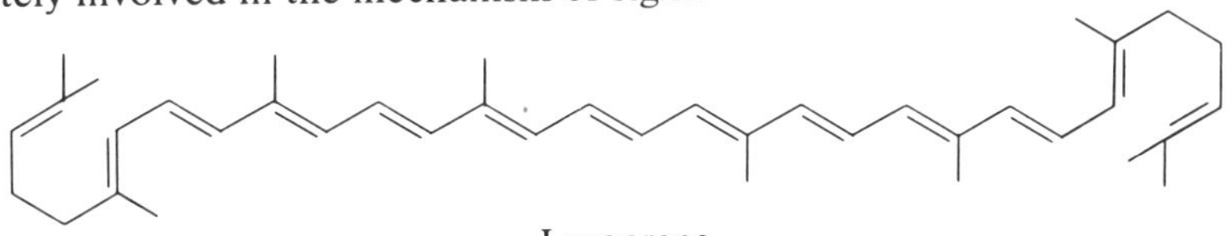

Lycoprene

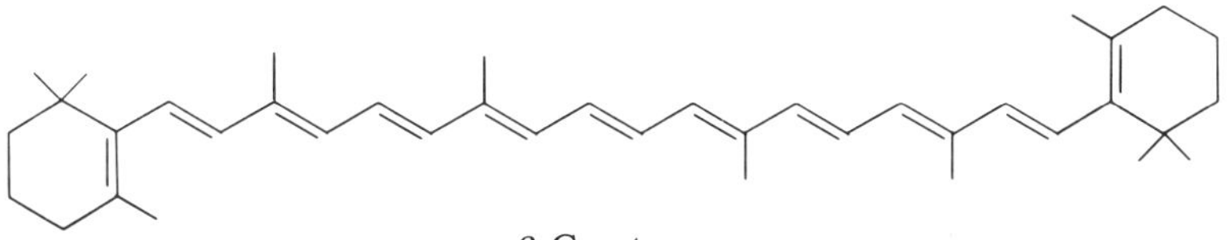

$\beta$-Carotene

Enzyme

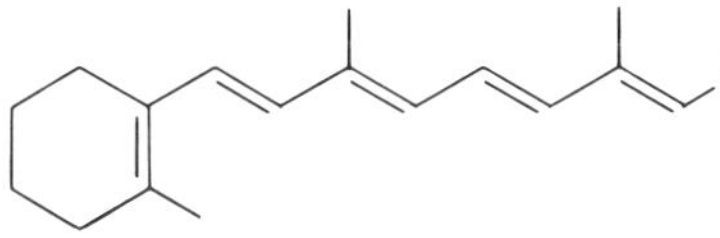

CHO   Trans-retinal

[R]    [O]

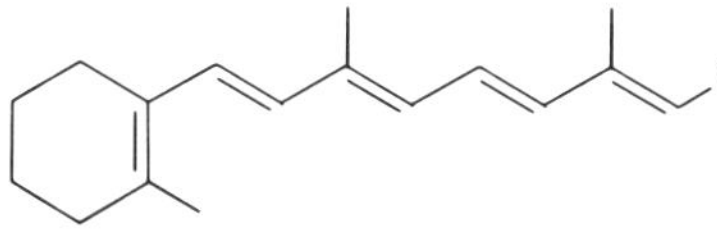

$CH_2OH$   Retinol (vitamin A)

## STEROIDS

Squalene, a triterpene, is the biochemical precursor of steroids.

Squalene  $\xrightarrow{\text{Several steps}}$

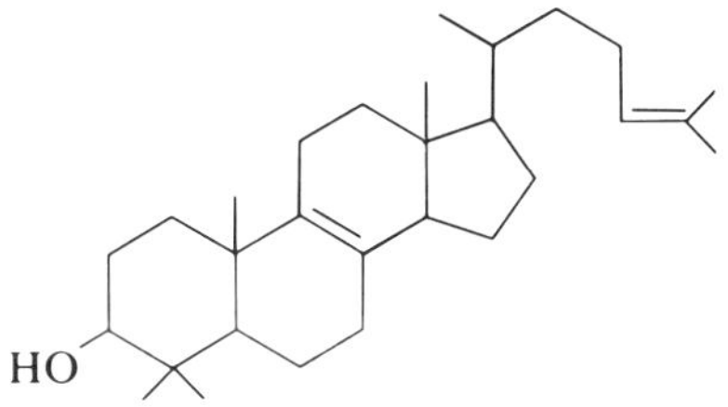

Lanosterol

HO

The four-ring structure is characteristic of most steroids, whereas the side chains and functional groups exhibit an almost infinite variety. We previously discussed the

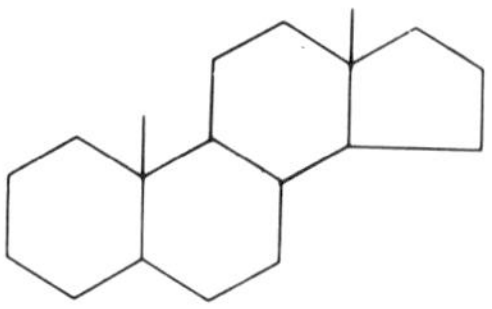

Characteristic steroid unit

importance and occurrence of sterols in Chapter I-8. However, many other important bodily functions are governed by steroids of various types. Bile acids, such as cholic acid, help hydrolyze and absorb fats in the lower digestive tract by acting as emulsifying agents.

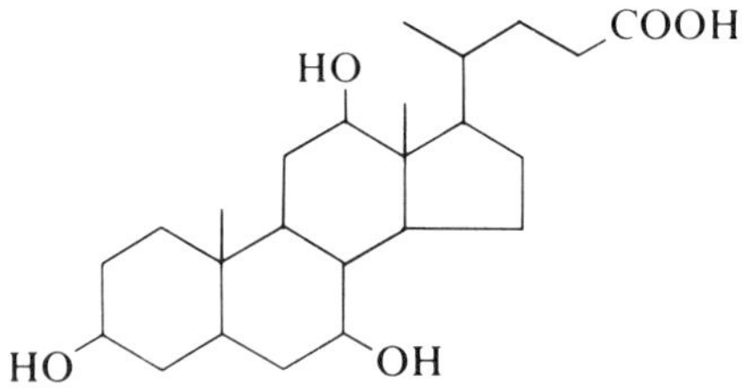

Cholic acid

Much research and interest has been devoted to the sex hormones over the past several decades. This research led to the development of several steroid analogs which mimic sex-hormone action related to ovulation in women, thus allowing predictable biochemical birth control. Norethynodrel and Norethyndrone both prevent ovulation in nonpregnant women and are now commonly used in "the Pill." Steroids are involved in many other bodily functions, but detailed description is beyond the scope of this text.

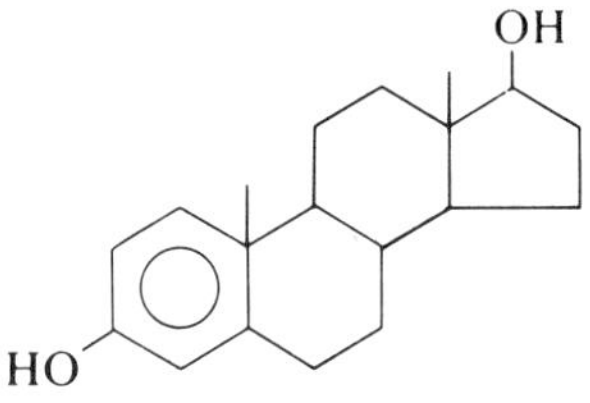

Estradiol

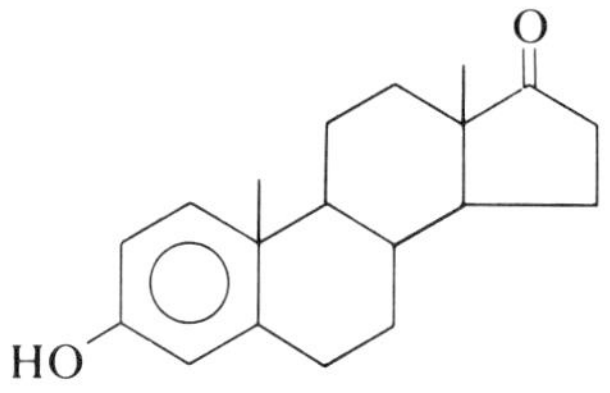

Estrone

Female sex hormones

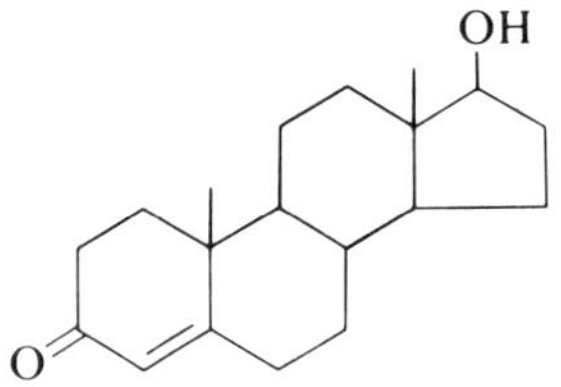

Testosterone

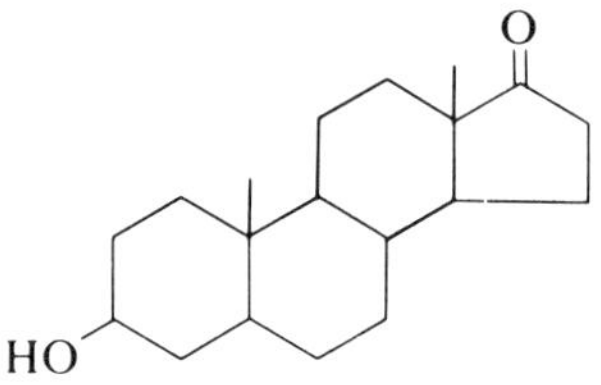

Androsterone

Male sex hormones

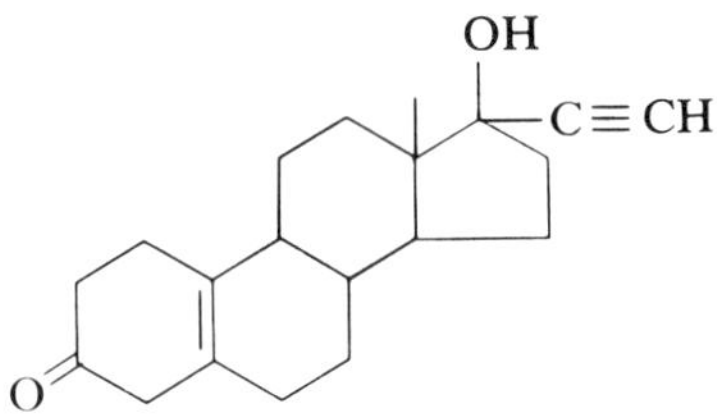

Norethynodrel

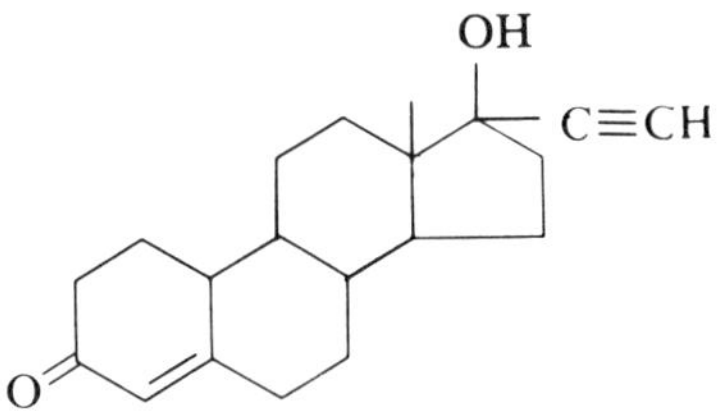

Norethyndrone (Norlutin)

Synthetic regulators of ovulation
(birth-control pills)

## FAT-SOLUBLE VITAMINS

Several terpenes are essential to human metabolism and must be included in our diet. They are referred to as *fat-soluble vitamins*. We previously discussed the conversion of $\beta$-carotene to retinol (vitamin A), which is essential for good vision. Vitamins D, E, and K also originate from terpenes, and their structures and functions are described briefly below.

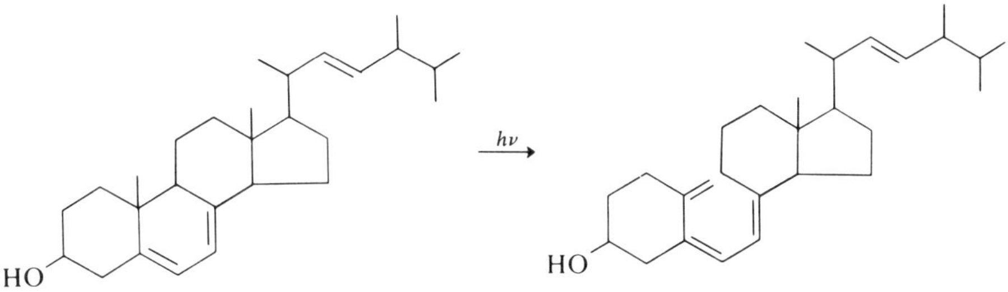

Ergosterol
(vitamin D precursor)

$h\nu$

Vitamin D$_2$
(calciferol)

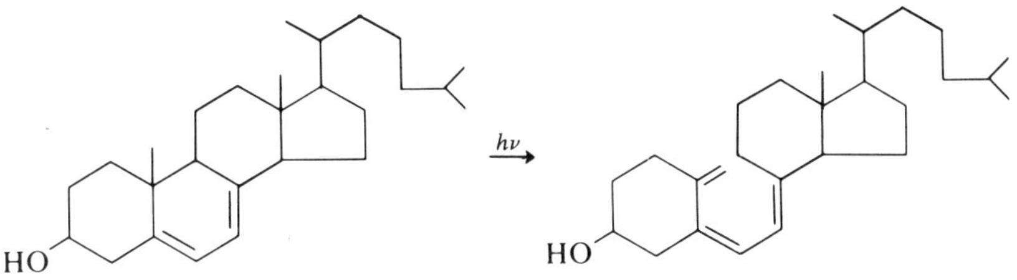

7-Dehydrocholesterol
(vitamin D$_3$ precursor)

$h\nu$

Vitamin D$_3$

D vitamins are essential for the prevention of rickets, a crippling bone disease.

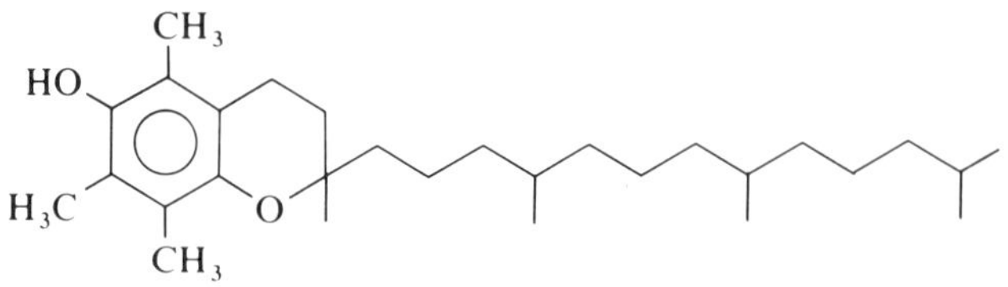

Vitamin E ($\alpha$-tocopherol)

Absence of vitamin E is associated with muscular dystrophy and liver diseases.

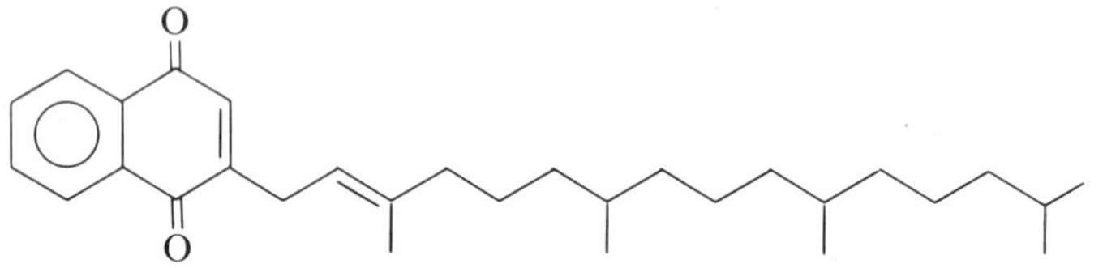

Vitamin $K_1$

Absence of vitamin $K_1$ is associated with the inability of the blood to clot.

## PROSTAGLANDINS

Prostaglandins exist in human semen and in animal seminal tissues in extremely small quantities, but they have high physiological activity in both. They have several interesting properties, among the more important of which is the ability to relax muscles and to sharply control blood pressure. In general, their structures are $C_{20}$ carboxylic acids, and their precursors are usually regarded to be polyunsaturated fatty acids. In humans, more than a dozen important prostaglandins have been identified, a few of which are shown below along with their fatty-acid precursor.

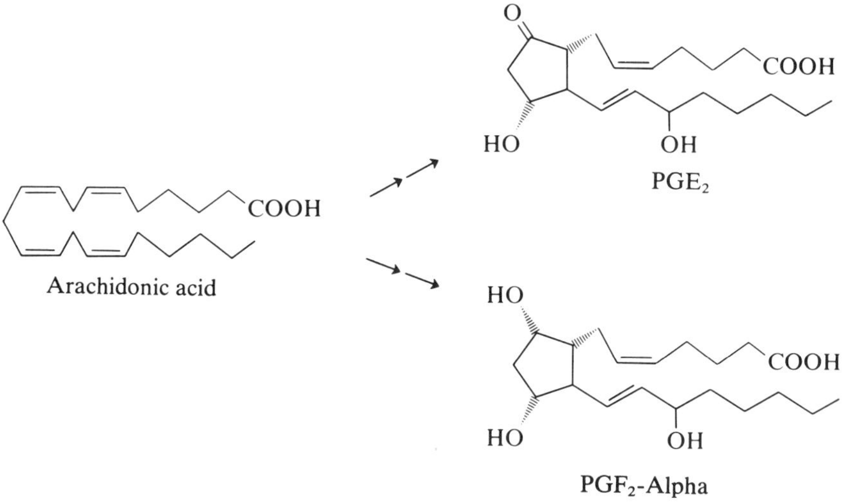

PGE$_2$

Arachidonic acid

PGF$_2$-Alpha

Individual prostaglandins have specific physiological effects in man. For example, $PGE_2$ lowers blood pressure, while $PGF_2$-Alpha raises blood pressure. For this reason there is much current interest in their pharmacological potential.

## REVIEW OF NEW TERMS AND CONCEPTS

Define each term and, if possible, give an example of each:

| | |
|---|---|
| Lipid | Fat |
| Oil | Wax |
| Soap | Saponification |
| Micelle | Emulsification |
| Detergent (syndet) | ABS |
| Phospholipid | Acetyl CoA |
| Terpene | Steroid |
| Fat-soluble vitamin | Prostaglandin |

## PROBLEMS: CHAPTER II-6

1. What characteristics distinguish a fat from an oil or wax? Give structures for compounds in each of these categories that are different from any discussed in this chapter.

2. An oil sample yields palmitic, oleic, stearic, and linoleic acids when saponified. How many different triglycerides could give rise to this product mixture if we assume that no more than two acids are present in each molecule?

3. Write a reaction for the saponification of two of the triglycerides you listed in Problem 2.

4. In this chapter we learned that alkyl sulfonates could function as synthetic detergents. Can you think of any other classes of organic acids, or other functionalities, that might function as syndets?

5. Suggest a structure that would qualify as a monoterpene (remember the isoprene rule).

6. The following compounds are all monoterpenes. How would you distinguish between them by chemical analysis?

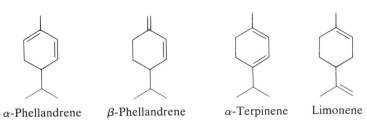

α-Phellandrene    β-Phellandrene    α-Terpinene    Limonene

**7.** Can you suggest *chemical* reagents that could interconvert trans-retinal and retinol (vitamin A)?

# 7

# Polymers:
## Backbone of Present-Day Life

The developments of modern technology have brought tremendous changes to our lifestyles in the past three decades. One of the most pervasive changes has been the gradual replacement of natural materials such as wood, wool, cotton, and metal with manmade synthetics. For better or worse we are becoming a "plastic" society. In this chapter we will learn how polymers are made and how they are used. We will also briefly examine how structure and function are related, and how polymers are able to replace dwindling naturally occurring substances.

## GENERAL CHARACTERISTICS OF POLYMERS

*Polymers* are molecules of high molecular weight whose structures are composed of a large number of simple repeating units. Polymer chains may also have identical composition but differing molecular weights. The basic polymer repeating unit is usually formed from low-molecular-weight compounds referred to as *monomers*, while the conversion process, monomer → polymer, is known as *polymerization*. Polymers may have many different types of structures. The monomer units may form a linear polymer whose structure is basically a straight chain. Polymer chains may also be *branched*, thus forming a network which extends in several directions.

$$-(A)_n-$$

$$\longleftarrow \quad -A-A-A-A-A-A-A- \quad \longrightarrow \quad \text{Linear polymer}$$

$$\begin{array}{c} | \\ A \\ | \\ A \\ | \\ -A-A-A-A-A-A-A- \\ | \\ A \\ | \\ A \\ | \end{array} \quad \begin{array}{l} \text{Branched-chain} \\ \text{polymer} \end{array}$$

Polymers may be formed from a single monomer or from two or more different monomers. The former are called *homopolymers*, the latter *copolymers*.

$$n\mathrm{A} \quad \longrightarrow \quad +A+_n \qquad \text{Homopolymer from one monomer}$$

$$n\mathrm{A} + n\mathrm{B} \quad \longrightarrow \quad +A-B+_n \qquad \text{Copolymer from two monomers}$$

Copolymer structure may be either *alternating*, in which a definite structural pattern may be observed, or *random*.

$$n\mathrm{A} + n\mathrm{B} \quad \longrightarrow \quad -A-B-A-B-A-B-A-B-A-B-A-B-A-B-A-$$
Alternating copolymer—
discernible pattern

$$n\mathrm{A} + n\mathrm{B} \quad \longrightarrow \quad -A-A-B-A-B-B-A-B-A-A-B-A-B-A-A-B-B-$$
Random copolymer—
no discernible order

## CLASSIFICATION OF POLYMERS

Polymers may generally be divided into two distinct categories: *addition polymers* and *condensation polymers*. Addition polymers are formed, as their name implies, by the addition of one monomer unit to another. In the final polymer structure, the repeating unit still incorporates all the atoms of the original monomer. In a condensation polymer, a small fragment, such as water, is eliminated during the polymerization process from reaction between the monomer functional groups. A few examples of these types of polymerization are shown below:

$$n \; \mathrm{CH_2}{=}\mathrm{CH} \quad \longrightarrow \quad +\mathrm{CH_2CH}+_n$$
$$\qquad\qquad\;\; | \qquad\qquad\qquad\qquad |$$
$$\qquad\qquad\;\; \mathrm{X} \qquad\qquad\qquad\qquad \mathrm{X}$$

Addition polymer

$$n \; \mathrm{HO(CH_2)_xOH} + n \; \mathrm{HOOC(CH_2)_yCOOH}$$

$$\searrow$$

$$\qquad\qquad\qquad\qquad\qquad\qquad\qquad \overset{\displaystyle O}{\overset{\displaystyle \|}{}} \qquad\quad \overset{\displaystyle O}{\overset{\displaystyle \|}{}}$$
$$\qquad\qquad\qquad\qquad\qquad\mathrm{HO}{+}(\mathrm{CH_2})_x\mathrm{O}\overset{}{C}(\mathrm{CH_2})_y{-}\overset{}{C}{+}_n + n \; \mathrm{H_2O}$$

Condensation polymer

Some polymers can be formed by either reaction:

$$n \; (CH_2)_x \underset{\underset{\displaystyle NH}{|}}{\overset{\overset{\displaystyle O}{\|}}{-C}} \longrightarrow \; +NH(CH_2)_x\overset{\overset{\displaystyle O}{\|}}{C}+_n \; \xleftarrow{-H_2O} \; n \; H_2N+CH_2+_x\overset{\overset{\displaystyle O}{\|}}{C}OH$$

Both addition and condensation polymerizations are extremely important in the industrial manufacture of polymers.

## ADDITION POLYMERS AND POLYMERIZATION

By far the most important class of addition polymers are formed from *vinyl monomers*, $CH_2{=}CHX$. Polymer formation may be initiated by free-radical, cationic, or anionic reagents, and the polymer chain grows by successive additions of monomer molecules. This process is known as *chain propagation*. The three types of addition are outlined below.

$$\text{Initiating molecule, M} \; \xrightarrow[\substack{or \\ irradiation}]{\substack{Chemical \\ reaction}} \; I^*$$

$$I^* = I\cdot \quad \text{(free radical)}$$
$$I^+ \quad \text{(cation)}$$
$$I{:}^- \quad \text{(anion)}$$

Free-radical polymerization mechanism

(a) Initiation:

$$I\cdot + CH_2{=}\underset{\underset{\displaystyle X}{|}}{CH} \longrightarrow ICH_2\underset{\underset{\displaystyle X}{|}}{\dot{C}H}$$

(b) Propagation step 1:

$$ICH_2\underset{\underset{\displaystyle X}{|}}{\dot{C}H} + CH_2{=}CHX \longrightarrow ICH_2\underset{\underset{\displaystyle X}{|}}{CH}CH_2\underset{\underset{\displaystyle X}{|}}{\dot{C}H}$$

Propagation step 2:

$$ICH_2\underset{\underset{\displaystyle X}{|}}{CH}CH_2\underset{\underset{\displaystyle X}{|}}{\dot{C}H} + CH_2{=}\underset{\underset{\displaystyle X}{|}}{CH} \longrightarrow ICH_2\underset{\underset{\displaystyle X}{|}}{CH}CH_2\underset{\underset{\displaystyle X}{|}}{CH}CH_2\underset{\underset{\displaystyle X}{|}}{\dot{C}H}$$

Etc.

(c) Chain termination reactions: The chain reaction can be interrupted when two radical chains combine, or by disproportionation.

$$2 \text{ R·} \longrightarrow \text{R}-\text{R}$$

$$\underset{\text{X}}{\text{RCH}_2\dot{\text{C}}\text{H}} + \underset{\text{X}}{\text{RCH}_2\dot{\text{C}}\text{H}} \longrightarrow \text{RCH}_2\text{CH}_2\text{X} + \text{RCH}{=}\text{CHX}$$

Cationic and anionic polymerization will be discussed later in this chapter.

Free-radical polymerization requires a free-radical initiator. Usually the initiatior is generated from a thermally unstable material such as a peroxide, which will decompose when heated, yielding reactive free radicals. Benzoyl and *t*-butyl peroxides are common reagents for free-radical polymerizations.

$$(\text{CH}_3)_3\text{C}-\text{O}-\text{O}-\text{C}(\text{CH}_3)_3 \xrightarrow{\Delta} 2\,(\text{CH}_3)_3\text{CO·} \longrightarrow (\text{CH}_3)_2\text{C}{=}\text{O} + \dot{\text{C}}\text{H}_3$$

· or CH$_3$·   then initiate free-radical polymerization.

---

**Exercise 7-1:**   Write a complete reaction sequence for the benzoyl-peroxide–induced polymerization of vinyl chloride, CH$_2$=CHCl. Show all initiation, propagation, and termination steps.

---

## VINYL ADDITION POLYMERS

Vinyl polymers are ubiquitous in our society, and we have grown so accustomed to their use that we don't even notice them. Vinyl polymers may be either homopolymers or random copolymers.

$$n\ \underset{\text{X}}{\text{CH}_2{=}\text{CH}} \longrightarrow \underset{\text{X}}{(\text{CH}_2\text{CH})_n}$$

Vinyl monomer        Vinyl homopolymer

$$n\ \underset{\text{X}}{\text{CH}_2{=}\text{CH}} + m\ \underset{\text{Y}}{\text{CH}_2{=}\text{CH}} \longrightarrow \underset{\text{X}\quad\text{Y}}{(\text{CH}_2\text{CHCH}_2\text{CH})_{n+m}}$$

Monomer X   Monomer Y        Vinyl copolymer— usually random

In Table 7-1 the most common vinyl homopolymers and copolymers are listed along with their common names and their more important uses.

Table 7-1   Vinyl Polymers and Their Uses

| Monomer | Polymer | Common trade name | Common uses |
|---|---|---|---|
| $CH_2=CH_2$ | $-(CH_2CH_2)_n$ | Polyethylene | Films, pipes, tubing |
| $CH_3CH=CH_2$ | $-(CH_2CH)_n$ <br>      $\vert$ <br>     $CH_3$ | Polypropylene | Fibers, molded objects |
| $CH_2=C(CH_3)_2$ | $-(CH_2C(CH_3)_2)_n$ | Butyl rubber | Tire innertubes, adhesive, insulation |
| $CH_2=CHCl$ | $-(CH_2CH)_n$ <br>      $\vert$ <br>     $Cl$ | Polyvinylchloride (PVC) | Pipes, panels, molded products, insulation, vinyl leather, transparent films (raincoats) |
| $CH_2=CHCl +$ <br> $CH_2=CCl_2$ | $Cl$ <br> $\vert$ <br> $-(CH_2CHCH_2C)_n$ <br>     $\vert$   $\vert$ <br>     $Cl$  $Cl$ | Saran | Plastic wrap, pipes, outdoor coverings, auto upholstery, food packaging |
| $\quad\quad O$ <br> $\quad\quad \|\|$ <br> $CH_2=C-COCH_3$ <br> $\quad\quad \vert$ <br> $\quad\quad CH_3$ | $COOMe$ <br> $\vert$ <br> $-(CH_2C)_n$ <br>     $\vert$ <br>     $Me$ | Polymethylmeth-acrylate, Plexiglass, Lucite, Perspec | Contact lenses, windows, clear sheets |
| $CH_2=CHCN$ | $-(CH_2CH)_n$ <br>      $\vert$ <br>     $CN$ | Polyacrilonitrile, Orlon, Acrilan, Acrylic | Fibers for clothing, carpets, blends with wool |
| $CH_2=CH-\langle\bigcirc\rangle$ | $-(CH_2CH)_n$ | Polystyrene | Packing materials, building products, synthetic rubber |
| $CF_2=CF_2$ | $-(CF_2CF_2)_n$ | Teflon | Nonstick linings, electrical insulation, inert linings for gaskets, valves, diaphragms |
| $\quad\quad\quad O$ <br> $\quad\quad\quad \|\|$ <br> $CH_2=CHOCCH_3$ | $-(CH_2CH)_n$ <br>      $\vert$ <br>     $OAc$ | Polyvinylacetate | Binder, adhesive coatings, chewing gum |
|  | $\downarrow$ <br> $-(CH_2CH)_n$ <br>      $\vert$ <br>     $OH$ | Polyvinyl alcohol | Binder, coatings for paper, leathers, tubing for oils |

Free-radical polymerization produces polymer chains which have little organized structure. Such random-chain structure polymers are said to be *amorphous*, and have little or no tendency to crystallize. In the early 1950s, however, Ziegler and Natta developed ionic catalysts, which enabled organic chemists to construct polymer chains with regular and predictable stereochemistry. The most common Ziegler–Natta catalyst is $AlEt_3 \cdot TiCl_4$. The catalyst has a complex structure which acts as a mold upon which the polymer forms. Since the structure of the catalyst does not change during the course of the reaction, stereoregular polymers are formed. For simple vinyl monomers, $CH_2{=}CHX$, two *stereoregular* chains are possible, one in which all the X groups are on the same side of the chain, and one in which the X groups alternate from one side of the chain to the other. The former are said to be *isotactic*, while the latter are referred to as *syndiotactic*. The random, amorphous polymers are referred to as *atactic*. These stereochemical distinctions are illustrated in Fig. 7-1.

Isotactic polypropylene

Syndiotactic polypropylene

Stereoregular polymers have more crystalline character, higher melting points, and higher tensile strength than do atactic polymers. Of particular importance to industry, they make much stronger fibers.

Cationic polymerizations can also be achieved with simple Lewis acids such as boron trifluoride. For example, $BF_3$ is used extensively for the polymerization of isobutylene in the preparation of butyl rubber.

(Small amount of isoprene added for cross-linking)

Anionic polymerization usually requires a strong base. Alkyl lithiums are used in the preparation of *cis*-1,4-polyisoprene (synthetic rubber) and also in some styrene, methylmethacrylate, and acrylonitrile polymerizations. In order for anionic polymerization to be effective, the carbionic intermediate should be stabilized by electron-withdrawing or delocalizing species.

402

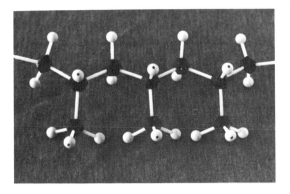
(a) Isotactic polypropylene

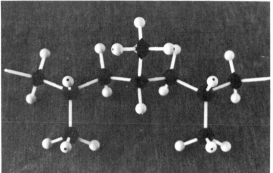
(b) Syndiotactic polypropylene

**Figure 7-1.** Stereochemical forms of polypropylene.

**Exercise 7-2:** Write complete polymerization reactions for the formation of poly-isobutylene from isobutene and $BF_3$, and also for the formation of polyacrylonitrile from $CH{=}CHCN$ and *n*-butyllithium.

## CONDENSATION POLYMERS

The most important classes of condensation polymers involve either ester or amide formation. Although the reactions appear to be quite simple in theory, in practice the reactants or reaction conditions must be carefully controlled in order to obtain polymers with useful properties.

$$HO{-}\boxed{A}{-}OH + \overset{(R)}{HOOC}{-}\boxed{B}{-}\overset{(R)}{COOH}$$

$$\downarrow \Delta$$

$$\overset{\displaystyle +}{}O{-}\boxed{A}{-}O{-}\overset{\displaystyle O}{\overset{\|}{C}}{-}\boxed{B}{-}\overset{\displaystyle O}{\overset{\|}{C}}\overset{\displaystyle +}{}_n$$

Condensation copolymer involving ester formation

$$H_2N{-}\boxed{A}{-}NH_2 + HOOC{-}\boxed{B}{-}COOH$$

$$\downarrow \Delta$$

$$\overset{\displaystyle +}{}NH{-}\boxed{A}{-}NH{-}\overset{\displaystyle O}{\overset{\|}{C}}{-}\boxed{B}{-}\overset{\displaystyle O}{\overset{\|}{C}}\overset{\displaystyle +}{}_n$$

Condensation copolymer involving amide formation

Polyethyleneterephthalate is by far the most important polyester commercially, and is known in the United States as *Dacron*, and in Europe as *Terylene*. It is formed from ethylene glycol and dimethylterephthalate by a transesterification process, followed by heating at 280° to effect polymerization. The polymer melt can then be spun into fibers and combined with such naturally-occurring fibers as cotton (65% cotton–35% polyester, 60/40 cloth). The combination of cotton and polyester yields a fabric that dries quickly without wrinkling, yet still retains the coolness and comfort of 100% cotton. Polyesters have truly revolutionized the clothing industry.

$$HOCH_2CH_2OH + \underset{\underset{COOCH_3}{|}}{\overset{\overset{COOCH_3}{|}}{\bigcirc}} \xrightarrow[200°]{Base} \underset{Intermediate}{HOCH_2CH_2O\overset{O}{\overset{||}{C}}-\bigcirc-\overset{O}{\overset{||}{C}}OCH_2CH_2OH} + CH_3OH$$

$$\downarrow \Delta$$

$$\underset{Dacron}{\left[ OCH_2CH_2O\overset{O}{\overset{||}{C}}-\bigcirc-\overset{O}{\overset{||}{C}} \right]_n} + HOCH_2CH_2OH$$

*Nylon-6,6* is the most important polyamide and is formed from 1,6-diamino-hexane (hexamethyleneamine) and hexanedioic acid (adipic acid). Equimolar mixtures of the two monomers are heated at 270° under pressure, and then heated at reduced pressure to remove as much water as possible. Fiber can be spun from the melt, or pellets can be obtained by extruding the melt.

$$HO\overset{O}{\overset{||}{C}}\underbrace{(CH_2)_4}\overset{O}{\overset{||}{C}}OH + H_2N\underbrace{(CH_2)_6}NH_2 \xrightarrow{\Delta} \left[ NH\overset{O}{\overset{||}{C}}\underbrace{(CH_2)_4}\overset{O}{\overset{||}{C}}NH\underbrace{(CH_2)_6} \right]_n$$

$$\underset{6 \text{ carbons}}{\phantom{xx}} \underset{6 \text{ carbons}}{\phantom{xx}} \underset{\underset{\text{Nylon-6,6}}{6 \text{ carbons} \quad 6 \text{ carbons}}}{\phantom{xx}}$$

---

**Exercise 7-3:**   Show how you would prepare Nylon-6,8.

---

A similar polyamide can be prepared from caprolactam, a cyclic amide. Nylon-6 is formed from this polymerization and is primarily used as a textile fiber. It is more flexible and has a lower melting point than Nylon-6,6.

$$\underset{NH}{\overset{\overset{O}{\overset{||}{C}}}{(CH_2)_5}} \xrightarrow[(H_2O)]{250°} \left[ NH(CH_2)_5\overset{O}{\overset{||}{C}} \right]_n \quad \text{Nylon-6}$$

## POLYURETHANES

Urethanefoam is widely used today as foam padding, particularly in furniture. A urethane is normally prepared by the addition of an alcohol to an isocyanate. As can be seen from its structure, a urethane combines aspects of both amides and esters and

$$R-N=C=O + R'OH \longrightarrow RNH\overset{\overset{\displaystyle O}{\|}}{C}OR'$$

An isocyanate        A urethan

thus should have interesting and unique polymer properties. In practice, 2,4-tolylene diisocyanate is added to a low-molecular-weight polyester (MW=1,000–2,000) formed from ethylene glycol and adipic acid. The foaming action is achieved

$$HOOC(CH_2)_4COOH + HOCH_2CH_2OH \longrightarrow HOCH_2CH_2O \left[ \overset{\overset{\displaystyle O}{\|}}{C}(CH_2)_4\overset{\overset{\displaystyle O}{\|}}{C}OCH_2CH_2O \right]_n H$$

by adding a small quantity of water during the diisocyanate polymerization process. The water reacts with an equally small quantity of the isocyanate, producing $CO_2$ which rapidly expands the polymer, forming a light, resilient material when cooled.

Diisocyanate      Dicarbamic acid—      Gas expands
                thermally unstable      polymer to
                                              form a foam

405

# DIENE POLYMERS AND SYNTHETIC RUBBER

Natural rubbers, such as gutta-percha and balata, are impure *trans*-1,4-polyisoprenes isolated from rubber trees in such places as Malaysia and Vietnam. Hevea rubber is all *cis*-1,4-polyisoprene, and is extracted in Brazil from *Hevea brasiliensis*,

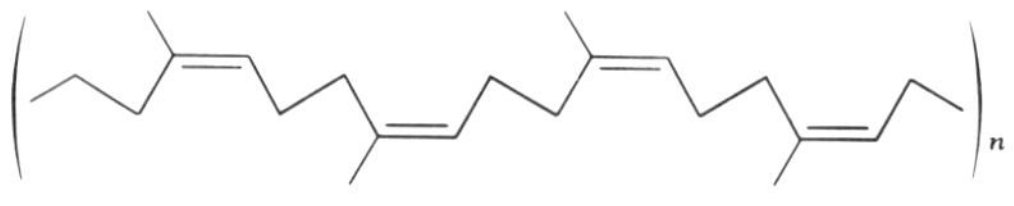

Hevea natural rubber

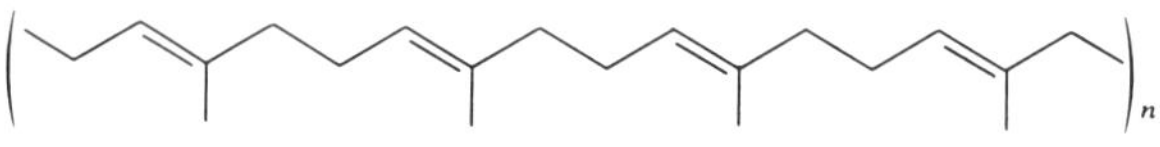

Gutta-percha and balata

the rubber tree. Disruption of the rubber trade during World War II led to the development of synthetic rubber, mainly by polymerization of chloroprene and isoprene.

$$\underset{\text{Chloroprene}}{CH_2{=}\overset{\overset{\displaystyle Cl}{|}}{C}CH{=}CH_2} \xrightarrow[\text{polymerization}]{\text{Free-radical}} \underset{\substack{\text{Polychloroprene}\\ \text{(Neoprene)}}}{\left( CH_2\overset{\overset{\displaystyle Cl}{|}}{C}{=}C{-}CH_2 \right)_n}$$

$$\underset{\text{Isoprene}}{CH_2{=}\overset{\overset{\displaystyle CH_3}{|}}{C}CH{=}CH_2} \xrightarrow[\substack{\text{or}\\ n\text{-BuLi}}]{\text{Ziegler catalyst}} \underset{\text{Polyisoprene—all } cis\text{-configuration}}{\left( CH_2\ \overset{\overset{\displaystyle CH_3}{|}}{C}{=}CCH_2 \right)_n}$$

Both natural and synthetic rubber can be made stronger and harder by heating with sulfur, which forms bridges between adjacent chains in a process known as *vulcanization.*

$$\begin{array}{c}
\sim CH_2\overset{\overset{\displaystyle CH_3}{|}}{C}{=}CHCH_2CH_2\overset{\overset{\displaystyle CH_3}{|}}{C}{=}CHCH_2\sim \\
\sim CH_2\underset{\underset{\displaystyle CH_3}{|}}{C}{=}CHCH_2CH_2\underset{\underset{\displaystyle CH_3}{|}}{C}{=}CHCH_2\sim \\
\downarrow \text{S, } \Delta\text{, catalyst} \\
\sim CH_2\overset{\overset{\displaystyle CH_3}{|}}{C}{=}CH{-}CH{-}CH_2\overset{\overset{\displaystyle CH_3}{|}}{C}{=}CH{-}CH_2\sim \\
| \\
S \\
| \\
\sim CH_2\underset{\underset{\displaystyle CH_3}{|}}{C}{=}CH{-}CH{-}CH_2\underset{\underset{\displaystyle CH_3}{|}}{C}{=}CH{-}CH_2\sim
\end{array}$$

## CLASSIFICATION OF POLYMER PROPERTIES

Polymers may be divided into three fairly distinct groups according to their properties: *elastomers*, *plastics*, and *fibers*. These three groups behave differently when deformed, when heated, and in their tendency to crystallize. When polymer chains are stretched, the material may be permanently deformed; however, cross-linking such as vulcanization allows the material to resist *permanent* deformation, and regain its original shape. Thus some polymers are *elastic*, while others are not. The three types of polymers and their differences are outlined below.

| | |
|---|---|
| Elastomers | Very elastic, material regains shape immediately; becomes more elastic when heated and brittle when cooled. Almost no tendency to crystallize. Examples: natural rubber, synthetic polychloroprene, polyisoprene. |
| Plastics | Moderate elasticity, but some permanent deformation when stretched. The elasticity is temperature dependent, but much less so than for elastomers. Moderate tendency to crystallize. Thermoplastics can be remelted many times with no change in properties. Examples: PVC, polyvinylacetate, polystyrene. |
| Fibers | Some elasticity and some permanent set when stretched. Elasticity independent of temperature. Very high tendency to crystallize. Examples: polyesters, polyamides. |

## REVIEW OF NEW TERMS AND CONCEPTS

Define each term and, if possible, give an example of each:

| | |
|---|---|
| Monomer | Polymer |
| Homopolymer | Copolymer |
| Alternating copolymer | Random copolymer |
| Addition polymer | Condensation polymer |
| Initiation | Chain propagation |
| Chain termination | Vinyl polymer |
| Vinyl copolymer | Atactic |
| Syndiotactic | Isotactic |
| Cationic polymerization | Anionic polymerization |
| Polyester | Polyamide |
| Polyurethane | Natural rubber |
| Synthetic rubber | Vulcanization |
| Elastomer | Plastic |
| Fiber | |

1. Make a trip to your local department store (J. C. Penney, Sears, Montgomery Ward, etc.) and look at either the men's or women's clothing section. Make a list of items that are 100% synthetic; a mixture of synthetic and natural fiber, such as wool or cotton; and last, those items that are 100% natural fiber. Calculate the percentage in each category. Are you surprised?

2. Investigate your room, apartment, or house. How many plastic and synthetic items can you identify? For each of these items, identify what natural substance it replaced (wood, metal, leather, etc.). Do you think it would be more or less expensive to have a "natural" house?

3. Write structures for the possible polymers that you would obtain from the following reactants.

   (a) $Cl_2C=CClF \xrightarrow[\text{peroxide} \atop \Delta]{\text{Benzoyl}}$

   (b)
   $$H_2N(CH_2)_4NH_2 + Cl\overset{O}{\overset{||}{C}}(CH_2)_4\overset{O}{\overset{||}{C}}Cl \longrightarrow$$

   (c)

   $+ HOCH_2CH_2OH \longrightarrow$

   (d)

   (e)

   $HOCH_2CH_2OH + O=C=N-\!\!\!\left\langle\bigcirc\right\rangle\!\!\!-N=C=O \longrightarrow$

   (f)

   $+ CH_2=CHCl \xrightarrow[\Delta]{\text{Peroxide}}$

4. A synthetic rubber can be made from 1,3-butadiene and styrene. Suggest possible structural units that might appear in this polymer. (*Hint:* Consider both 1,2- and 1,4-addition.)

5. Write structures for the possible stereoregular polymers produced by Ziegler–Natta catalysis for the following monomers:

   (a) $CH_2=CHCl$

   (b) $CH_3CH=CHCH_3$

6. Automobile tires show tiny surface cracks after exposure to air for extended periods. These cracks appear more quickly and are more serious in large, industrial cities. Can you suggest a chemical explanation for this phenomenon?

7. Hevea rubber is much more elastic and deformable than either gutta-percha or balata. Can you suggest a possible explanation for this fact?

8. Why is it necessary to prepare polyvinylalcohol from polyvinyl acetate instead of a direct polymerization?

# 8

# Drugs:
## Boon and Bane

The use of medicines and drugs probably distinguishes our society from those gone before it more than any other single factor. Their use has lengthened our lives, increased our health and vitality even into old age, and more recently, blown our minds. For better or worse we now live in a drug-oriented culture, though few people will admit it even to themselves. Of course, we must realize that many everyday items are really drugs: caffeine, aspirin, alcohol, and nicotine as well as the more readily identified tranquilizers and barbiturates. Drugs can be both used and abused. In this chapter we will learn what categories of drugs exist and how they can cure our ills. We will also learn how drugs are currently abused in the United States, from overuse of over-the-counter and prescription drugs to the search for the illusive perfect high.

## CLASSIFICATION OF DRUGS

Drugs are more readily classified by their effect upon human physiology than by structural features. Most common prescription drugs are chemically complex and their preparations (for the most part) and reactions are beyond the scope of this text. In fact, the chemotherapeutic mechanisms of many of these drugs are not well understood, even by physicians. Therefore in this section we will briefly examine the

410

types of diseases that drugs can control and the desirable physiological effects that can be derived from drug action.

## Drugs that Affect the Cardiovascular System

Diseases of the cardiovascular system, which include both the heart and the blood vessels (veins and arteries) are quite common in modern man. In fact, heart disease, high blood pressure, and stroke (as a group) constitute the number-one killer in the United States at the present time. This should not be any great surprise, because of the increase in our life expectancy over the past three decades. Some of the more common cardiovascular problems, and drugs that offer some relief, are discussed below:

Excessively high blood pressure, or *hypertension*, results when the blood vessels contract and blood flow is restricted. Several drugs may be used to dilate the vessels, allowing blood to flow more normally and thus lowering pressure. Useful drugs to combat hypertension include sodium chlorothiazide (Diuril), methyldopa (Aldomet), and hydrochlorothiazide (Dyazide, Hydrodiuril).

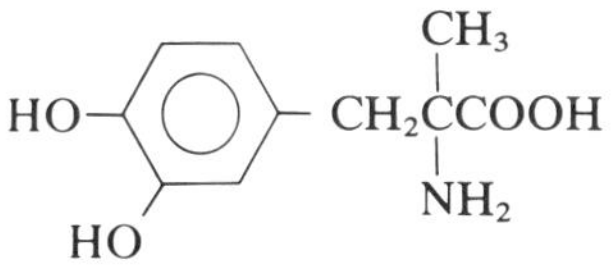

$$HO-\text{\Large\bigcirc}-CH_2\overset{\overset{\displaystyle CH_3}{|}}{\underset{\underset{\displaystyle NH_2}{|}}{C}}COOH \qquad \alpha\text{-Methyldopa}$$

*Angina pectoris* is severe and disabling heart pain that results when blood flow to heart muscle is restricted. *Nitroglycerine* can relieve this pain in minutes by increasing blood flow.

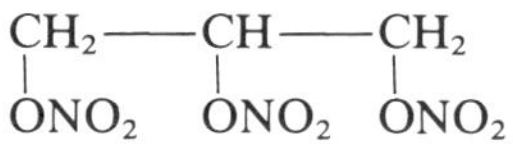

$$\underset{\underset{\displaystyle ONO_2}{|}}{CH_2}-\underset{\underset{\displaystyle ONO_2}{|}}{CH}-\underset{\underset{\displaystyle ONO_2}{|}}{CH_2} \qquad \text{Nitroglycerine}$$

Irregular heart beat or flutter (*arrhythmia*) can be relieved by several drugs such as procaine or procaine amide. Severe arrhythmia due to heart failure can be arrested by *Digoxin*, which is related to the cardiac glycoside digitalis.

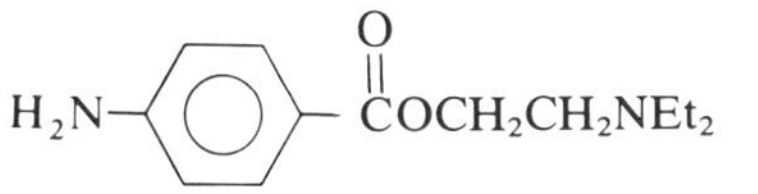

$$H_2N-\text{\Large\bigcirc}-\overset{\overset{\displaystyle O}{\|}}{C}OCH_2CH_2NEt_2 \qquad \text{Procaine}$$

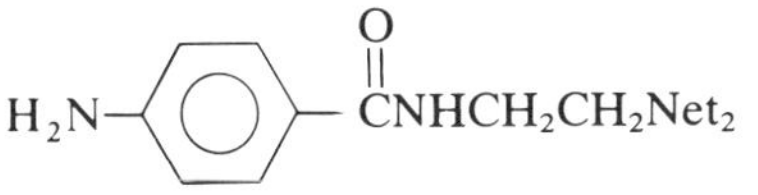

$$H_2N-\text{\Large\bigcirc}-\overset{\overset{\displaystyle O}{\|}}{C}NHCH_2CH_2Net_2 \qquad \text{Procaine amide}$$

Body tissue sometimes accumulates a great deal of excess water. *Diuretic* drugs cause such excess water to be excreted, thus reducing blood pressure and easing strain on the heart. Some common diuretic drugs include chlorthiazide (Diuril), furosemide (Lasix), hydrochlorothiazide (Hydrodiuril), triamterene (Dyazide), and ethacrynic acid.

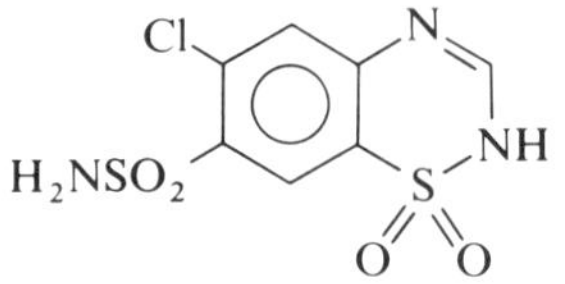

Chlorthiazide

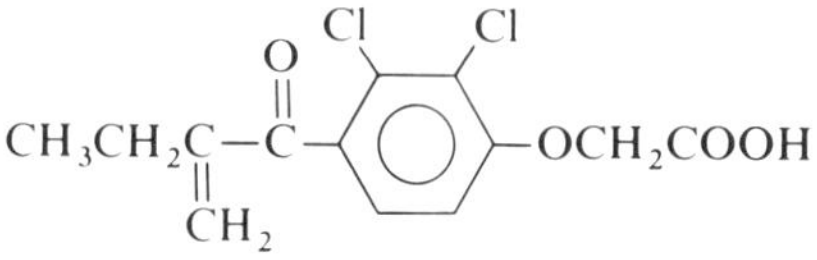

Ethacrynic acid

### Drugs that Affect the Central Nervous System

Our society relies on drugs that affect the central nervous system in an alarming fashion. Drug use in this category has increased dramatically over the past two decades, and many people rely on these drugs regularly to "get through the day" or to reduce nervous tension in order to be able to sleep. These drugs can be divided into several categories:

**Tranquilizing drugs.**   Tranquilizing drugs are taken to reduce anxiety or to modify psychotic behavior without inducing sleep. They may be either very strong or mild in their effect and thus modify behavior over a wide range. The central nervous system has many natural modifying chemicals that sometimes can get out of balance. Two of these chemicals are serotonin and dopamine. Tranquilizing drugs imitate, interact with, or counteract these normal agents.

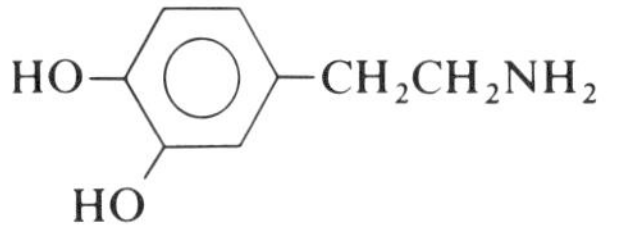

Dopamine

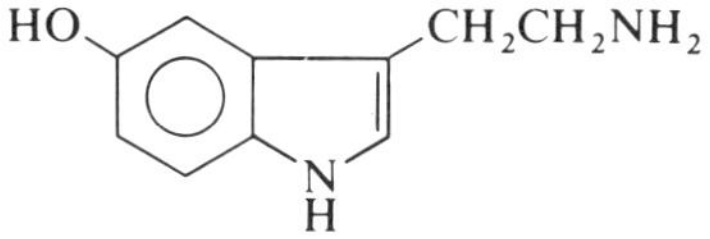

Serotonin

*Strong tranquilizers.*   Phenothorazine and chlorpromazine and their derivatives are used in the treatment of severe psychosis.

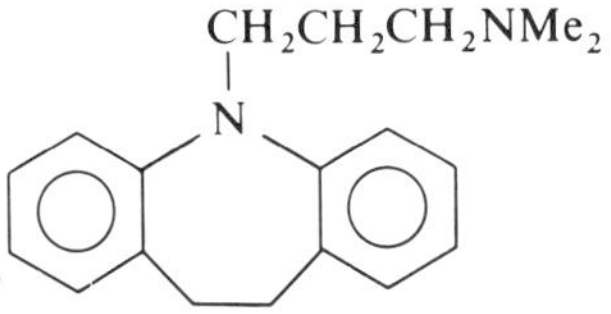

Imipramine

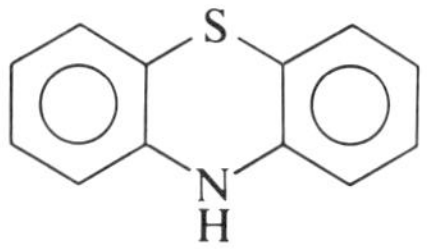

Phenothiazine

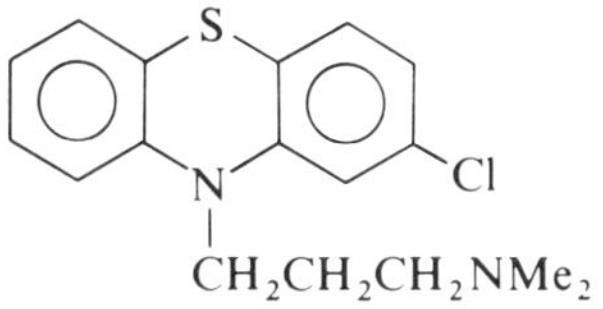

Chlorpromazine
(Thorazine)

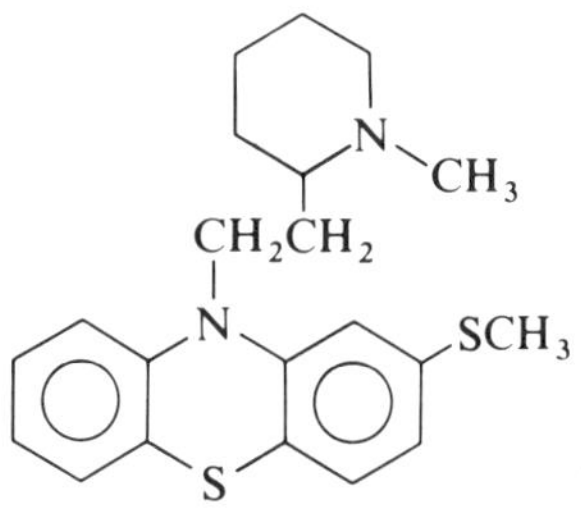

Thioridazine
(Mellaril)

*Mild tranquilizers.* Modern society seems to generate a great deal of anxiety in a significant portion of the population. Thus we have experienced greater use of mild tranquilizers, which reduce this anxiety. *Valium* and *Librium* are by far the most commonly prescribed drugs in the United States, the "mother's little helper" referred to in the song by the Rolling Stones rock group.

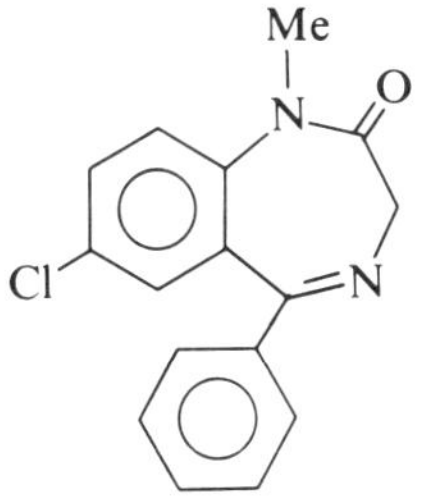

Diazepam
(Valium)

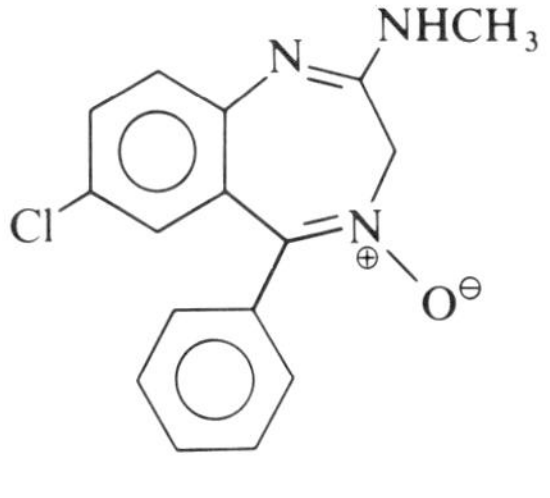

Chlordiazepoxide
(Librium)

*Meprobamate* (Equanil or Miltown) is another mild tranquilizer with sales in the millions of dollars per year.

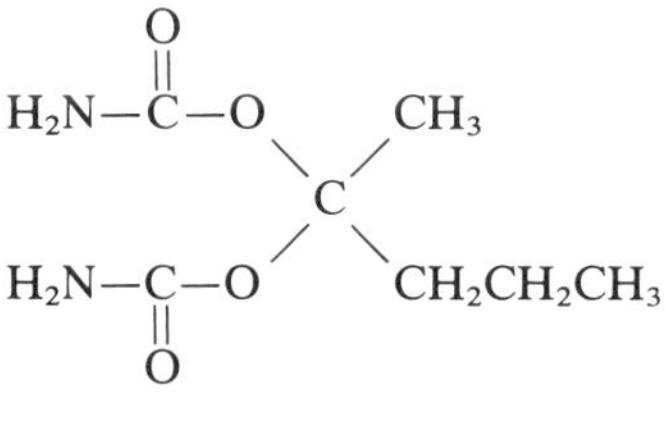

Meprobamate
(Equanil, Miltown)

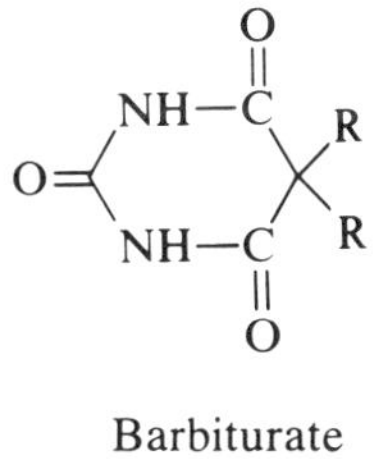

Barbiturate
Note similarity of meprobamate functional
groups compared with those of barbiturate

Many other products (Cope, Compoz, Vanquish, etc.) that *claim* to have tranquilizing effects have appeared on the market in recent years, but they are not true tranquilizers. Usually these products contain aspirin plus an antihistamine, which makes many people drowsy. Consumers equate this sleepiness with the tranquilizing effect of Valium or Librium, but the drugs are different in their physiological activity.

**Hypnotic or sleep-inducing drugs.**  Some people are often troubled with an inability to sleep, particularly if they are affected by nervous anxiety. Barbiturates, obtained by condensation reaction from malonic esters and urea, induce a deep sleep that may last several hours. Long-term use of barbiturates often leads to physical dependence. The R groups of the substituted barbiturates are easy to vary, and they affect both the duration of sleep and the length of time needed to take effect. The commonly used barbiturates are listed below.

| | Urea | Substituted malonic ester | Substituted barbituric acid (barbiturate) |

| $R_1$ | $R_2$ | Name |
| --- | --- | --- |
| Et | Et | Barbital (Veronal) |
| Et | $-CHCH_2CH_2CH_3$ <br> $\quad CH_3$ | Pentobarbital (Nembutal) |
| $CH_2=CHCH_2-$ | $-CHCH_2CH_2CH_3$ <br> $\quad CH_3$ | Secobarbital (Seconal) |
| Et | (phenyl ring) | Phenobarbital (Luminal) |
| Et | $-CH_2CH_2CH(CH_3)_2$ | Amobarbital (Amytal) |

**Stimulants.**  When people realize that they are in a dangerous situation or when a fear response is generated, adrenalin is released into their systems, increasing blood pressure, metabolic rate, and oxygen absorption through the lungs. The normal chemical responsible for transmitting nerve impulses in the body is noradrenalin. Adrenalin release also triggers increased levels of noradrenalin. Thus our whole system is on "alert" when adrenalin and noradrenalin are flowing, and we feel stimulated and energetic.

$$HO-\bigcirc-\underset{\underset{OH}{|}}{C}HCH_2NH_2$$
$$HO$$

Noradrenalin
(norepinephrine)

$$HO-\bigcirc-\underset{\underset{OH}{|}}{C}HCH_2NHCH_3$$
$$HO$$

Adrenalin
(epinephrine)

Synthetic stimulants have been around for years. Because they stimulate the sympathetic nervous system, they are referred to as *sympathomimetic* drugs. Almost everyone is familiar with the effects of *caffeine*, normally from ingesting large quantities of coffee, tea, Coca-Cola, or No-Doz. Caffeine is a mild stimulant at best, but many more powerful synthetic stimulants have been developed that more closely imitate adrenalin. Probably the most important of these are the *amphetamines*, which increase alertness and mask fatigue. They are often used by people who need to stay awake for extended periods of time, such as long-distance truck drivers and college students cramming for exams. Because a loss of appetite often accompanies

Caffeine

Amphetamine
(benzedrine)

Methedrine
(methamphetamine)
("speed")

steady use, they are often used as diet pills, but extended use of amphetamines often produces unwanted side effects ranging from restlessness and unusual excitability to hallucinations and psychotic behavior. Methedrine is particularly dangerous, and people addicted to its use often inject it directly in their veins for greater effect. It is not uncommon for users to spend 3 to 4 days on "speed" trips, only to "crash" in total exhaustion when the drug is no longer effective. This behavior pattern often leads to debilitating illness and death ("speed kills").

**Analgesic drugs—pain killers.**   Pain-killing drugs have been a boon to humans since the earliest recorded history. In fact, until the present century physicians could only treat the *pain* associated with many illnesses because they could not treat the *cause*. Originally, the most useful pain killers were derived from plants. For years alkaloids such as crude opium (laudanum) were administered to relieve pain or induce sleep for those who were seriously ill. Unfortunately, alkaloids produce both euphoria and addiction, and they became one of the first classes of drugs to be abused.

*Plant alkaloids.* When partially ripe opium poppy pods are slit, they exude a sticky fluid, which dries to a rubbery solid that can be collected and compressed into solid blocks. This crude opium contains more than 20 different alkaloids, of which *morphine* is the most important. Morphine is a powerful pain killer that also produces a euphoric sense of pleasure, well-being, and drowsiness. Unfortunately it is also extremely addictive—the body soon becomes tolerant of the drug and ever-increasing doses are required to achieve the same physical effects. If morphine usage is discontinued, withdrawal symptoms, which include intense pain and nausea, occur.

Morphine and derivatives:
(a) morphine  $R_1 = CH_3, R_2 = R_3 = H$
(b) codeine  $R_1 = R_2 = CH_3, R_3 = H$

(c) heroin  $R_1 = CH_3, R_2 = R_3 = CH_3\overset{\overset{\displaystyle O}{\|}}{C} -$

*Codeine*, once a common additive for cough syrups, is also found in crude opium. It is not as effective as morphine in its analgesic properties, but neither is it as addictive. *Heroin*, produced by acetylating the free OH groups of morphine, is more addictive than morphine. Curiously, heroin was originally prescribed as a cure for morphine addiction before its effects were really understood. Heroin is not only a potent pain killer but also produces a much more euphoric high than morphine. Heroin overdose, in which the central nervous system becomes completely paralyzed, frequently occurs because street samples, of unknown and widely varying purity, are used. The addict never knows how much heroin he is injecting in his vein, and sometimes the effect is fatal. Obviously, heroin addiction in the United States today is one of society's great unresolved problems.

*Synthetic analgesics.* Several synthetic analgesics have been developed over the years that mimic the pain-killing properties of the plant alkaloids but are much less addictive and often noneuphoric. *Methadone* is one such product and is currently used in treatment of heroin addiction. If the addict uses methadone in plance of heroin, his withdrawal symptoms are less severe.

Methadone

*Meperidine* is less effective than morphine as a pain killer, but it can be taken by people who suffer from severe nausea when treated with morphine. Unfortunately, meperidine (Demerol) is also addictive. In fact almost all drugs that have pain-killing

Meperidine (Demerol)

properties seem to be addictive in one way or another. It is possible that these two effects are linked and incapable of dissociation from one another.

*Aspirin, the universal pain killer.* Aspirin, *acetylsalicylic acid*, is the most widely used drug in the world, and probably the safest. It can relieve an amazing variety of discomforts, including minor aches and pains, fever produced by viral and bacterial agents (anti-inflammatory), and pain caused by rheumatoid arthritis. It also inhibits undesirable clotting in arteries. Its greatest use, however, is as a headache remedy.

Aspirin
(acetylsalicylic acid)

Prolonged use, or overuse, of aspirin can often irritate the stomach. In addition, some people are allergic to aspirin and may suffer rashes or asthmatic attacks. For this reason, aspirin substitutes have been developed, the most important of which is *acetaminophen*, marketed under such names as Tylenol or Datril. *Phenacetin* has also been used as an aspirin substitute or in combination with aspirin, but its use has been linked to kidney damage and it has been dropped from formulations such as Anacin and Exedrin. Acetaminophen has replaced phenacetin in Excedrin.

Phenacetin
(acetophenetidin)

Acetaminophen

Although many "combination-of-ingredients" products are currently on the market, none have been shown to be more effective in relieving minor pain or fever than plain aspirin.

## Drugs that Attack the Common Cold
## and Hay Fever

No two people can agree on exactly what causes a common cold. Conventional scientific wisdom suggests that most colds are caused by viruses, and thus are immune to antibiotics aimed at bacterial agents. Most people *do* agree on the symptoms of a common cold—stuffy nose, watery eyes, headache, minor body aches, postnasal drip, and sore throat. Most, if not all, of the cold medicines on the market today treat these symptoms and let the cause take care of itself. The following discussion of drugstore remedies, although by no means complete, will give you some indication of the spectrum of compounds available.

**Antihistamines.** When the body is attacked by foreign agents such as viruses, the body's normal defense mechanisms may trigger the release of histamine, which causes headache and difficulty in breathing. Many *antihistamines* that counteract the body's reaction to the presence of histamine are available. Histamine generation is also characteristic of allergic reactions such as hay fever and sinus attacks. Two of the more potent antihistamines are chlorpheniramine (Ornade) and diphenhydramine.

Chlorpheniramine
(Chlor-trimeton)

Diphenhydramine
(Benadryl)

Nasal congestion may be relieved by any of the following reagents, which are usually combined with antihistamines such as chlorpheniramine.

Phenylephrine
(Neo-Synephrine)

Phenylpropanolamine
(Propadrine)

Coughing can be controlled by a number of different compounds. In older cough formulations codeine, an opium alkaloid, was used, and although it is effective, it is habit forming. Chloroform had been used until recently when questions arose about its possible carcinogenic (cancer causing) properties. The best cough suppressant now available seems to be dextromethorphan, whose activity rivals that of codeine, but which is not addictive.

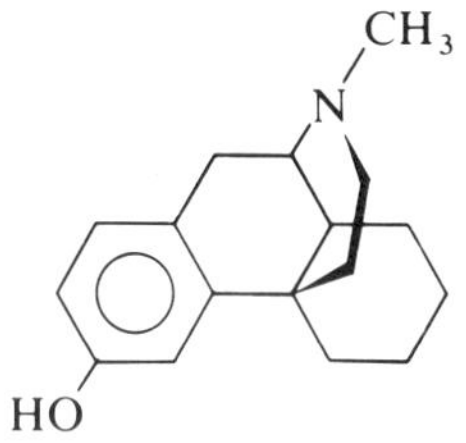

Dextromethorphan

### Drugs that Affect the Mind—Hallucinogens

Earlier in this chapter attention was directed to the natural chemicals that are involved with the transmission of nerve impulses and sensory data, dopamine and serotonin. Hallucinogenic drugs are similar in structure to these chemicals and

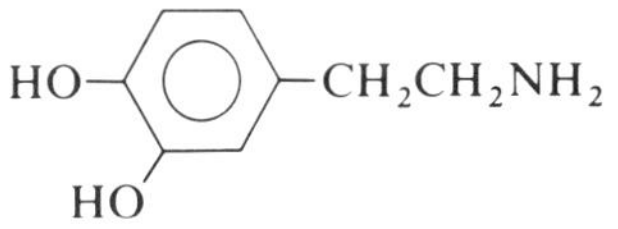

Dopamine

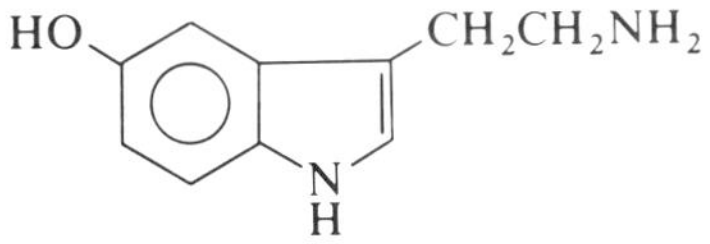

Serotonin

totally distort our natural perceptions of space, time, color, sound, and the appearance of objects. Not only do our perceptions of real things appear to be altered by the drug, but fantastic and often fearsome images from our subconscious can surface and appear to be *very* real.

Some hallucinogenic drugs have been used by humans for centuries. Marijuana use has been described in ancient Indian writings dating to thousands of years before the birth of Christ, and mescaline and psilocybin are important components of religious ceremonies of several North American Indian tribes.

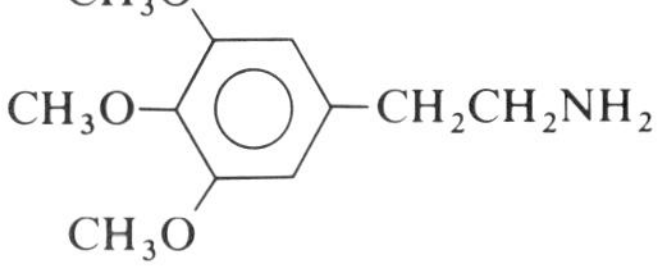

Mescaline

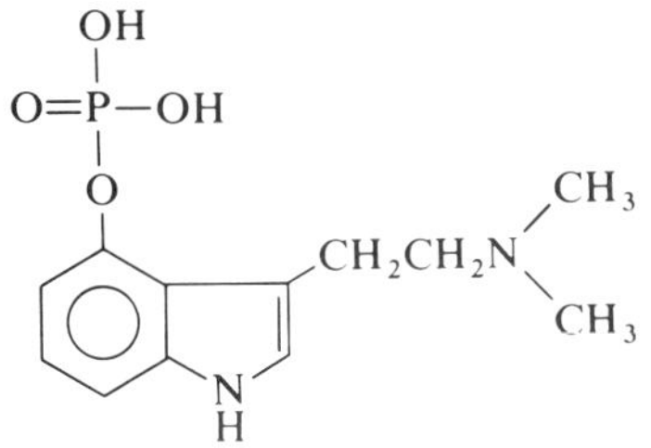

Psilocybin

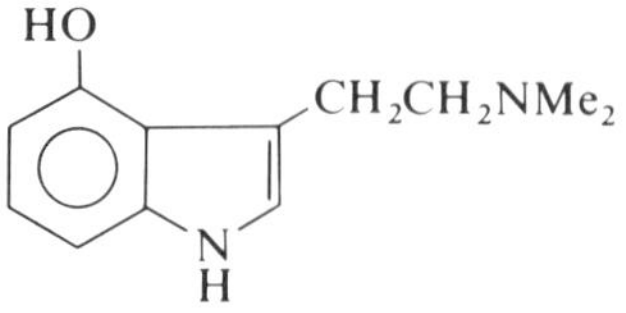

Psilocin

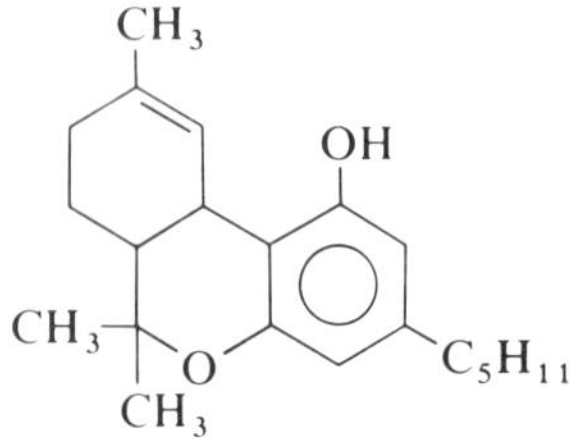

Tetrahydrocannabinol
(THC)

Major hallucinogenic component of
marijuana—also known as bhang, ganja,
and charas in India, as well as grass,
tea, pot, dope, weed, MJ, maryjane, and
many other street names.

Modern hallucinogens are mainly synthetic. By far the most widely known is LSD, or
lysergic acid diethylamide. LSD is a dangerous drug. Extended use has been
tentatively linked with chromosomal damage, and spontaneous abortion rates in

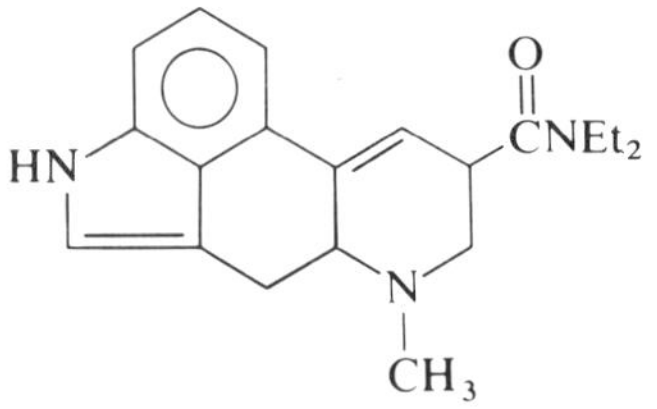

LSD—synthetic amide prepared from
naturally-occurring lysergic acid

pregnant women seem to be high. In high doses, LSD distorts reality beyond the
control of the user. He may think he can fly and so jump out out a window, or think
he's so strong he can stop a car with his bare hands. What is more frightening,
however, is the return of hallucinations (flashbacks) that can occur weeks or months
later with little or no warning. People with unstable or psychotic personalities have
been known to completely lose contact with reality on a "bad trip" and never regain
it. Even though these facts have been well publicized, LSD use still continues, but
less than in the late 1960s.

In recent years, "mescaline," "psilobycin," and "THC" have been offered for
sale by street drug dealers. Many of these samples have been analyzed, and only one

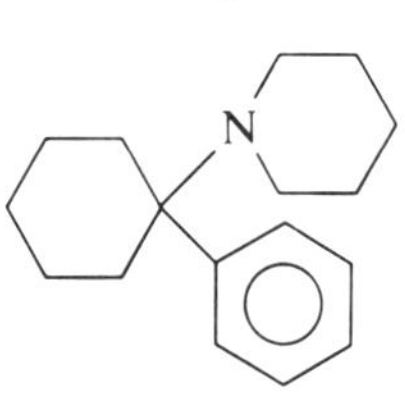

Phencyclidine (PCP)

out of a hundred contains even a little of what was purported to be. Instead, most of these street samples contain phencyclidine (PCP, or angel dust), a potent animal tranquilizer. This substance is extremely dangerous, and large doses may lead to death, particularly when mixed with alcohol.

## FROM LABORATORY TO DRUGSTORE

In this chapter we have described various drug classifications and the effects they have on our bodies. But how are drugs discovered or designed, and how are they brought to the marketplace? At best, it is a long, tedious process. In general, there are two possible routes that may lead to the discovery of new drugs: (1) random screening of compounds isolated from natural sources (usually plants) against specific diseases and (2) modification of existing structures whose therapeutic properties are known. Currently, the federal government operates a cancer screening program that tests new compounds and plant extracts against tumors or leukemic cells and is probably the biggest example of the first route. Pharmaceutical

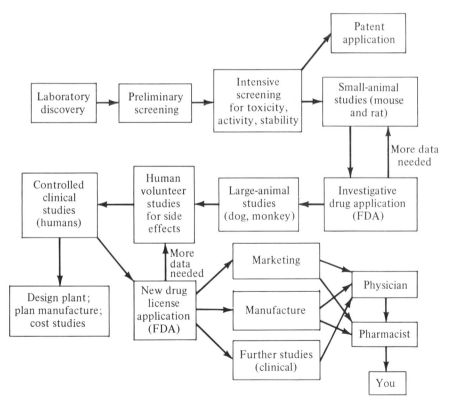

**Figure 8-1.**   Development of a new drug.

companies (Merck, Searle, Abbot, Wyeth, Parke Davis, etc.) are more intimately involved in the second route.

Random screening is a hit-or-miss affair and few compounds become marketable. It is not enough that a compound have activity against a specific disease or disease symptom: it must also be nontoxic and have minimal side-effects. It does no good to cure the disease yet kill the patient. Of every 20,000 compounds tested by random screening, only *one* is likely to be usable!

The approach of the pharmaceutical companies is more scientific and much more successful. In fact, 90% of all new drugs developed during the past several decades comes from the basic research groups of these companies. Expenditures for this research are staggering. The Food and Drug Administration (FDA) requires evidence that a new drug is safe before they will allow it to be marketed. Conservatively, from the time a potentially useful drug is discovered in the lab until the time it is released for sale to the public with FDA approval, development and testing may cost as much as $25 million. It should not be surprising, then, that the number of new drugs being released has declined drastically over the last decade. In Fig. 8-1 the procedures involved in new drug development are outlined. The time from discovery to marketing is now approximately 7 to 10 years.

## REVIEW OF NEW TERMS AND CONCEPTS

Define each term and, if possible, give an example of each:

| | |
|---|---|
| Anti-hypertension drug | Anti-arrhythmia drug |
| Diuretic | Tranquilizer |
| Hypnotic drug | Barbiturate |
| Stimulant | Amphetamine |
| Analgesic | Plant alkaloid |
| Antihistamine | Hallucinogen |

## PROBLEMS: CHAPTER II-8

1. Give examples for each of the following:

   (a) a drug to combat hypertension
   (b) a drug used to treat irregular heartbeat
   (c) a drug to calm a psychotic person
   (d) a drug to relieve common nervous tension
   (e) a drug to combat sleeplessness
   (f) a drug to promote alertness
   (g) a drug to counteract severe pain
   (h) a drug to treat mild aches and pains

     (i) a drug to combat an allergic reaction

     (j) a cough suppressant

     (k) a hallucinogenic drug

**2.** Go to your medicine chest or local drugstore and look at the list of ingredients in several nonprescription drugs. Can you identify the ingredients and explain what each one does? Choose two out of the following list of suggestions.

     (a) cough syrups—Vick's Formula 44, Robitussin

     (b) headache remedy—Anacin, Bufferin, Vanquish, Excedrin

     (c) decongestant—Dristan Mist, Coricidin D

     (d) cold capsules—Contac, Sinutabs

**3.** Several of the drugs discussed in this chapter can be readily synthesized by techniques learned in this text. Prepare the following from benzene, toluene, and any organic compounds containing three carbons or fewer:

     (a) amphetamine           (b) methedrine

     (c) aspirin               (d) phenacetin

     (e) Benadryl            (f) phenylpropanol amine

     (g) phencyclidine (PCP)

**4.** In some hallucinogenic drugs, it is easy to see the relationship between the drug and a natural chemical involved with the central nervous system. Thus psilocybin is an obvious competitor for serotonin in the human system. Can you identify any structural components in LSD that might mimic a natural substance in the human system?

# Index

425

429

431

Williamson synthesis, 152–53
Wittig reagent, 129
Wöhler, Friedrich, 1
Wood alcohol, 145
Wurtz reaction, 47–48, 127

## X

D-(+)-Xylose, 239

## Z

Z configuration, 82
Ziegler catalyst, 402, 406
Zwitterion, 283

## COMMON CHARACTERISTIC IR GROUP
## STRETCHING VIBRATIONS

| Group | Type | Group frequency range (cm$^{-1}$) |
|---|---|---|
| O—H | Alcohols[a] | 3200–3600 |
| | Acids[b] | 3000–3500 |
| N—H | Amines | 3300–3500 |
| C—H | Alkane | 2850–3000 |
| | Alkene | 3020–3080 |
| | Aromatic | 3000–3100 |
| | Alkyne | 3300 |
| | Aldehyde | 2700–2800 |
| C≡N[c] | Nitrile | 2210–2260 |
| C≡C[c] | Alkyne | 2100–2250 |
| C=C | Alkene | 1600–1680 |
| | Aromatic | 1400–1600 |
| C=O[d] | Aldehyde | 1720–1740 |
| | Ketone | 1705–1725 |
| | Carboxylic acid | 1700–1725 |
| | Ester | 1730–1750 |
| | Amide | 1640–1700 |
| C—X | Halogen | 600–800 (Cl) |
| | | < 600 (Br) |

[a] If hydrogen-bonded, these peaks will be broad.

[b] Acids are hydrogen bonded and the absorption will be broad.

[c] When present, these absorptions are weak.

[d] When present, these absorptions will be the strongest stretching frequencies in the spectrum.